applied time series analysis for the social sciences

applied time series analysis for the social sciences

Richard McCleary
Richard A. Hay, Jr.

with Errol E. Meidinger
and David McDowall

Foreword by Kenneth C. Land

SAGE PUBLICATIONS Beverly Hills London

For information address:

SAGE Publications, Inc.
275 South Beverly Drive
Beverly Hills, California 90212

SAGE Publications Ltd.
28 Banner Street
London EC1Y 8QE, England

Printed in the United States of America
Library of Congress Cataloging in Publication Data

McCleary, Richard.
Applied time series analysis for the social sciences.

Bibliography: p.
Includes index.
1. Time-series analysis. 2. Prediction theory. 3. Social sciences—Statistical methods. I. Hay, Richard, joint author. II. Title.

HA30.3.M32 300'.1'51955 79-27873
ISBN 0-8039-1205-6
ISBN 0-8039-1206-4 pbk.

FIRST PRINTING

To Donald T. Campbell

CONTENTS

List of Figures 9
Foreword 13
Acknowledgments 15

1. Statistical Models for Time Series Analysis 17
1.1 Caveat: The Limits of Time Series Analysis 20
1.2 An Outline of the Volume 24

2. Univariate ARIMA Models 29
2.1 Trend and Drift 31
2.2 The Random Walk and Other Integrated Processes 36
2.3 The Backward Shift Operator 45
2.4 Variance Stationarity 53
2.5 Autoregressive Processes of Order p: ARIMA (p,0,0) Models 61
2.6 Moving Average Processes of Order q: ARIMA (0,0,q) Models 64
2.7 The General ARIMA (p,d,q) Model 66
2.8 The Autocorrelation Function (ACF) 79
2.9 Seasonality 87
2.10 Identifying a Seasonal Model 91
2.11 Model Building 103
2.12 Example Analyses
2.12.1 Sutter County Workforce 104
2.12.2 Boston Armed Robberies 110
2.12.3 Swedish Harvest Index 121
2.12.4 Hyde Park Purse Snatchings: Outliers 128
2.13 Conclusion 132
Appendix: Expected Values 136

3. ARIMA Impact Assessment 141
3.1 The Zero-Order Transfer Function 145
3.2 The First-Order Transfer Function 154
3.2.1 Chlorpromazine Impacts on Perceptual Speed 160
3.2.2 Sutter County Workforce: Temporary Impacts 164
3.2.3 Testing Rival Impact Hypotheses 168
3.3 Interpreting Impact Parameters in the Natural Log Metric 171
3.4 Higher Order and Compound Intervention Components 185

3.5 U.S. Suicides 191
3.6 A Final Note on the Outlier Problem 199
3.7 Conclusion 201

4. Univariate ARIMA Forecasts 205
4.1 Point and Interval Forecasts 206
4.2 ARIMA Forecast Profiles 213
4.3 Conclusion: The Use of Forecasting 219

5. Multivariate ARIMA Models 227
5.1 The Cross-Correlation Function 229
5.2 A Forecasting Example 236
5.3 Prewhitening 243
5.4 Multivariate Population Growth Model 257
5.5 Conclusion and Recommendations 270

6. Nonlinear Estimation 275
6.1 Maximum Likelihood Estimation 276
6.2 Numerical Solutions to the S $(\hat{\beta})$ Minimum 285
6.2.1. The Grid Search Method 288
6.2.2. Marquardt's Algorithm 293
6.3 Monitoring and Controlling the Algorithm 301
6.4 Time Series Software 305
Appendix: ARIMA Software Distributors 310

Appendix A: Symbols and Conventions 311
Appendix B: Data Sets 313
Bibliography 321
Index 327
About the Authors 331

LIST OF FIGURES

1.2 A Time Series 22

2.1(a) German Immigration to the United States (in Thousands), 1870–1914 33
2.1(b) Closing Price of IBM Common Stock (Series B from Box and Jenkins) 34
2.2(a) Coin Flip Gamble: A Random Walk 38
2.2(b) Coin Flip Gamble: A Random Walk, Differenced 41
2.4(a) Nonfatal Disabling Mine Injuries, 1930–1978 50
2.4(b) Nonfatal Disabling Mine Injuries, Differenced 50
2.4(c) Nonfatal Disabling Mine Injuries, Logged 51
2.4(d) Nonfatal Disabling Mine Injuries, Logged and Differenced 52
2.8(a) Expected ACFs for Several ARIMA Processes 68
2.8(b) Expected ACFs and PACFs for Several ARIMA Processes 78
2.10(a) Expected ACFs for Several ARIMA $(P,D,Q)_{12}$ Processes 88
2.10(b) Expected ACF or PACF for Several ARIMA (p,d,q) $(P,D,Q)_{12}$ Processes 90
2.11(a) The ARIMA Model-Building Strategy 92
2.11(b) Expected ACF for the Process $Y_t = .5y_{t-1} + a_t$; to aid comparison with figure 2.11(c), standard errors have been calculated with T=100 observations 95
2.11(c) Estimated ACFs for the Process 96
2.12.1(a) Sutter County Workforce 105
2.12.1(b) ACF and PACF for the Raw Series 106
2.12.1(c) ACF and PACF for the Regularly Differenced Series 107
2.12.1(d) ACF and PACF for the Regularly *and* Seasonally Differenced Series 108
2.12.1(e) *Diagnosis:* ACF and PACF for the Model Residuals 109
2.12.2(a) Boston Armed Robberies 111
2.12.2(b) ACF and PACF for the Raw Series 112
2.12.2(c) ACF and PACF for the Regularly Differenced Series 113
2.12.2(d) *Diagnosis:* ACF and PACF for the Model Residuals 114
2.12.2(e) Boston Armed Robberies (Logged) 115
2.12.2(f) ACF and PACF for the Raw (Logged) Series 117
2.12.2(g) ACF and PACF for the Regularly Differenced (Logged) Series 118
2.12.2(h) *Diagnosis:* ACF and PACF for the Model Residuals (Logged) 119
2.12.2(i) The Residuals of the Unlogged Model Indicate that the Generating Process is Nonstationary in Variance 122
2.12.3(a) Swedish Harvest Index 124
2.12.3(b) ACF and PACF for the Raw Series 125
2.12.3(c) *Diagnosis:* ACF and PACF for the Model Residuals 126
2.12.4(a) Hyde Park Purse Snatchings 128
2.12.4(b) ACF and PACF for the Raw Series (Outlier Included) 129
2.12.4(c) ACF and PACF for the Raw Series (Outlier Excluded) 130
2.13(a) An Input-Output Representation of the ARIMA Model 133
2.13(b) An Input-Output Representation of the ARIMA Model-Building Strategy 133

3.1(a) Directory Assistance, Monthly Average Calls per Day 147
3.1(b) ACF and PACF for the Raw Preintervention Series 148
3.1(c) ACF and PACF for the Regularly Differenced Preintervention Series 149
3.1(d) ACF and PACF for the Regularly *and* Seasonally Differenced Preintervention Series 150

3.1(e) *Diagnosis:* ACF and PACF for the Model Residuals (Preintervention Series Only) 151
3.1(f) *Diagnosis:* ACF for the Model Residuals 152
3.2(a) Pattern of Impact Expected of a First-Order Transfer Function 157
3.2(b) Pattern of Impact Expected for Several Values of δ_1 158
3.2.1(a) Daily Perceptual Speed Scores for a Schizophrenic Patient 161
3.2.1(b) ACF and PACF for the Raw Preintervention Series 162
3.2.2 Pattern of Impact Expected for Several Values of δ_1 167
3.2.3 A Simple Theory of Impact 172
3.3(a) Minneapolis Public Drunkenness Intakes 176
3.3(b) Minneapolis Public Drunkenness Intakes (Logged) 176
3.3(c) ACF and PACF for the Raw (Logged) Preintervention Series 177
3.3(d) ACF and PACF for the Seasonally Differenced (Logged) Preintervention Series 178
3.3(e) *Diagnosis:* ACF and PACF for the Residuals of an ARIMA $(0,0,0)\ (0,1,1)_{12}$ Model and Intervention Component 179
3.3(f) *Diagnosis:* ACF and PACF for the Residuals of an ARIMA $(0,1,0)\ (0,0,1)_{12}$ Model and Intervention Component 180
3.4 Pattern of Impact Expected of a Compound Intervention Component 187
3.5(a) U.S. Suicide Rate, 1920–1969 191
3.5(b) ACF and PACF for the Raw Series 192
3.5(c) ACF and PACF for the Regularly Differenced Series 193
3.5(d) ACF and PACF for the Raw Series (1943–1969 Only) 194

4.2(a) *Forecast Profile:* Series B 216
4.2(b) *Forecast Profile:* Hyde Park Purse Snatchings 217
4.2(c) *Forecast Profile:* Swedish Harvest Index 219
4.2(d) *Forecast Profile:* Sutter County Workforce 220

5.1 Expected CCFs for Several Bivariate Relationships 233
5.2(a) Paris and New York IBM Common Stock Prices 236
5.2(b) *Identification:* CCF Estimated from the Differenced Series 237
5.2(c) *Identification:* ACF for the Residuals of the Model $Y_t = .767\ X_{t-2} + \dfrac{a_t}{1 - B}$ 238
5.2(d) *Diagnosis:* ACF for the Residuals of the Model $Y_t = .987\ X_{t-2} + \dfrac{1 - .88B}{1 - B}\, a_t$ 239
5.2(e) *Diagnosis:* CCF for the Differenced X_t Series and the Model Residuals 240
5.3(a) Swedish Population Rates, 1750–1849 244
5.3(b) *Identification:* CCF for the h_t and p_t Series 247
5.3(c) The Bivariate ARIMA Model-Building Strategy 251
5.3(d) *Identification:* CCF for the Prewhitened p_t and h_t Series 253
5.3(e) *Identification:* ACF for the Residuals of the Model $p_t = .866\ h_{t-1} + a_t$ 254
5.3(f) *Diagnosis:* ACF for the Residuals of the Model $p_t = .828\ h_{t-1} + \dfrac{a_t}{1 - .459B}$ 255
5.3(g) *Diagnosis:* CCF for Prewhitened h_t Series and the Model Residuals 256
5.4(a) Swedish Fertility Rates, 1750–1849 257
5.4(b) *Identification:* CCF for the Prewhitened p_t and h_t Series 258

5.4(c)	*Identification:* ACF for the Residuals of the Model $p_t = .239\,f_t + a_t$	259
5.4(d)	*Identification:* ACF for the Residuals of the Model $p_t = .239\,f_t - .01\,h_{t-1} + a_t$	260
5.4(e)	*Identification:* CCF for the Prewhitened f_t and h_t Series	263
5.4(f)	*Identification:* ACF for the Residuals of the Model $f_t = \frac{3.498}{1 - .372B} h_{t-1} + a_t$	264
5.4(g)	*Identification:* CCF for the f_t^* and p_t Series	265
5.4(h)	An Input-Output Diagram of the Multivariate Model	268
6.1(a)	Log-Likelihood Function for a Coin-Flip Experiment: 6 Heads in 10 Flips	278
6.1(b)	S $(\hat{\beta})$ for an ARIMA (0,1,1) Model Fit to Series B	284
6.1(c)	S $(\hat{\beta})$ for the Directory Assistance Example of Section 3.1	285
6.2(a)	The Precision Function, ϵ	289
6.2(b)	Two Hypothetical Sum of Squares Functions Differing Only in Steepness	295
6.2(c)	A Flowchart Diagram of Marquardt's Algorithm	298
6.2(d)	S $(\hat{\beta})$ for the Sutter County Workforce Example of Section 2.12.1: Marquardt's Algorithm "Finds" the Minimum in Four Steps	300
6.3	S $(\hat{\beta})$ for an ARIMA (1,0,1) Process, $\phi_1 \approx \theta_1$	304

FOREWORD

Social scientists have been confronted with revolutionary developments in a variety of areas of statistical theory and methodology during the 1970s, including Bayesian statistical inference, exploratory data analysis, log-linear and related models for categorial data, robust statistics, structural equation models in unobservable variables, and time series analysis. Two characteristics of these developments are noteworthy. First, they make the repertoire of available statistical tools considerably more adaptable to the substantive concerns and empirical data limitations encountered in social science research. But, second, this increase in adaptability typically is bought at the expense of increased computational complexity—in the form of iterative estimation algorithms that are manageable only with the assistance of modern electronic computers.

One consequence of these two qualities is that the "cookbook" approach of introductory statistical methods courses (in which statistical recipes are stated in closed form and applied to simplified, often artificial, data sets) no longer is a feasible mode of instruction. Rather, the emphasis must be on the application of sophisticated computer algorithms to real data sets. While the algorithms may remain essentially "black boxes" to the student, the fact that they are computerized facilitates their application to a large and diverse array of empirical data sets. In this way, the student can obtain an intuitive "feel" for the types of problems likely to be encountered in applications.

This new didactic style is exemplified in the present volume by Professors McCleary, Hay, Meidinger, and McDowall. Their subject matter is the synthesis of time series analysis and forecasting methods brought together in 1970 by George E. P. Box and Gwilyn M. Jenkins *(Time Series Analysis: Forecasting and Control,* San Francisco: Holden-Day; 1976, revised edition). While these *A*uto*R*egressive *I*ntegrated *M*oving *A*verage (ARIMA) models (popularly called "Box-Jenkins" models) have been widely applied in engineering, economics, and business for nearly a decade, social science applications outside of economics still are relatively rare. One obvious reason for this is the fact that available textbooks typically assume a mathematical and statistical maturity greater than appropriate for social science audiences outside of economics (which has a long tradition of econometric analyses of time series).

It is precisely this void in the statistical time series textbook literature that McCleary and his coauthors seek to fill. Assuming no training in statistics beyond intermediate statistical methods (at, say, the level of H.M. Blalock, Jr., 1979, *Social Statistics,* revised second edition, New York: McGraw-

Hill), the authors take the reader through a gentle introduction to univariate ARIMA models (emphasizing the Box-Jenkins iterative cycle of model identification, estimation, and diagnosis), impact assessments, and forecasts. This is followed by chapters on multivariate ARIMA models and ARIMA estimation algorithms. The text is noteworthy for its clear, concise exposition punctuated by numerous analyses of real time series. These pertain to a diverse array of noneconomic topics. By including listings of the time series and references to the available ARIMA software, the authors encourage readers to develop "hands on" experience in the didactic mode described above. For the highly motivated student, the authors also include annotated bibliographic references to more advanced literature.

In addition to its value as a didactic aid, this book is a timely and welcome addition to the professional social science literature for two reasons.

First, there is a wide and growing interest in the study of social change via the analysis of historical time series. This has been stimulated, in part, by the "social indicators movement" and the increased availability of time series data on noneconomic social conditions resulting therefrom. (See, for example, Kenneth C. Land and Seymour Spilerman, eds., 1975, *Social Indicator Models,* New York: Russell Sage Foundation, for a statement of some of the analytic problems created by social indicator time series.) Although annual social indicators time series often are too short for the application of ARIMA models and methods, this is less likely to be the case for social indicator time series collected at quarterly, monthly, weekly, or daily intervals. In any case, for sufficiently long series in areas in which there exists little prior theory or research, the ARIMA models described in this book provide powerful analytic tools.

Second, as the authors note, there has been a convergence in recent years between the "statistical time series" and "dynamic structural equation models" literatures (popularly called the "Box-Jenkins" and "econometric" literatures, respectively). For dynamic structural equation modelers, this convergence has taken the form of a greater sensitivity to the stochastic properties of the error or disturbance terms in time series models. Conversely, time series analysts have come to recognize that lag structures need not always be identified *de novo* from the time series to be modeled in areas in which substantial prior theory and/or research exists. The present volume should give nonspecialist readers the background in statistical time series analysis necessary to appreciate more fully the nature of this convergence.

—Kenneth C. Land
University of Illinois at Urbana-Champaign

ACKNOWLEDGMENTS

Our interest in time series analysis was first stimulated by the pioneering work of Donald T. Campbell. He has remained a constant source of inspiration and support over the years. We are no less indebted to Richard A. Berk who first introduced us to the practical and methodological exigencies of applied social research. He has also generously supported development of the software used in this volume. Berk, Clifford C. Clogg, Bruce Foster, Kenneth C. Land, Michael C. Musheno, and J. Richard Zelonka all read and commented on various drafts of the manuscript; the academic computing centers of Arizona State University and Northwestern University provided computer time and support for the analyses; and our collaborators, Errol E. Meidinger and David McDowall, made invaluable contributions to the organization and content of the volume. Finally, thanks are due to our colleagues at the Center of Criminal Justice, Arizona State University, and the Vogelback Computing Center, Northwestern University, for implicit and explicit support of this project.

—Richard McCleary
Tempe, AZ

—Richard A. Hay, Jr.
Evanston, IL

1 Statistical Models for Time Series Analysis

Monographs on time series analysis ordinarily address *either* theoretical issues (e.g., Doob, 1953; Feller, 1971; Anderson, 1975) *or* practical issues (e.g., Nelson, 1973; Glass et al., 1975; Pindyck and Rubinfeld, 1976; Makridakis and Wheelwright, 1978). This volume is of the latter type. We will be concerned largely with the practical or applied aspects of time series analysis and especially with applications of interest to economists, political scientists, psychologists, and sociologists.

Time series analysis can similarly be divided into two methodological areas: harmonic analysis and regression analysis. These two methods are sometimes called *analysis in the frequency domain* and *analysis in the time domain,* respectively. We will not cover harmonic methods at all in this volume. While there are no practical limitations on the use of harmonic methods with social science data (see, e.g., Mayer and Arney, 1974), this type of analysis ordinarily requires a background in the calculus and algebra of complex variables (imaginary numbers). We suspect that most readers will lack this mathematical preparation. Regression approaches to time series analysis, in contrast, have been widely used in the social sciences. All of our readers will have had some training in multiple regression methods. We have consciously addressed our development of time series analysis to this level of understanding.

Readers with broad backgrounds in regression methods may nevertheless find this volume novel. The time series models developed here are not the models ordinarily developed in econometrics texts, but rather are *stochastic process* models. The particular class of stochastic process models to be developed are the *A*uto*R*egressive *I*ntegrated *M*oving *A*verage (ARIMA) models of George E. P. Box and Gwilym M. Jenkins (1976; Box and Tiao, 1965, 1975). Although elements of ARIMA modeling can be traced back some 50 years, Box and Jenkins (and George C. Tiao) must be credited with integrating the elements into a comprehensive theory, extending it greatly, and popularizing it. ARIMA modeling has rightly been called the "Box-Jenkins approach to time series analysis."

It will be instructive at this early point to make the tenets of the Box-Jenkins approach explicit. ARIMA models posit a random shock, a_t, as the driving force of a time series process, Y_t. As an analogy, consider a coffee-brewing machine of the sort widely used in university departments at present. To brew a pot of coffee, 12 cups of cold water are fed into one end of the machine. A few moments later, 12 cups of coffee are delivered from the other end. Of course, we do not always wish to brew 12 cups of coffee. Sometimes we feed only 6 cups of cold water into the machine with the expectation that only 6 cups will be delivered at the other end of the machine. And of course, sometimes we expect to receive 12 cups of coffee from the machine, but for some reason receive only 11.5 cups. We can diagram the coffee brewing process as an input-output process:

$$\text{(Cold Water) } a_t \rightarrow \square \rightarrow Y_t \text{ (Hot Coffee),}$$

Both the a_t input and the Y_t output are measured in cup units. To some extent, there is a prescribed relationship between the size of each input and the size of each output. Inputting 6 cups of cold water, for example, we would be surprised if 12 cups of coffee were delivered at the other end of the machine. On the other hand, we might be equally surprised if *exactly* 6 cups of coffee were delivered. Inside the coffee machine, the unobserved and often mysterious brewing process is at work, sometimes delivering slightly more than 6 cups of coffee and sometimes delivering slightly less.

If we were interested in this mysterious internal process, we could perform an experiment with the aim of unraveling the relationship between a_t and Y_t. We would first hire a graduate student to do nothing but brew coffee under scientific conditions. Every 15 minutes, the graduate student would input a precisely measured amount of cold water, and after a few minutes brewing time would receive an amount of hot coffee from the machine to be measured precisely. To ensure experimental control, each input to the machine would be randomly determined. Our graduate student would consult a table of Normal (Gaussian) random numbers to determine how many cups of cold water to feed into the machine for each trial.

After many trials (say, 500 pots of coffee), we could analyze the Y_t output series to draw inferences about the brewing process. We would no doubt discover the following:

(1) The most important determinant of Y_t is a_t. Other things being equal, the more cold water input, the more hot coffee output.
(2) To a lesser extent, Y_t may be also be determined by a_{t-1}, the previous input. A small percentage of each input, for example, may remain inside the machine to be delivered in the next output. The larger the input on one trial, the larger the residual remaining inside the machine to be delivered in the next output.
(3) To a lesser extent, Y_t may be also be determined by Y_{t-1}, the previous output. A particularly large output, for example, as a result of a particularly large input, may somehow reduce or increase the efficiency of the brewing process. This change in efficiency will show up in the next output.
(4) To a much lesser extent, and for the same reasons, Y_t may be determined by inputs and outputs further removed in time, such as a_{t-2} and Y_{t-2}.

These are the basic tenets of the Box-Jenkins approach to time series analysis, or ARIMA modeling. In general, we may say that a pot of coffee has its size determined by the few immediately preceding outputs and inputs, that is,

$$Y_t = \phi_1 Y_{t-1} + \phi_2 Y_{t-2} + a_t - \theta_1 a_{t-1} - \theta_2 a_{t-2}\,.$$

Inputs and outputs further removed from the present, a_{t-3} and Y_{t-3}, for example, may also play some role in the process but their influence will be so small as to be statistically insignificant. In practice, of course, the number of past inputs and outputs required in the model will be determined empirically, but in almost every case, no more than two prior inputs and outputs will have a statistically significant influence on the present output.

The most important tenet of the ARIMA model is that the present input, a_t, will have a greater impact on the present output than any earlier input. This means that the parameter θ_1 must be a fraction. The parameter θ_2, also a fraction, will ordinarily be smaller than θ_1, so in general

$$1 > \Theta_1 > \Theta_2 > \ldots > \Theta_q.$$

This is the most important principle of the ARIMA model. The influence of a past event (or input) on present events diminishes as time passes. This same principle applies to the influence of past outputs:

$$1 > \phi_1 > \phi_2 > \ldots > \phi_p.$$

But if there is a single guiding principle of the Box-Jenkins approach, it is *parsimony*. This principle reflects not only a view of nature but also a view of the relationship between a time series model and nature. In almost all cases, a social science time series process can be modeled as a probabilistic function of a few past inputs (random shocks) and outputs (time series observations). As we develop the algebra of ARIMA modeling in the next chapter, it will become clear that there is a real difference between parsimony and simplicity. While parsimonious, univariate ARIMA models also give a surprisingly sophisticated representation of nature.

The reader who is familiar with the more widely used regression approaches to time series analysis (structural equation or econometric models) should not assume that ARIMA models are substantially different than regression models. While ARIMA models require the novel input-output explanation, the two approaches are in fact identical. The only real difference between ARIMA and regression approaches to time series analysis is a practical one. Whereas regression models can be built on the basis of prior research and/or theory, ARIMA models must be built empirically from the data. Because ARIMA models must be identified from the data to be modeled, relatively long time series are required. *No time series that we analyze in this volume is shorter than 50 observations long*. The reader may use this rule of thumb when deciding whether to analyze time series data from an ARIMA or regression approach. When relatively long time series are available, an empirical ARIMA approach will ordinarily give the best results. But when relatively long series are not available, regression approaches informed by prior research and/or theory will give the best results.

1.1 Caveat: The Limits of Time Series Analysis

If our experiences as teachers are typical, there is a danger to learning any sophisticated statistical method. The power of the method may desensitize the student to the more fundamental questions of interpretation. Though absorbing, the statistical problems of analyzing time series data are generally less important than the problems of interpreting the results of an analysis. Lacking an easy interpretation, the time series analysis has failed.

To emphasize this point, we note first that, while the statistical methods of time series analysis are relatively new, the *logic* of time series analysis is not. In his classic investigation into the causes of suicide, for example, Emile Durkheim wrote:

> It is a well known fact that economic crises have an aggravating effect on the suicidal tendency. . . . In Vienna, in 1873 a financial crisis occurred which reached its height in 1874; the number of suicides immediately rose. . . . What

> proves this catastrophe to have been the sole cause of the increase is the special prominence of the increase when the crisis was acute, or during the first four months of 1874 [1951:241].

We would like to believe that Durkheim actually plotted the annual suicide rates of European cities, searched for peaks and valleys in this time series plot, and then compared the peaks and valleys with the economic histories of the cities. In the most general sense, this is a time series analysis.

Aside from an increase in methodological sophistication, nothing has changed. Contemporary economists, political scientists, psychologists, and sociologists share an interest with the early social philosophers in the change of social phenomena over time. At the most pragmatic level, this traditional preoccupation with the temporal ordering of things can be explained as a function of "cause." When causal relationships are an issue, social scientists have traditionally resorted to longitudinal research designs. Time series *analysis* is a statistical *method* for interpreting the results of certain longitudinal research designs. When used appropriately, time series analysis brings a powerful inferential logic to bear on questions of social cause. When used inappropriately, however, the relative weaknesses of time series analysis far outweigh its strengths.

Assuming that time series analysis is the appropriate method for addressing a particular research question, the analyst must pay some attention to defining the time series. On the face of it, *a time series is a set of N time-ordered observations of a process*. Each observation should be an interval level measurement of the process and the time separating successive observations should be constant. Minimal violations of these requirements are acceptable. Monthly time series observations, for example, are sometimes separated by 28 days and sometimes by 30 or 31 days. This minimal departure from the ideal presents no real problems for the analysis, however.

By this definition, a time series is a *discrete* data set. Figure 1.2 shows a plotted time series. The observations of this series, the data that is, are the equally spaced symbols (o) strung out along the time dimension. It is often more realistic (and aesthetically pleasing) to connect the symbols with a broken line as we have done in Figure 1.2. The analyst should nevertheless be conscious that the time series is actually a *discrete* data set.

To be sure, the discrete time series may be a measure of some underlying *continuous* process. Stock market time series, for example, are usually reported as daily closing prices. The price of a stock fluctuates more or less continuously throughout the trading day. *Closing* price is the value of the stock at 3:00 P.M. when the New York Stock Exchange "closes" or ends its business day. We might instead choose to record the daily opening or noon price but the principle is still the same. So long as the continuous process is

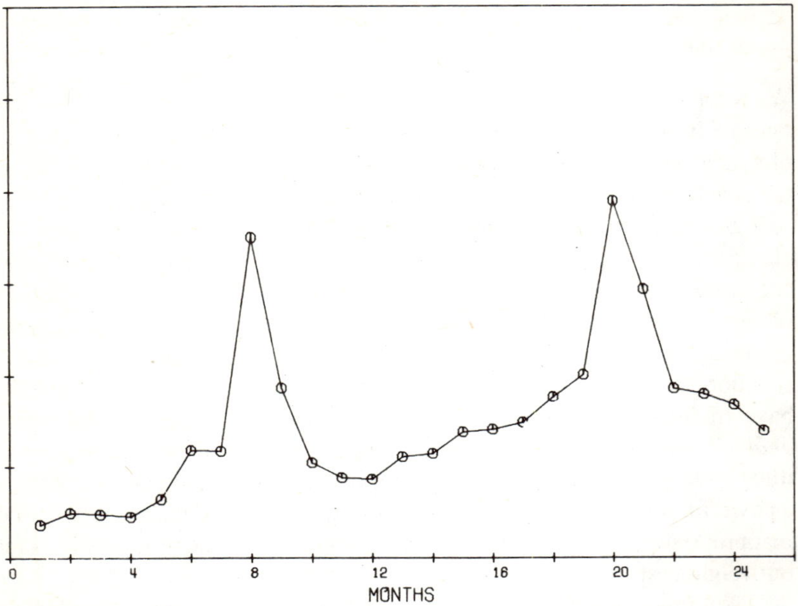

FIGURE 1.2 A Time Series

recorded consistently, a discrete time series is likely to give a good approximation to the continuous process.

Many other time series will approximate processes that are actually discrete. For example, a discrete (but rare) event process can be *aggregated* into a time series. An illustration of this would be a time series of monthly traffic fatalities. Traffic fatalities are discrete, rare events and thus might be best analyzed as Poisson outcomes. If the data are aggregated into monthly totals, however, the resulting time series will ordinarily capture the essence of the rare event process. The average "waiting time" between fatalities in a given month will be roughly proportional to the total number of fatalities in that month.

While there are many other ways in which continuous or discrete social processes can be represented by a time series, the principles of logic are the same. The social scientist is interested in making inferences about the process *underlying* the time series. The time series must thus always give an adequate representation of the true social process. Lacking this quality, inferences based on a time series analysis will be invalid.

But assuming that a time series analysis is the optimal research method, and assuming that an adequate data set is available, some attention must be paid to the limits of logical inference. For all practical purposes, the time series analyst is interested in "predicting the future" of a social process, that is, in measuring the past "change" in a social process to extrapolate that "change" into the future. Extrapolation cannot be accommodated outside a well-defined axiomatic foundation, however: a theory of the process. Discussing the measurement of economic growth, for example, Simon Kuznets has noted that:

> The difficulties in measuring economic growth, supply of empirical data apart, lie precisely in this point: modern economic growth implies major structural changes and correspondingly large modifications in social and institutional conditions under which the greatly increased product per capita is attained. Yet for purposes of measurement, the changing components of the structure must be reduced to a common denominator; otherwise it would be impossible to compare the product of the economy of the United States with that of China, or the product of an advanced country today with its output a century ago before many of the goods and industries that loom so large today were known [1959:15].

Kuznets's point here is that we cannot compare apples with oranges unless we have some theoretically sound dimension on which they are comparable. To measure change in the output of any process, we must work from a set of axioms that allows a comparison of today's social and economic structures with yesterday's social and economic structures. Gross national product, for example, must be defined in such a way that its meaning is not appreciably time-bound.

In many substantive areas, the interpretation of measured change in a time series presents no significant problems. Even when serious problems arise, however, interpretability can be increased through appropriate changes in operational definitions. There is ordinarily a smaller problem interpreting change over a few months or years, for example, than over a few decades or centuries. Interpretability can thus sometimes be increased by narrowing the time frame of the analysis. Of course, there is a trade-off here: Making the time series more interpretable may also make it more trivial. The limitations of time series analysis in this sense must be clearly understood. Time series analysis is generally more appropriate for gauging incremental change than for gauging structural change. In some cases, meaningful indicators of structural change can be found and, in these cases, the analytical power of time series analysis can be brought to bear. But meaningful indicators of structural change are sometimes not available and, in these cases, any time series analysis will be absurd.

It is also ordinarily easier to understand change in relatively concrete time series than in relatively abstract time series. It would be easier to measure change in "work-related accidental fatalities," for example, than in "unemployment." Our definition of the first indicator has changed little over a long period of time—although it has changed. Our definition of unemployment, on the other hand, changes constantly as a result of changing social norms. What would have been an unthinkably high rate of unemployment during one era may be seen as an absurdly low rate of unemployment during another era. While the absolute meaning of an unemployment rate may not change at all over time, the social meaning of unemployment changes constantly.

We may say the same thing of "crime." Social scientists cannot agree among themselves whether rates of crime and unemployment are higher or lower today than they were 50 years ago. The crux of this disagreement can be eliminated by simply conducting analyses of concretely defined time series. But again, to insist upon concrete, objective indicators may sacrifice substantive importance. We are not arguing that the analyst should ignore a time series problem when the available data are not obviously concrete. But the analyst must always be conscious of and acknowledge the limitations of an analysis.

We have intended this short essay to be a *caveat,* warning the reader that the problems of analyzing time series data are relatively unimportant compared to the problems of interpreting analytic results. Without interpretability, there is nothing. More specifically, the reader has been warned (1) to use time series methods only when those methods are appropriate to the research question; (2) to be sure that a time series gives an adequate representation to the underlying process; and (3) to recognize the limits of each analysis.

1.2 An Outline of the Volume

We might have called this section "How To Use Our Book." Here we explain our motives, describing as best as possible what material will be presented in subsequent chapters, in what order this material will be presented, and why. Some readers may become anxious and confused in subsequent chapters, perhaps understanding the material presented, but questioning its relevance to the overall scheme of time series analysis. A careful reading of this outline may reduce anxiety levels, permitting the reader to concentrate more fully on the elements of time series analysis.

ARIMA models and Box-Jenkins time series analysis are not necessarily more difficult a topic than other statistical models and methods used in the social sciences. ARIMA models are nonetheless "different" than most other statistical models and methods in terms of their underlying principles, statistical properties, and applications. One major difference which we have already noted is that ARIMA models are not arbitrarily fit to data, but rather are built empirically. The analyst selects a particular model for a given time series from the general class of ARIMA models. The decision is based on empirically derived characteristics of the series. To make a wise decision, however, the analyst must be aware of the statistical properties of ARIMA models and must be adept at relating these properties to information about the series.

These two concerns, statistical knowledge or understanding and practical application, have determined the orientation of this volume. In each chapter, we first develop the statistical properties of the time series models. We have attempted to do this in a manner that is accurate and thorough and yet lucid to a reader with only an elementary statistical background. When some esoteric or mathematical point must be made, we use footnotes. For all practical purposes, these footnotes may be disregarded in the first reading of a chapter. After developing statistical properties, we apply the time series models to several time series which, in our opinion, are typical of those encountered in the social sciences. Each of these example series has been carefully selected to illustrate a variety of characteristics. Each of these series is listed in an Appendix to this volume and our intention is that the reader will replicate our analyses.

The volume is divided into six chapters, each covering a distinct topic in time series analysis and each building on material developed in preceding chapters. Chapter 2 is the core of the volume and an understanding of the material presented there is *absolutely* essential. Chapters 3, 4, and 5 extend the basic ARIMA model developed in Chapter 2 to a particular application. These chapters may be read out of order, although we do not recommend it. Chapter 6 deals with several practical issues of parameter estimation which will be of most interest to those readers who are about to start a time series research project.

In Chapter 2, we present the basic concepts of univariate Box-Jenkins time series analysis. The univariate ARIMA model is the baseline "building block" which we use in subsequent chapters for impact assessment, forecasting, and causal modeling. Separate and distinct components of the ARIMA model (the autoregressive, integrated, and moving average components) are developed in sequence and then integrated into the general ARIMA model. Once developed, the general ARIMA model is used in an analysis of four example time series.

Chapter 3 presents a general impact assessment model for the analysis of an "interrupted time series quasi-experiment." This model has been widely used to analyze or assess the effects of planned and unplanned interventions on social systems. The impact assessment model consists of a noise model, as developed in Chapter 2, coupled to an impact model. After developing the algebra of a general impact model, we illustrate its use by analyzing several example time series.

In Chapter 4, we develop the use of ARIMA models for forecasting future values of a time series. Our treatment of this topic is intentionally brief. Most of the books written about applied time series analysis are oriented exclusively to forecasting applications and we would have little original thought to add to this body of literature. Readers still may gain some understanding of ARIMA models and algebra from our treatment of forecasting. After presenting the forecast profiles for several time series, we conclude this chapter with a discussion of the uses (both proper and improper, in our opinion) of forecasting in social science research.

Chapter 5 extends the Box-Jenkins approach to multivariate time series analysis. One or more independent variable time series (inputs) may be used to explain the stochastic behavior of a dependent variable time series (output). Multivariate ARIMA time series models are a rather novel concept in social science research although, in our opinion, they have a great potential. Once again, we present the underlying statistical concepts of multivariate ARIMA models, develop a model-building strategy, and conduct several example analyses.

As in many other areas of quantitative social research, Box-Jenkins time series analysis depends upon sophisticated computer software. ARIMA models are nonlinear, for example, so parameters must be estimated with numerical routines. In Chapter 6, we derive the likelihood function, illustrate the solution procedure, and discuss related topics that may affect the analysis. Several available software packages for the analysis of time series data are reviewed and the use of interactive software is discussed. It may seem unusual to relegate a chapter on "estimation" to the the end of the book. Yet most social science graduate students are able to use multiple regression software packages without actually understanding how parameter estimates are generated inside the computer. The same principle holds for Box-Jenkins parameter estimation. While most readers will have no trouble conducting time series analyses without understanding the mechanics of parameter estimation, the reader who intends to go further in this area must at least understand the general principles of nonlinear estimation. For all readers, Chapter 6 is likely to be insightful. For readers who intend to do major work with time series analysis, Chapter 6 is necessary reading.

A Note to the Instructor

A graduate seminar in time series can finish the material presented here in 8 to 12 weeks, depending upon many obvious factors. Our seminars have typically included no more than 12 students drawn from several graduate social science departments. To be admitted to the seminar, students have been required to have a set of time series data and a short research proposal. Grades are based on an article-length research report. The time series data listed in Appendix B, and analyzed in the volume, provide an excellent practicum experience for the seminar. After analyzing these data, however, students must generalize the experience to their own time series data. Of course, the seminar requires access to an appropriate software package, preferably one that is interactive. In Section 6.4, we describe several time series software packages that are available at almost all academic computing centers.

2 Univariate ARIMA Models

If this volume has a single most important chapter, it is this one. Here we develop a general ARIMA model-building strategy for a single time series. In subsequent chapters, we develop methods for applying the univariate ARIMA model to problems of social research. An understanding of these applications will require an understanding of the material developed here.

We have divided this chapter into two distinct parts. The first part (Sections 2.1 to 2.9) deals mainly with abstract issues, especially with the statistical properties of univariate ARIMA time series models. This material (and, indeed, the rest of the volume) presumes a knowledge of fundamental statistical concepts and a familiarity with algebra. In general, the reader should have a working knowledge of the material ordinarily presented in a first-year social statistics course: measures of central tendency, variance, covariance, correlation, expected values, the Normal distribution, ordinary least-squares (OLS) regression, and so forth. A short appendix at the end of this chapter summarizes the rules for applying expectation operators to random variables. The reader who is unfamiliar with the concept of expected values may read this appendix before starting the chapter.

We try to present the material in the first half of this chapter in an illustrative manner that is both intuitively plausible and technically correct. Readers are urged not to rely on our algebra, but to work through each demonstration or derivation. Although this may become tedious at times, an understanding of abstract concepts will open the door to an understanding of the general class of ARIMA models. More important, the derivation of model algebra is in its own right an intellectually challenging and satisfying exercise which should not be missed.

In the second part of this chapter (Sections 2.10 to 2.13), we describe the concrete procedures used to build univariate ARIMA models for given sets of time series data. After developing a general model-building strategy, we apply it step by step to the analysis of four time series, each typical of those encountered in social science research. These analyses illustrate how such problems as nonstationarity, seasonality, outliers, and ambiguous identification information can be handled within the general ARIMA model-building strategy.

As an introduction to the first part of this chapter, we now return to the most crucial of all definitions, *a time series is a set of ordered observations:*

$$Y_1, Y_2, Y_3, \ldots, Y_{t-1}, Y_t, Y_{t+1} \ldots.$$

In cross-sectional analyses, the order of observation is not of any great consequence and may even be undefined. If the analyst is measuring the performance of subjects in an experiment, for example, it usually makes no difference which subject was tested first and which subject was tested last. Indeed, it is often the case that all subjects are tested simultaneously, so no order of testing is defined.

In longitudinal analyses, on the other hand, the order of an observation is crucial. If the analyst is measuring the improvement in performance from test to test, for example, the order of an observation is as important as the observation itself. The order of an observation is conventionally denoted by a subscript. The general observation is written as Y_t, meaning the tth observation of a time series. This implies that the preceding observation is Y_{t-1} and the subsequent observation is Y_{t+1}.

We will make a distinction throughout this chapter between *process* and *realization*. An observed time series is a realization of some underlying stochastic process. In this sense, the relationship between realization and process in time series analysis is analogous to the relationship between *sample* and *population* in cross-sectional analysis.

A related and equally important concept is the *model*. A realization, or time series, is used to build a *model* of the process which generated the series. The procedures used to build this model are broadly referred to as time series *analysis,* which implies that the series is being "picked apart" or decomposed into its components. ARIMA models are built around three process components: the autoregressive, integrated, and moving average components. The first component to be considered is the integrated component which is closely related to the concept of trend.

2.1 Trend and Drift

Our discussion of ARIMA algebra begins with the concept of *trend*. When we think of the component parts of a time series process, we tend not to think of ARIMA structures, but rather of more fundamental, commonsensical component parts. Trend is one such component. Even those of us who have no experience with longitudinal analysis have a fundamental understanding of what a trend is. A trend is motion in a specific direction, usually (to simplify matters) upward or downward. We can say thus that the trend in government during this century has been *away* from the state level and *toward* the federal level. Not surprisingly, our notion of trend in terms of ARIMA structures is almost identical to this commonsense notion.

More specifically, we define trend as *any systematic change in the level of a time series*. While this definition lacks the mathematical rigor we might like, it is the best definition we (and mathematicians) can construct. A time series that is trend*less* can ordinarily be represented by the model

$$Y_t = b_0 + N_t,$$

where the parameter b_0 is interpreted as the *level* of the time series. N_t is a "noise" component or a stochastic process. The most important type of stochastic process in time series analysis is "white noise," which we represent as a_t. Assuming a white noise process, we can rewrite the model for a trendless time series process as

$$Y_t = b_0 + a_t.$$

A white noise process has the statistical property[1]

$$a_t \sim NID(0,\sigma_a^2),$$

that is, white noise consists of a series of random shocks, each distributed Normally and independently about a zero mean with constant variance, σ_a^2. As the mean of white noise is zero, the expected value of a trendless process is:

$$EY_t = b_0 + Ea_t = b_0.$$

The parameter b_0 is the level about which the realized time series fluctuates. As the level of the time series process is constant throughout its course, the time series is trendless with a "flat" appearance.

If a time series process is trendless, the parameter b_0 is the arithmetic mean of the series, estimated as

$$\hat{b}_0 = \bar{Y} = 1/N \sum_{i=1}^{N} Y_i,$$

for a time series of N observations.

Unfortunately, most social science time series processes are not well represented by this simple model. A time series process following a linear trend, for example, requires a model with an extra parameter:

$$Y_t = b_0 + b_1 t + a_t.$$

For this model, the expected value of Y_t is:

$$EY_t = b_0 + b_1 t + Ea_t = b_0 + b_1 t,$$

a regression of Y_t on the time series t ($t = 0, 1, \ldots, N$). The level of the time series process is thus expected to change systematically throughout its course. Starting at $t = 0$,

$$\begin{aligned} EY_0 &= b_0 \\ EY_1 &= b_0 + b_1 \\ EY_2 &= b_0 + 2b_1 \\ &\vdots \\ EY_N &= b_0 + Nb_1. \end{aligned}$$

The parameter b_0 is the *intercept* of the model, that is, the expected level of the process when $t = 0$. The parameter b_1 is the *slope* of the model, that is, the expected change in level from one observation to the next.

Unfortunately, few social science time series processes appear to be trendless and this presents a problem for the time series analyst. Trend must be removed or modeled. One common (but almost always inappropriate) method of detrending a time series is to use a linear regression model for the trend. With this method, the analyst defines the order of each observation ($t = 1, 2, \ldots, N$) as an independent variable. Then, using the time series itself as a dependent variable, the parameters b_0 and b_1 are estimated with OLS formulae. For a linear trend, this method yields the detrended series:

$$\hat{y}_t = Y_t - \hat{Y}_t.$$

The detrended series, $\hat{y}_t$, is then analyzed.

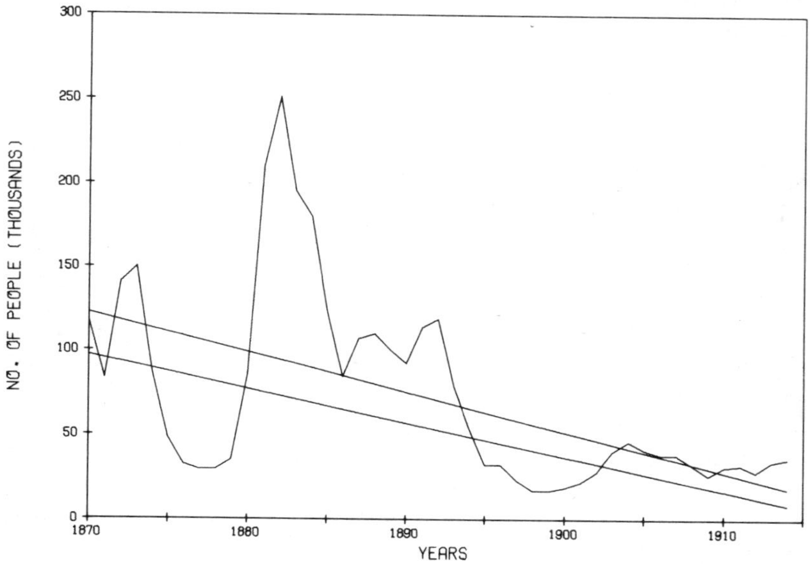

FIGURE 2.1(a) German Immigration to the United States (in Thousands), 1870–1914

As there are a number of problems with this method, it is not generally recommended. One major problem with OLS trend models is that the parameters b_0 and b_1 cannot be estimated with any accuracy. In Figure 2.1(a), for example, we show a time series of total immigration (in thousands) from Germany to the United States during 1870–1914 as reported by Fried (1969). Does this series follow a trend? Reasonable people might agree that total immigration decreased steadily during this period. The OLS trend line for this series is:

$$\hat{Y}_t = 124.78 - 2.37t.$$

But note the abnormal spike in this series starting in 1881. If we ignore this spike, the OLS trend line becomes:

$$\hat{Y}_t = 98,82 - 2.01t.$$

We have superimposed these trend lines over the time series in Figure 2.1(a) to illustrate the estimation problem: *OLS trend estimates are sensitive to outliers*.

The underlying problem is that the OLS linear regression model of trend depends upon the sum of squares function which, for the German immigration time series, is:

$$\sum_{t=1}^{N}(Y_t - \hat{Y}_t)^2 = \sum_{t=1}^{N}[b_0 - \hat{b}_0 + (b_1 - \hat{b}_1)t]^2.$$

OLS estimates of b_0 and b_1 are derived by minimizing this sum of squares function. As the independent variable, t, increases monotonically, however, some of the time series observations are more important than others in the sum of squares function. As a rule, the first and last observations of the time series (Y_1 and Y_N) usually make the greatest contribution to the sum of squares. In an extreme case, the OLS estimates of b_0 and b_1 are derived so that the OLS trend line passes through Y_1 and Y_N regardless of how well the middle observations ($Y_2, Y_3, \ldots, Y_{N-1}$) are fit.

Intuitively, we would like our estimate of trend to be *dynamic*. In practical terms, this means, first, that each observation of the time series should have more or less the same influence in determining the trend line and, second, that the trend line should fit the start, middle, and end of the series equally well. OLS trend estimates are *static*, not dynamic. An OLS estimate of trend is largely determined by the first and last observations of the series.

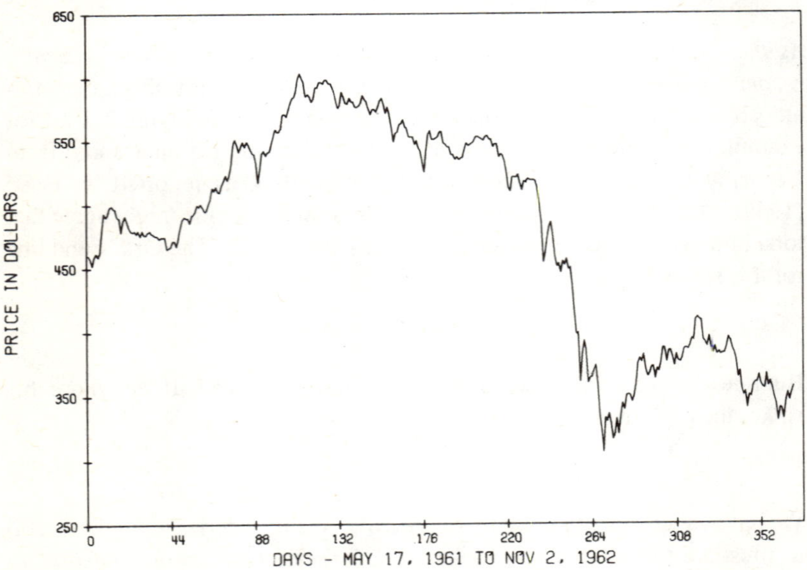

FIGURE 2.1(b) Closing Price of IBM Common Stock (Series B from Box and Jenkins)

But there is a more basic problem with OLS detrending methods. Figure 2.1(b) shows what is perhaps the most famous of all time series: "Series B," from Box and Jenkins (1976), 369 consecutive daily closing prices of IBM common stock. Does this series follow a trend? Examining only the first half of this series, reasonable people would see an upward trend. Examining only the second half, the same people would see a downward trend. Brown (1962) demonstrated that this time series could be detrended adequately with a quadratic polynomial of the sort

$$Y_t = b_0 + b_1 t + b_2 t^2.$$

Commenting on this procedure, however, Box and Jenkins note that:

> One of the deficiencies in the analysis of time series in the past has been the confusion between *fitting* a series and forecasting it. For example, suppose that a time series has shown a tendency to increase over a particular period. . . . A common method of analysis is to decompose the series arbitrarily into three components—a "trend," a "seasonal component," and a "random component." The trend might be fitted by a polynomial and the seasonal component by a Fourier series. . . . Such methods can give extremely misleading results. . . . Now, it is true that short lengths of Series B do look as if they might be fitted by quadratic curves. This simply reflects the fact that a sum of random deviates can sometimes have this appearance [1976: 300].

In fact, the IBM series does *not* follow a trend. For want of a better word, we say that this series "drifts," first upward and then downward.

The real difference between trend and drift is that trend is *deterministic* behavior while drift is *stochastic* behavior. Deterministic behavior can be expressed as a systematic or fixed function of time. Stochastic behavior, on the other hand, can be expressed only as the outcome of a process operating through time. Whereas future values of a deterministic process are fixed by the definition of the function, future values of a stochastic process are free to vary in a probabilistic manner. The real problem with regard to time series analysis is that, given an observed realization of finite length, it is extremely difficult to tell whether a progressive change in the level of the series is due to deterministic trend or to stochastic drift. If we model it deterministically, when in fact it is stochastic, then we may make disastrous errors in the prediction of future values or in assessing the magnitude of an exogenous intervention on the series behavior.

We do not mean to imply that trend does not exist. Indeed, many social science time series processes increase or decrease systematically. Demographic time series processes, for example, may trend due to population growth; economic time series processes may trend due to inflation; and time series measures of human performance may trend due to learning. In each case, a few known causal forces underlie the trend. In Chapter 5, we will develop bivariate and multivariate methods that can account for trends of this sort. Our point here is that a time series can drift upward or downward for extremely long periods of time due only to random forces. Unless there is a strong theoretical basis (or empirical evidence) for assuming that a time series process trends deterministically, there are great advantages to be gained by modeling it stochastically.

As Box and Jenkins note, the issue of trend versus drift is really the issue of *fitting* versus *modeling* a time series. OLS detrending methods always require an assumption that change in the realized process is due to the constant, deterministic effects of a few causal forces. If the analyst can make this assumption, then it will be best to include these exogenous forces in the time series model directly rather than attempting to exclude them indirectly through detrending.

The alternative to detrending a time series is to build a dynamic model which accounts for what appears to be (or may actually be) trend. In the next section, we develop difference equation models for drift and trend. A difference equation model accounts for both trend and drift without requiring an a priori distinction between the two. Unlike OLS detrending parameters, the parameters of a difference equation are easily estimated. But most important of all, difference equation models are *dynamic* models. Trend or drift is determined by the entire set of observations, not by the first or last observations.

2.2 The Random Walk and Other Integrated Processes

The IBM stock series can be thought of as the result of a random walk. A random walk process is a stochastic process wherein successive random shocks accumulate or *integrate* over time and thus a random walk is called an *integrated* process.

To illustrate the random walk, suppose a gambler bets on the flip of a fair coin. When a coin flip results in heads, the gambler wins one dollar; for tails the gambler loses one dollar. We can then define

$$\begin{aligned} a_t &= +\$1 \text{ if the } t^{th} \text{ coin flip results in a head} \\ &= -\$1 \text{ if the } t^{th} \text{ coin flip results in a tail.} \end{aligned}$$

Because the flip of a fair coin results in heads as often as tails, the series of coin flips is a series of binomial experiments. This means that

$$P(\text{heads}) = P(\text{tails}) = \tfrac{1}{2}$$

$$Ea_t = \tfrac{1}{2}(+\$1) + \tfrac{1}{2}(-\$1) = 0,$$

in the long run, the gambler expects to break even, winning exactly as much money as he or she has lost. Also,

$$Ea_t^2 = (½)(½) = ¼$$

$$Ea_t a_{t+k} = 0.$$

The last equation expresses the notion that successive flips of the coin are expected to have independent outcomes, that is, a_t and a_{t+k} are not related. The outcome of successive coin flips thus approximates white noise. Now define the total amount of money won or lost by the gambler after the tth coin flip as

Y_t = total money won (or lost) at the end of t coin flips.

With this definition, we have a random walk process. At the end of the first coin flip,

$$Y_1 = a_1,$$

that is, the total money won (or lost) consists of the money won (or lost) on the first coin flip. At the end of the second coin flip,

$$Y_2 = a_1 + a_2,$$

that is, the total money won (or lost) consists of the money won (or lost) on the first flip plus the money won (or lost) on the second flip. At the end of the third coin flip,

$$Y_3 = a_1 + a_2 + a_3,$$

and at the end of the fourth coin flip,

$$Y_4 = a_1 + a_2 + a_3 + a_4,$$

and at the end of the tth coin flip,

$$Y_t = a_1 + a_2 + \ldots + a_{t-1} + a_t,$$

which is the random walk. Y_t is the sum of the money won (or lost) on t flips of the coin. As each random shock is *expected* to be zero (even though no single random shock can be zero), the sum of t random shocks is also expected to be zero, that is,

$$EY_t = Ea_1 + Ea_2 + \ldots + Ea_{t-1} + Ea_t = 0.$$

If we had to guess the value of the gambler's holding at the end of t flips, we would guess that the gambler had broken even. This guess would be correct in one sense but incorrect (or at least misleading) in another sense. While the expected value of Y_t is zero, the process will almost always drift high above or below its expected value. In fact, it is quite unlikely that the gambler will break even by the tth flip.

This random walk gambling example illustrates a major difference between cross-sectional and longitudinal stochastic processes. If 1000 people each flip a coin simultaneously, we expect to observe 500 heads and 500 tails. It is unlikely that we would observe this exact outcome, of course, but the observed frequencies will ordinarily be quite close to the expected. On the other hand, if one person flips a coin 1000 times, the short-run gain (Y_t) is *expected* to be zero also, that is, we would expect 500 heads and 500 tails. However, the observed frequencies are quite likely to be much different than the expected. Heads or tails is likely to be in the lead throughout most of the 1000 flips. If two gamblers each have only a finite amount of cash, say $25, one of the gamblers is likely to win all of the other's money long before the 1000th flip. This phenomenon, "gambler's ruin," is a surprising property of the random walk process.[2]

In Figure 2.2(a), we show the results of a coin flip gamble. This result is typical of a random walk process. The random variate (Y_t) makes wide swings away from its expected level. *It drifts,* and if we had only a short

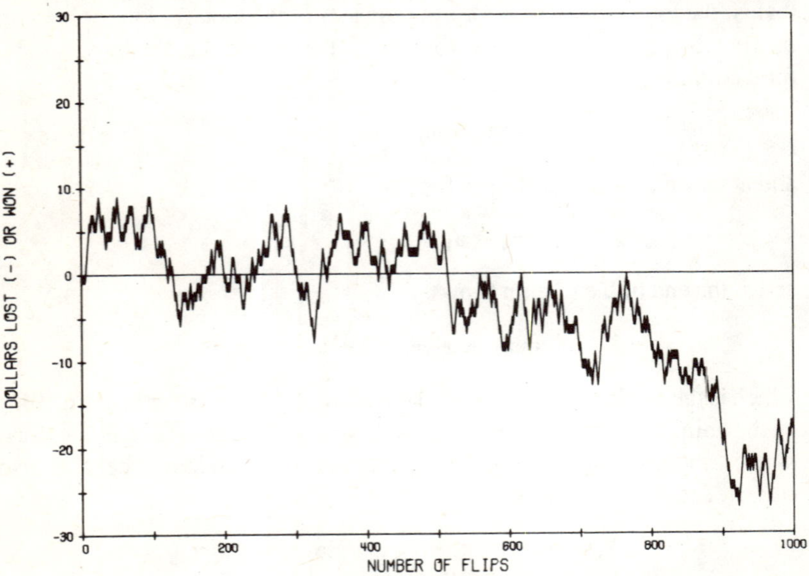

FIGURE 2.2(a) Coin Flip Gamble: A Random Walk

realization of the process, we might conclude that the process follows a trend. The probabilistic forces underlying this drift have a straightforward explanation. If positive and negative shocks were roughly equal in the short run, we would see no drift whatsoever. Relatively long runs of positive or negative shocks can occur, however. In coin flip gambling, the probability that a run of eight heads will be realized is:

$$\text{P(eight consecutive heads)} = (\tfrac{1}{2})^8 = .0039063,$$

which is a relatively slight probability. Yet when a run of eight heads does occur, as it inevitably will, the Y_t process drifts far above its expected value of zero. The level of the process stays at this zenith until a run of eight tails drives it back down. Of course, we would expect to wait a long time before realizing a counterbalancing run of eight tails.

Processes that can be thought of as random walks are frequently encountered in the social sciences. A major difference between the random walks we encounter in social science processes, however, and the random walk illustrated by the coin flip gamble is the *size* of each random shock. In coin flip gambling, random shocks are equal in absolute value because a shock must be ±\$1. In the typical social science process, on the other hand, random shocks vary in size as well as in sign. There are large shocks, medium-size shocks, and small shocks in the typical social science random walk.

As an example, consider this simple model for the population of a small town. Let

Y_t = the town population at the end of the tth month

a_t = number of births minus the number of deaths in the tth month.

Then, for some "starting point," Y_0, we have the random walk process:

$$\begin{aligned}
&Y_0 = \text{the town population at the end of month zero}\\
&Y_1 = Y_0 + a_1\\
&Y_2 = Y_0 + a_1 + a_2\\
&Y_3 = Y_0 + a_1 + a_2 + a_3\\
&\vdots\\
&Y_t = Y_0 + a_1 + a_2 + \ldots + a_{t-1} + a_t,
\end{aligned}$$

which describes how the population of this small town changes from month to month.

If birth rates and death rates are equal in the long run, then the expected value of each random shock is zero. Actually, the number of births and the number of deaths in a month will seldom be exactly equal, so most random shocks will not be exactly zero. A plausible assumption is that the random shocks will be normally and independently distributed about the mean zero with constant variance, that is,

$$a_t \sim \text{NID}(0, \sigma_a^2).$$

So the random shocks are a white noise process. If this model of the town population is realistic, the expected population after *t* months is[3]:

$$\begin{aligned} EY_t &= Y_0 + Ea_1 + Ea_2 + \ldots + Ea_{t-1} + Ea_t, \\ &= Y_0 \end{aligned}$$

which was the population at the "starting point." However, we would expect to see the town population drift upward or downward for long periods of time. This feature of the realized time series is only drift, however, not trend.

It might be valuable at this point to explain exactly what a random shock is. In the town population model, random shocks are the thousands of variables which "cause" birth and death. These many factors vary across time and interact in complex and complicated patterns which we call "random." None of these factors alone could explain the birth or death rates which make up an observation of the process. But jointly, the effects of these many factors are aptly described as white noise.

Because a random walk observation is the sum of all past random shocks, there is a rather simple way to model the random walk. The series is simply *differenced*. Differencing a time series amounts to subtracting the first observation from the second, the second from the third, and so forth:

$$\begin{aligned} z_1 &= Y_1 - Y_0 \\ &= (Y_0 + a_1) - Y_0 = a_1 \\ z_2 &= Y_2 - Y_1 \\ &= (Y_0 + a_1 + a_2) - (Y_0 + a_1) = a_2 \\ z_3 &= Y_3 - Y_2 \\ &= (Y_0 + a_1 + a_2 + a_3) - (Y_0 + a_1 + a_2) = a_3 \\ &\vdots \\ z_t &= Y_t - Y_{t-1} \\ &= (Y_0 + a_1 + a_2 + \ldots + a_{t-1} + a_t) \\ &\qquad - (Y_0 + a_1 + a_2 + \ldots + a_{t-2} + a_{t-1}) = a_t. \end{aligned}$$

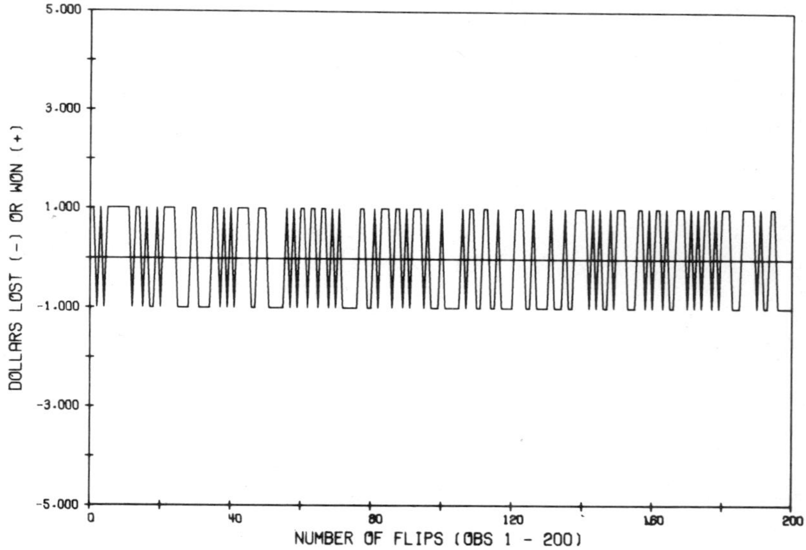

FIGURE 2.2(b) Coin Flip Gamble: A Random Walk Differenced

In the remainder of the volume, except where explicitly noted, we will represent the differenced Y_t series by z_t.

By differencing the random walk process, we obtain a time series whose observations are the contemporaneous random shocks, $a_1, a_2, \ldots, a_t$. In other words, differencing transforms the random walk into a white noise process. In Figure 2.2(b), we show the coin flip gambling series after it has been differenced. The differenced series no longer drifts away from its expected value, but instead fluctuates about its mean level.

A random walk is an example of an *integrated* process in which "integration" means "addition." If a time series is the realization of an integrated process, it can be modeled by simple differencing. The random walk, for example, is well represented by an ARIMA (0,1,0) model where d = 1 is the number of differences required to make the series stationary.

So far, the random shocks or white noise process has been assumed to have a zero mean. Suppose now that the white noise process has a nonzero mean. In the coin flip gambling example, this implies that heads and tails have different payoffs. In the town population example, this implies that birth and death rates are not equal in the long run. Representing the nonzero level of the white noise process by the constant Θ_0, then,

$$Ea_t = \Theta_0.$$

Successive realizations of the integrated process are now expected to be:

$$
\begin{aligned}
Y_0 &= Y_0 \\
EY_1 &= Y_0 + Ea_1 = Y_0 + \Theta_0 \\
EY_2 &= Y_0 + Ea_1 + Ea_2 = Y_0 + 2\Theta_0 \\
EY_3 &= Y_0 + Ea_1 + Ea_2 + Ea_3 = Y_0 + 3\Theta_0 \\
&\vdots \\
EY_t &= Y_0 + Ea_1 + \ldots + Ea_t = Y_0 + t\Theta_0.
\end{aligned}
$$

So when the white noise process has a nonzero mean, a random walk follows a linear trend. What happens when the Y_t process is differenced?

$$
\begin{aligned}
Ez_1 &= E(Y_1 - Y_0) = Ea_1 = \Theta_0 \\
Ez_2 &= E(Y_2 - Y_1) = Ea_2 = \Theta_0 \\
Ez_3 &= E(Y_3 - Y_2) = Ea_3 = \Theta_0 \\
&\vdots \\
Ez_t &= E(Y_t - Y_{t-1}) = Ea_t = \Theta_0.
\end{aligned}
$$

The differenced series, z_t, is expected to equal a nonzero constant, Θ_0. This leads to the difference equation model of linear trend:

$$
\begin{aligned}
Y_t - Y_{t-1} &= \Theta_0 \\
Y_t &= Y_{t-1} + \Theta_0.
\end{aligned}
$$

To illustrate the difference equation model of linear trend, consider the sequence of integers:

$$1, 2, 3, 4, \ldots, t.$$

This sequence has a perfect linear trend which we can represent as

$$Y_t = Y_0 + \Theta_0 t,$$

where $Y_0 = 0$ and $\Theta_0 = 1$. But if we difference this sequence,

$$
\begin{array}{llll}
2 - 1, & 3 - 2, & 4 - 3, & \ldots, t - (t-1) \\
1, & 1, & 1, & \ldots, 1
\end{array}
$$

the differenced series is equal to the constant, $\Theta_0 = 1$. We may thus write a difference equation model for this sequence of integers as

$$
\begin{aligned}
Y_t - Y_{t-1} &= 1 \\
Y_t &= Y_{t-1} + 1.
\end{aligned}
$$

It might be useful here to draw an analogy between the OLS trend equations which we discussed in Section 2.1 and difference equation models of trend. In the OLS model, we solved the equation

$$Y_t = \hat{b}_0 + \hat{b}_1 t + a_t,$$

using t as an independent variable. The parameter b_1 is interpreted as the slope or linear trend of the Y_t process. As noted, it is practically impossible to derive a satisfactory estimate of b_1. In the difference equation model, on the other hand, we use the value of Y_{t-1} (rather than t) as an independent variable. The constant of the difference equation, Θ_0, is interpreted in the same way that b_1 of the OLS equation is interpreted: as the slope or linear trend of the process. But while the OLS trend equation

$$\hat{Y}_t = \hat{b}_0 + \hat{b}_1 t$$

and the difference equation

$$\hat{Y}_t = Y_{t-1} + \hat{\Theta}_0,$$

describe exactly the same deterministic trend (when $\hat{b}_1 = \hat{\Theta}_0$, that is), there a major *practical* difference between these two equations.[4] There is no satisfactory method of estimating the parameter b_1 in the OLS trend equation but the analogous parameter of the difference equation, Θ_0, is easily estimated as

$$\hat{\Theta}_0 = \bar{z} = 1/N \sum_{t=1}^{N} z_t,$$

that is, Θ_0, is estimated as the mean of the differenced series. Beyond this practical point, there are substantive issues which make the difference equation the preferred formulation. We will address these issues at a later point in the chapter.

A time series process that does not require differencing (because it neither drifts nor trends) is said to be *stationary in the homogeneous sense*. A time series process that *does* require differencing (because it drifts or trends) is nonstationary in the homogeneous sense. As noted, a time series that follows a random walk or that follows a linear trend may be best represented by an ARIMA (0,1,0) model written as

$$Y_t = Y_{t-1} + \Theta_0 + a_t.$$

The sense of this ARIMA (0,1,0) model is quite simple. *The best prediction of the current time series observation* (Y_t) *comes from the preceding observation* (Y_{t-1}) *and a constant*. To distinguish between drift and trend, the analyst may examine the estimated value of Θ_0 with the null hypothesis:

$$H_o: \hat{\Theta}_0 = 0.$$

The hypothesis may be tested with a t statistic. If $\hat{\Theta}_0$ is not statistically different from zero, the analyst must conclude that the process is drifting, not trending.

In the general case, an ARIMA (0,d,0) model implies that a time series is white noise after being differenced d times. There are two types of integrated processes which will be well represented by ARIMA (0,d,0) models, d^{th}-order random walks and time series processes with d^{th}-order polynomial trends. A second-order random walk, for example, would be:

$$Y_t = a_t + 2a_{t-1} + 3a_{t-2} + 4a_{t-3} + \ldots.$$

And differencing this process,

$$\begin{aligned} Y_t - Y_{t-1} &= a_t + 2a_{t-1} + 3a_{t-2} + 4a_{t-3} + \ldots \\ &\quad - a_{t-1} - 2a_{t-2} - 3a_{t-3} - \ldots \\ &= a_t + a_{t-1} + a_{t-2} + a_{t-3} + \ldots. \end{aligned}$$

And differencing this process,

$$\begin{aligned} (Y_t - Y_{t-1}) - (Y_{t-1} - Y_{t-2}) &= a_t + a_{t-1} + a_{t-2} + a_{t-3} + \ldots \\ &\quad - a_{t-1} - a_{t-2} - a_{t-3} - \ldots \\ &= a_t. \end{aligned}$$

Thus, a second-order random walk will be well represented by an ARIMA (0,2,0) model. To demonstrate how a dth-order polynomial trend can be accommodated by an ARIMA (0,2,0) model, consider the sequence of squared integers

$$1, 4, 9, 16, 25, \ldots, t^2$$

If we difference this sequence,

$$\begin{array}{l} 4-1,\ 9-4,\ 16-9,\ 25-16,\ \ldots,\ t^2 - (t-1)^2 \\ 3, \quad 5, \quad 7, \quad 9, \quad \ldots, 2t-1 \end{array}$$

and if we difference it again,

$$\begin{array}{l} 5-3,\ 7-5,\ 9-7,\ \ldots,\ (2t-1) - (2t-3) \\ 2, \quad 2, \quad 2, \quad \ldots, 2 \end{array}$$

we obtain a sequence of constants. Again, to distinguish d^{th}-order drift from d^{th}-order trend, the analyst need only test the statistical significance of $\hat{\Theta}_0$. In the social sciences, a time series will ordinarily have to be differenced only once. While ARIMA (0,1,0) processes are quite common, higher order random walks and higher order polynomial trends are rare.

We will close this discussion of integrated processes by applying an input-output analogy to the ARIMA (0,1,0) process. White noise is conceptualized as the "driving force" of a time series process. We may view an ARIMA (0,1,0) model as a "black box" in which the *input* is white noise and the *output* is a time series. For processes that drift or trend, the ARIMA (0,1,0) black box *integrates* random shocks. Once a shock enters the black box, it remains inside, influencing all future outputs. We represent this as

$$a_t \dashrightarrow \boxed{\sum_{i=1}^{\infty} a_{t-i}} \dashrightarrow Y_t.$$

All future outputs of the ARIMA (0,1,0) black box will contain the random shock a_t as well as all prior random shocks back into the infinitely distant past. To unlock this black box, we simply difference the time series.

2.3 The Backward Shift Operator

At this point, we introduce the backward shift operator, B, such that

$$B(Y_t) = Y_{t-1}.$$

This expression does not mean "B multiplies Y_t," but rather, means that "B operates on Y_t to shift it backward one point in time." B is thus similar to other logical operators such as the derivative or integral operators in calculus or the natural logarithm operator in algebra.

The properties of the backward shift operator are:

$$B^n(Y_t) = Y_{t-n}$$

and

$$B^nB^m(Y_t) = B^{n+m}(Y_t) = Y_{t-n-m}.$$

The operator obeys all the laws of exponents that we routinely use in polynomial algebra. We will make immediate use of the backward shift operator to difference a time series. For this purpose, we use the operator expression

$$(1 - B)Y_t = (1)Y_t - (B)Y_t = Y_t - Y_{t-1} = z_t.$$

Second differencing is accomplished by

$$(1 - B)^2 Y_t = (1 - 2B + B^2)Y_t = Y_t - 2Y_{t-1} + Y_{t-2},$$

which is the same result we obtain by differencing a time series and then differencing it again. To demonstrate this identity, we difference the time series, obtaining

$$\begin{aligned} z_1 &= Y_1 - Y_0 \\ z_2 &= Y_2 - Y_1 \\ &\vdots \\ z_t &= Y_t - Y_{t-1}. \end{aligned}$$

We then difference this series, obtaining

$$\begin{aligned} z_1^* &= z_1 - z_0 \\ z_2^* &= z_2 - z_1 = (Y_2 - Y_1) - (Y_1 - Y_0) \\ z_3^* &= z_3 - z_2 = (Y_3 - Y_2) - (Y_2 - Y_1) \\ &\vdots \\ z_t^* &= z_t - z_{t-1} = (Y_t - Y_{t-1}) - (Y_{t-1} - Y_{t-2}). \end{aligned}$$

We see that the general term z_t^* is:

$$\begin{aligned} z_t^* &= (Y_t - Y_{t-1}) - (Y_{t-1} - Y_{t-2}) \\ &= Y_t - Y_{t-1} - Y_{t-1} + Y_{t-2} = Y_t - 2Y_{t-1} + Y_{t-2} \\ &= (1 - 2B + B^2)Y_t \\ &= (1 - B)^2 Y_t, \end{aligned}$$

which is the identity we wanted to demonstrate. In the general case, to difference a series d times we apply a dth-order differencing operator, $(1 - B)^d$, to the series.

Another useful property of the backward shift operator is *invertibility*. The property of invertibility allows us to move forward and backward in time by applying the operator and its inverse to a time series. We see an analogous relationship between the natural logarithm operator, Ln, and exponentiation. For example,

$$\mathrm{Ln}(K) = x$$

then

$$e^x = K.$$

We will be using the natural logarithm operator at a later point. For now, we introduce it only to illustrate the principle of an inverse operator. The inverse of the backward shift operator, B, is the *forward* shift operator, F. The forward shift operator has the same properties as the backward shift operator, that is,

$$F^n(Y_t) = Y_{t+n}$$

and

$$F^nF^m(Y_t) = F^{n+m}(Y_t) = Y_{t+n+m}.$$

The inverse relationship between B and F is defined by

$$(B)(F) = (F)(B) = 1,$$

$$(B)(F)\,Y_t = (B)Y_{t+1} = Y_t,$$

and

$$B^nF^mY_t = F^{m-n}Y_t = Y_{t+m-n}.$$

In other words, "F undoes what B has done."

In the remainder of this volume, we will use only the backward shift operator, B. To denote the forward shift operator, F, we will use B^{-1}, and thus,

$$B^{-1}B = 1.$$

When the backward shift operator is used to difference a time series, we denote the inverse operation by $(1 - B)^{-1}$, and thus

$$(1 - B)Y_t = z_t$$

$$(1 - B)^{-1}z_t = Y_t,$$

so

$$(1 - B)^{-1}(1 - B)Y_t = Y_t.$$

In practice, the inverse of the differencing operation implies division,

$$(1 - B)^{-1}z_t = z_t/(1-B)$$

and division by the backward shift operator is a difficult concept to grasp. The inverse operator can be evaluated with a Taylor series expansion, however, as the infinite series

$$(1 - B)^{-1} = (1 + B + B^2 + \ldots + B^{n-1} + B^n + \ldots).$$

We will not go into the derivation of this identity but instead direct the interested reader to any introductory calculus text. However, we will demonstrate the identity by showing that the differencing operator, (1 − B), and the infinite series are inverses. First, differencing a Y_t process, we obtain

$$(1 - B)Y_t = Y_t - Y_{t-1} = z_t.$$

Then, applying the inverse operator to the z_t process, we obtain

$$\begin{aligned}(1 - B)^{-1}z_t &= z_t(1 + B + B^2 + \ldots B^{n-1} + B^n + \ldots) \\ &= z_t + z_{t-1} + z_{t-2} + \ldots + z_{t-n+1} + z_{t-n} + \ldots \\ &= (Y_t - Y_{t-1}) + (Y_{t-1} - Y_{t-2}) + (Y_{t-2} - Y_{t-3}) + \ldots \\ &\qquad + (Y_{t-n+1} - Y_{t-n}) + (Y_{t-n} - Y_{t-n-1}) + \ldots \\ &= Y_t + (Y_{t-1} - Y_{t-1}) + (Y_{t-2} - Y_{t-2}) + \ldots \\ &\qquad + (Y_{t-n} - Y_{t-n}) + \ldots \\ &= Y_t,\end{aligned}$$

which demonstrates the identity of $(1 - B)^{-1}$ and the infinite series.

In the following sections of this chapter and in the following chapters, we will routinely use the backward shift operator to describe time series models. The operator greatly simplifies our algebra and, indeed, some of the higher order ARIMA models cannot be written economically without the operator convention.

2.4 Variance Stationarity[5]

As noted in Section 2.2, ARIMA models require a time series process to be stationary in the homogeneous sense. While homogeneous sense stationarity is a *necessary* condition of an ARIMA model, however, it is not a *sufficient* condition. Before we can properly represent a time series with an ARIMA model, the time series must also have a stationary *variance*.

Our working definition of homogeneous sense stationarity is based on the process *level*. If a time series process is stationary in the homogeneous sense, then that process has a single constant level throughout its course, that is,

$$EY_t = \Theta_0$$

for all t. By this working definition, if a process drifts or trends, it is *not* stationary in the homogeneous sense. This presents no real problem, however, because when the process is differenced,

$$z_t = (1 - B)^d Y_t$$

and

$$Ez_t = \Theta_0$$

for all t. As the differenced process has a single constant level throughout its course, it is stationary in the homogeneous sense. From this working definition, we might say that a process that is stationary in the homogeneous sense is *stationary in its level*. Finite realizations of such processes appear "flat" or trendless. If we divide the series into two segments of equal length, the first segment will have the same level as the second segment.

A process that is stationary in the homogeneous sense (or one that has been made so by differencing) need not be stationary in its *variance*, however. A process that is stationary in variance will have a single constant variance throughout its course, that is,

$$E(Y_t - \Theta_0)^2 = \sigma_a^2$$

for all t or

$$E(z_t - \Theta_0)^2 = \sigma_a^2$$

for all t. Because the expression for variance is based on the process level, Θ_0, it follows that any process that is stationary in variance is also stationary in level (or has been made so by differencing). The converse is not true, however. A process that is stationary in level need not be stationary in variance.

In Figure 2.4(a), we show a time series of nonfatal disabling mine injuries for the United States during the period 1930–1977. Nonfatal disabling injuries decreased systematically during this period, trending downward. We can guess (correctly) that this series is nonstationary in the homogeneous sense and must be differenced. In Figure 2.4(b), we show the first-differenced time series. Differencing has effectively detrended the series, that is, has made the series stationary in the homogeneous sense. *Note, however, that in both Figures 2.4(a) and Figures 2.4(b), the process variance decreases steadily throughout the length of the series.* Year-to-year fluctuations are much larger in the first half of the series than in the second half. We can guess (correctly) that this series, even after differencing, is not stationary in its variance.

If our experiences are typical, most of the time series that social scientists will be interested in are either stationary or else nonstationary only in the

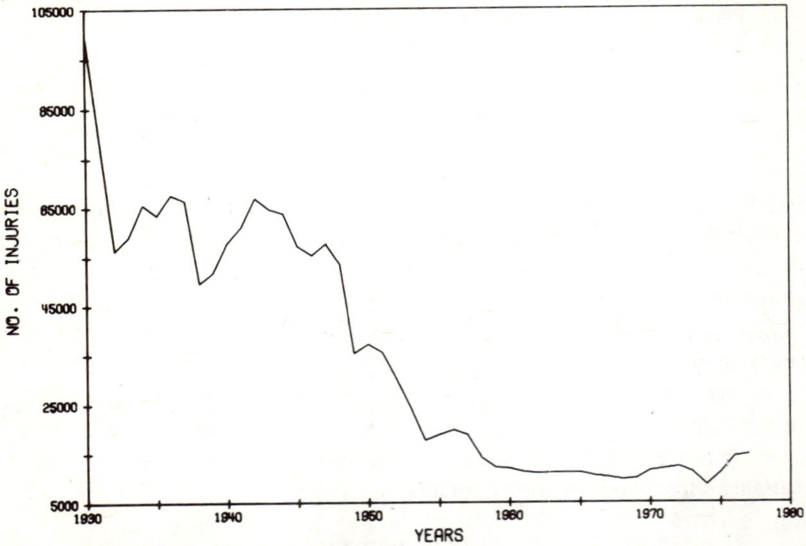

FIGURE 2.4(a) Nonfatal Disabling Mine Injuries, 1930–1978

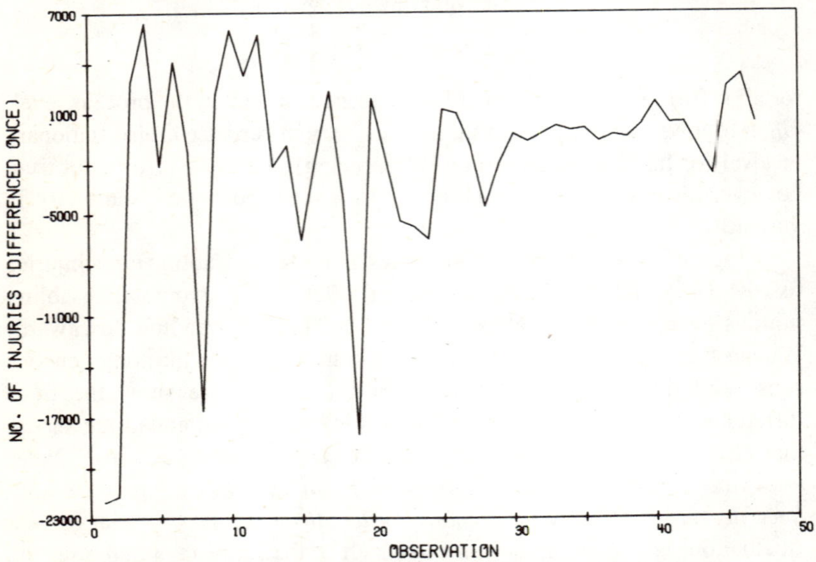

FIGURE 2.4(b) Nonfatal Disabling Mine Injuries, Differenced

homogeneous sense. In practice, then, the analyst will find that most time series can be made stationary in the larger sense (in level and variance) by differencing. Time series of the sort shown in Figure 2.4(a) are not totally uncommon, however. Many social processes have naturally defined "floors" which constrain the stochastic behavior of the process. As the process approaches its floor, process variance is constrained. In the case of the mine injury series, process variance is roughly proportional to the process level, a relationship which we express as

$$EY_t = \Theta_0 t$$

and

$$\sigma_a^2 \propto \Theta_0 t.$$

What this means simply is that, first, the level of the process decreases from observation to observation by the quantity Θ_0. Second, as the level decreases, process variance decreases proportionally.

Fortunately, there is a rather simple transformation which may be applied to such processes to make them stationary in the larger sense. In Figure 2.4(c), we show a time series of natural logarithms of the nonfatal disabling injuries. The log-transformed series still follows a downward trend, that is, is still nonstationary in the homogeneous sense. However, log-transformation has made the year-to-year fluctuations roughly the same in both halves of the series. In Figure 2.4(d), we show the first-differenced log-

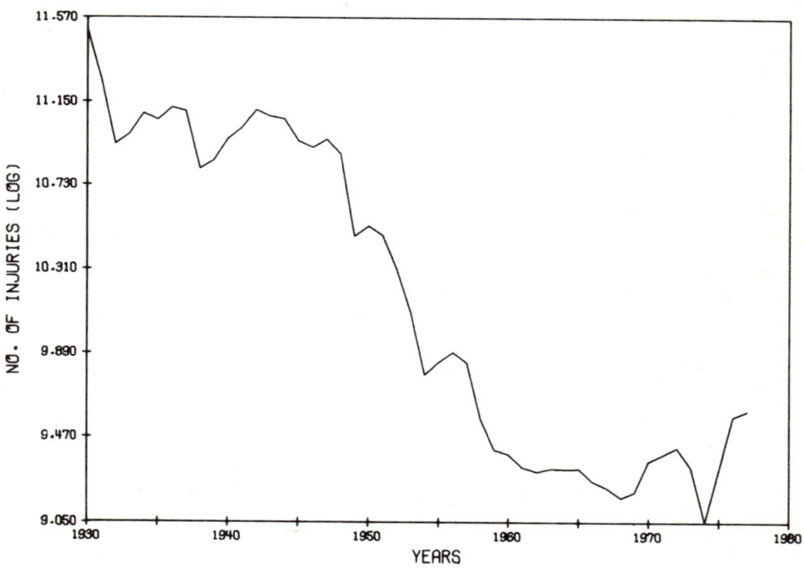

FIGURE 2.4(c) Nonfatal Disabling Mine Injuries, Logged

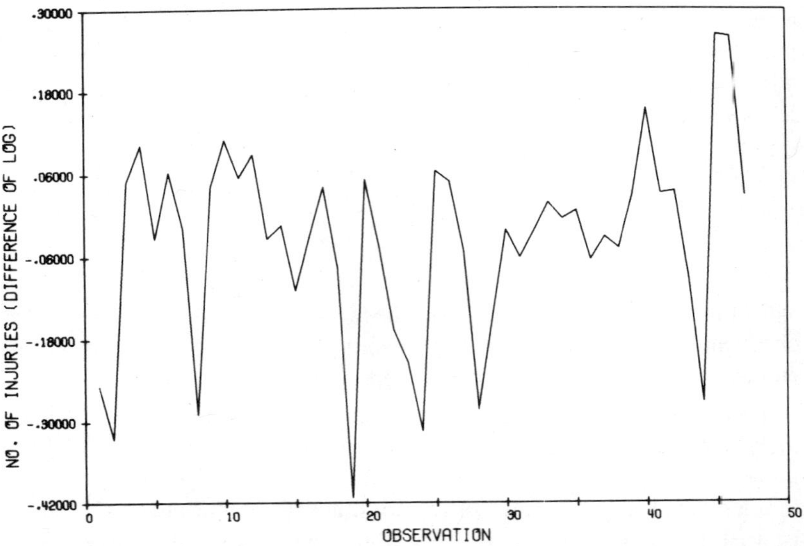

FIGURE 2.4(d) Nonfatal Disabling Mine Injuries, Logged and Differenced

transformed series. After differencing, the series has a single level and single variance throughout its course. We may guess (correctly) that the first-differenced log-transformation series is stationary in the larger sense.

In general, whenever the variance of a time series decreases (or increases) as the level of the series decreases (or increases), the series can be made stationary in the larger sense (in both level and variance) by log-transformation and differencing. Log-transformation is effective in such cases because the absolute value of successive random shocks changes systematically. Thus,

$$a_t^2 = Ka_{t-1}^2$$

for some constant of proportionality K. From this we see that

$$\text{Ln}(a_t^2) = \text{Ln}(a_{t-1}^2) + \text{Ln}(K)$$
$$\text{Ln}(a_t^2) - \text{Ln}(a_{t-1}^2) = \text{Ln}(K).$$

Log-transformation and differencing results in a constant variance. Readers who are familiar with variance stabilizing transformations in the context of regression models (e.g., Draper and Smith, 1966: 131–134) will recognize similarities in the approach taken here.

We note in conclusion that the analyst must always be able to assume that a time series process is stationary in both its level *and* variance. A series that is nonstationary in variance must be transformed prior to analysis. If the analyst ignores this problem (and if the assumption of larger sense stationarity is not satisfied), the analysis may lead to incorrect inferences about the time series process. These assumptions are similar in kind to the assumptions typically required of a regression analysis. Homoskedasticity of disturbance terms must be assumed to derive least-squares estimates of regression model parameters, for example, and ARIMA models are not different in this respect.

2.5 Autoregressive Processes of Order p: ARIMA (p,0,0) Models

Autoregressive processes of order p may be modeled by using p lagged observations of the series to predict the current observation, that is,

$$y_t = \phi_1 y_{t-1} + \phi_2 y_{t-2} + \ldots + \phi_p y_{t-p} + a_t,$$

which, using the backward shift operator, may be rewritten as

$$(1 - \phi_1 B - \phi_2 B^2 - \ldots - \phi_p B^p) y_t = a_t.$$

As a convention, we will use y_t to denote a *deviate* time series, that is,

$$y_t = Y_t - \Theta_0$$

for the stationary process, Y_t. Autoregressive processes are always stationary processes and ARIMA (p,0,0) models are always stationary models. Should a process be nonstationary, however, it may happen that its difference will be well represented by an ARIMA (p,0,0) model. Thus, for the nonstationary process, Y_t,

$$\begin{aligned} z_t &= (1 - B)^d Y_t \\ y_t &= z_t - \Theta_0 \\ &= \phi_1 y_{t-1} + \phi_2 y_{t-2} + \ldots + \phi_p y_{t-p} + a_t. \end{aligned}$$

The process is well represented by an ARIMA (p,d,0) model.

There is no essential difference in the algebras of ARIMA (p,0,0) and ARIMA (p,d,0) models except, of course, that ARIMA (p,d,0) models imply a nonstationary process. To simplify our discussion, then, we will deal only with ARIMA (p,0,0) models and only with deviate time series.

When a process is nonstationary, of course, our argument will apply to the differenced process.

The most commonly encountered autoregressive processes in the social sciences are first-order autoregressive processes. First-order autoregression is well represented by an ARIMA (1,0,0) model:

$$y_t = \phi_1 y_{t-1} + a_t$$
$$(1 - \phi_1 B)y_t = a_t.$$

It may be instructive to think of the ARIMA (1,0,0) model as an OLS regression model in which the current time series observation is regressed on the preceding time series observation. Unlike the OLS regression model, however, the parameter ϕ_1 must be constrained to the interval

$$-1 < \phi_1 < +1.$$

The purpose of these constraints will be soon made clear.

Tracking a random shock through time, the ARIMA (1,0,0) model exhibits a distinctive pattern of stochastic behavior called *autoregression*. Let

$$y_0 = a_0,$$

then,

$$\begin{aligned}
y_1 &= \phi_1 y_0 + a_1 \\
&= \phi_1 a_0 + a_1 \\
y_2 &= \phi_1 y_1 + a_2 \\
&= \phi_1(\phi_1 a_0 + a_1) + a_2 \\
&= \phi_1^2 a_0 + \phi_1 a_1 + a_2 \\
y_3 &= \phi_1 y_2 + a_3 \\
&= \phi_1(\phi_1^2 a_0 + \phi_1 a_1 + a_2) + a_3 \\
&= \phi_1^3 a_0 + \phi_1^2 a_1 + \phi_1 a_2 + a_3 \\
&\vdots \\
y_t &= \phi_1^t a_0 + \phi_1^{t-1} a_1 + \ldots + \phi_1 a_{t-1} + a_t.
\end{aligned}$$

Recalling that ϕ_1 is a fraction, successive powers of ϕ_1 converge to zero. If $\phi_1 = .5$, for example, then

$$
\begin{aligned}
\phi_1^2 &= .25 \\
\phi_1^3 &= .125 \\
\phi_1^4 &= .0625 \\
&\vdots \\
\phi_1^t &\simeq 0
\end{aligned}
$$

for t large. So while the initial random shock stays in the process indefinitely, its impact diminishes exponentially. After one observation, the impact of a_0 is only a fraction of its initial impact. By time t, the impact of a_0 is so small that we may think of it as zero. Returning to the black box input-output analogy used in Section 2.2 to describe integrated processes, we see that a portion of the random shock "leaks" out of the autoregressive black box as time passes:

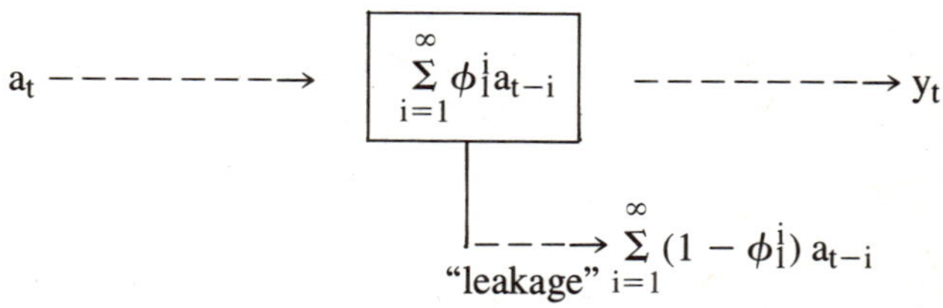

The autoregressive black box accumulates random shocks in the same manner as the integrated black box accumulates random shocks. There is a significant difference between these two black boxes, however. Random shocks leak out of the autoregressive black box over time. For an initial random shock, a_0, the portion remaining in the black box at successive points in time is:

Time	Portion Remaining	Portion Lost Through Leakage
$t = 0$	a_0	. . .
$t = 1$	$\phi_1 a_0$	$(1 - \phi_1)a_0$
$t = 2$	$\phi_1^2 a_0$	$(1 - \phi_1^2)a_0$
$\vdots$		
$t = t$	$\phi_1^t a_0 \approx 0$	$(1 - \phi_1^t)a_0 \approx a_0$.

After t moments, the portion of a_0 remaining in the black box is so small that we may think of it as zero. The portion of a_0 lost through leakage is so large that we may think of it as 100% of a_0.

To be perfectly correct in this analogy, the y_t process must then be passed through another black box:

$$y_t \dashrightarrow \boxed{+\Theta_0} \dashrightarrow Y_t,$$

which adds a constant to each observation, thus transforming the y_t process into the Y_t process. Of course, for an ARIMA (p,d,0) model, a nonstationary process is implied and the sequence of black boxes is:

$$y_t \dashrightarrow \boxed{+\Theta_0} \dashrightarrow z_t,$$

and then,

$$z_t \dashrightarrow \boxed{\sum_{i=1}^{\infty} z_{t-i}} \dashrightarrow Y_t.$$

Thus, while we have confined our discussion to ARIMA (p,0,0) models and to deviate time series, it is clear that any argument can be generalized to ARIMA (p,d,0) models and to the raw Y_t process by simply defining an appropriate black box filter.

Autoregression refers to a stochastic behavior in which a random shock has an exponentially diminishing impact over time. While it may not be obvious, this aspect of autoregression is determined by the bounds placed on the parameter ϕ_1:

$$-1 < \phi_1 < +1.$$

These are called the *bounds of stationarity for autoregressive parameters*. If an ARIMA (1,0,0) model is written as

$$(1 - \phi_1 B)y_t = a_t,$$

the implications of these bounds became apparent. If $\phi_1 = 1$,

$$(1 - B)y_t = a_t,$$

the ARIMA (1,0,0) model becomes an ARIMA (0,1,0) model which reflects a nonstationary process. Nonstationary processes result from an integration of random shocks over time:

$$Y_t = Y_0 + a_1 + a_2 + \ldots + a_{t-1} + a_t.$$

The impact of shocks from the distant past do not diminish over time and this is not consistent with autoregressive behavior.

This implication may be demonstrated again by introducing an important property of the ARIMA (1,0,0) model: The ARIMA (1,0,0) model can be expressed identically as *the infinite sum of exponentially weighted past random shocks*. To deduce this important property, write the ARIMA (1,0,0) model as

$$y_t = \phi_1 y_{t-1} + a_t$$

and

$$y_{t-1} = \phi_1 y_{t-2} + a_{t-1}.$$

Substituting for y_{t-1},

$$\begin{aligned} y_t &= \phi_1(\phi_1 y_{t-2} + a_{t-1}) + a_t \\ &= \phi_1^2 y_{t-2} + \phi_1 a_{t-1} + a_t. \end{aligned}$$

Similarly,

$$y_{t-2} = \phi_1 y_{t-3} + a_{t-2}$$

which we may substitute into the expression for y_t to obtain

$$\begin{aligned} y_t &= \phi_1^2(\phi_1 y_{t-3} + a_{t-2}) + \phi_1 a_{t-1} + a_t \\ &= \phi_1^3 y_{t-3} + \phi_1^2 a_{t-2} + \phi_1 a_{t-1} + a_t. \end{aligned}$$

Continuing the substitution process back into the infinite past,

$$y_t = \sum_{i=0}^{\infty} \phi_1^i a_{t-i},$$

which demonstrates that an ARIMA (1,0,0) process is identical to the infinite sum of exponentially weighted past shocks.

This is a rather important point. In Section 2.3, we demonstrated that the inverse difference operator was identical with the infinite series

$$\begin{aligned} (1 - B)^{-1} &= 1 + B + B^2 + \ldots + B^k + \ldots \\ &= \sum_{k=0}^{\infty} B^k. \end{aligned}$$

Writing the ARIMA (1,0,0) model as an infinite sum of exponentially weighted past random shocks, it is clear that the inverse first-order autoregressive factor is identical with the infinite series

$$(1 - \phi_1 B)^{-1} = 1 + \phi_1 B + \phi_1^2 B^2 + \ldots + \phi_1^k B^k + \ldots$$
$$= \sum_{k=0}^{\infty} \phi_1^k B^k.$$

This identity gives a "solution" of the ARIMA (1,0,0) model as

$$(1 + \phi_1 B) y_t = a_t$$
$$y_t = (1 - \phi_1 B)^{-1} a_t$$
$$= (1 + \phi_1 B + \phi_1^2 B^2 + \ldots + \phi_1^k B^k + \ldots) a_t$$
$$= a_t + \phi_1 a_{t-1} + \phi_1^2 a_{t-2} + \ldots + \phi_1^k a_{t-k} + \ldots .$$

Moreover, as the infinite series formulation of the inverse autoregressive operator converges to zero, it may be truncated after a few terms without appreciably changing an evaluation of the solution. For example, if $\phi_1 = .5$,

$$y_t = a_t + .5a_{t-1} + .25a_{t-2} + .0625a_{t-3} + .03125a_{t-4}$$
$$+ \ldots + (.5)^k a_{t-k} + \ldots .$$

After a few terms, the value of ϕ_1^k is approximately zero. All successive terms of the series could be ignored.

While the infinite series identity is important in and of itself, it gives a crucial insight into the nature of process stationarity. Whenever $\phi_1 = 1$, an ARIMA (0,1,0) model is implied but when $\phi_1 > 1$, *a growth process is implied*. Suppose, for example, that $\phi_1 = 1.5$. The ARIMA (1,0,0) process is then

$$(1 - 1.5B) y_t = a_t$$
$$y_t = (1 + 1.5B)^{-1} a_t$$
$$= (1 + 1.5B + 2.25B^2 + \ldots + (1.5)^k B^k + \ldots) a_t$$
$$= a_t + 1.5a_{t-1} + 2.25a_{t-2} + \ldots (1.5)^k a_{t-k} + \ldots .$$

Past random shocks have larger weights and, thus, are more important to the y_t process than the current random shock. As time passes, the random shock becomes more important.

As an example of such a process, consider the hypothetical situation in which a savings account earns 5% monthly compounded interest. If the monthly deposit to the account is essentially random, that is,

$$a_t = \text{deposit in the } t^{th} \text{ month}$$

and

$$y_t = \text{savings and accrued interest in the } t^{th} \text{ month,}$$

then the level of the savings account is given by

$$y_t = 1.05y_{t-1} + a_t.$$

In other words, the level of the savings account in this month is equal to 105% of the previous month's level (y_{t-1}) plus the current month's deposit (a_t). The current level of the account can be expressed identically in terms of past deposits only.

$$y_t = a_t + 1.05a_{t-1} + 1.1025a_{t-2} + 1.157625a_{t-3} + \ldots + (1.05)^k a_{t-k} + \ldots .$$

When expressed in this form, it is apparent that *the most important determinant of the current level of the savings account is the first deposit*. This is contrary to the principles of autoregression, and to avoid this situation, the value of ϕ_1 must be constrained to the bounds of stationarity for autoregressive parameters.

The principles of autoregression can be generalized to higher order processes. An ARIMA (2,0,0) model, for example, is:

$$y_t = \phi_1 y_{t-1} + \phi_2 y_{t-2} + a_t$$

or

$$(1 - \phi_1 B - \phi_2 B^2) y_t = a_t.$$

For p = 2, the current series observation is equal to a portion of the two preceding observations. In the general case, an ARIMA (p,0,0) model reflects a type of stochastic behavior in which the present observation is equal to portions of the p preceding observations. In practice, however, we have found that most autoregressive social science processes are well represented by ARIMA (1,0,0) models. ARIMA (2,0,0) processes are less common and higher order ARIMA (p,0,0) processes are quite rare. Indeed, in the next section, we will describe a class of moving average ARIMA models which will parsimoniously represent higher order autoregressive processes.

To demonstrate the principles of autoregression for an ARIMA (2,0,0) process, we begin with the expression for y_t and then substitute backward into time:

$$
\begin{aligned}
y_t &= \phi_1 y_{t-1} + \phi_2 y_{t-2} + a_t \\
&= \phi_1(\phi_1 y_{t-2} + \phi_2 y_{t-3} + a_{t-1}) \\
&\qquad + \phi_2(\phi_1 y_{t-3} + \phi_2 y_{t-4} + a_{t-2}) \\
&= a_t + \phi_1 a_{t-1} + \phi_2 a_{t-2} \\
&\qquad + \phi_1^2 y_{t-2} + 2\phi_1\phi_2 y_{t-3} + \phi_2^2 y_{t-4} \\
&= a_t + \phi_1 a_{t-1} + \phi_2 a_{t-2} \\
&\qquad + \phi_1^2(\phi_1 y_{t-3} + \phi_2 y_{t-4} + a_{t-2}) \\
&\qquad + 2\phi_1\phi_2(\phi_1 y_{t-4} + \phi_2 y_{t-5} + a_{t-3}) \\
&\qquad + \phi_2^2(\phi_1 y_{t-5} + \phi_2 y_{t-6} + a_{t-4}) \\
&= a_t + \phi_1 a_{t-1} + (\phi_2 + \phi_1^2)a_{t-2} + 2\phi_1\phi_2 a_{t-3} + \phi_2^2 a_{t-4} \\
&\qquad + \phi_1^3 y_{t-3} + 3\phi_1^2\phi_2 y_{t-4} + 3\phi_1\phi_2^2 y_{t-5} + \phi_2^3 y_{t-6}
\end{aligned}
$$

and so forth and so on. Clearly, with enough arithmetic, we can write the ARIMA (2,0,0) process as the infinite sum of exponentially weighted past shocks. Unlike the ARIMA (1,0,0) process, however, the infinite series is not a simple function of ϕ_1 and ϕ_2.

Of course, like the ARIMA (1,0,0) process, the parameters ϕ_1 and ϕ_2 must be constrained to the bounds of stationarity for autoregressive parameters. On commonsense grounds, one might think that the bounds of stationarity for an ARIMA (2,0,0) model should be:

$$-1 < \phi_1, \phi_2 < +1.$$

But there are a number of $\phi_1\phi_2$ interaction terms in the infinite series, so these simple bounds will not ensure stationarity. While we will not do so here, it can be demonstrated that the bounds of stationarity for an ARIMA (2,0,0) model are:[6]

$$
\begin{aligned}
-1 < \phi_2 &< +1 \\
\phi_1 + \phi_2 &\leq +1 \\
\phi_2 - \phi_1 &< +1.
\end{aligned}
$$

So long as the values of ϕ_1 and ϕ_2 satisfy these bounds, a stationary process is implied. To demonstrate this, let

$$\phi_1 + \phi_2 = 1$$

or

$$\phi_2 = 1 - \phi_1,$$

which violates the bounds of stationarity. The ARIMA (2,0,0) process then becomes:

$$(1 - \phi_1 B - \phi_2 B^2)y_t = a_t$$
$$(1 - \phi_1 B - (1 - \phi_1)B^2)y_t = a_t$$
$$(1 - \phi_1 B + \phi_1 B^2 - B^2)y_t = a_t$$
$$y_t - y_{t-2} = \phi_1(y_{t-1} - y_{t-2}) + a_t,$$

which is a difference equation representation of a nonstationary process.

ARIMA (p,0,0) models describe a type of stochastic behavior in which the current observation is a weighted sum of p preceding observations. In general, ARIMA (p,0,0) processes can be written as an infinite sum of exponentially weighted past shocks. As noted, most autoregressive social processes can be well represented by ARIMA (1,0,0) models. ARIMA (2,0,0) processes are less common. Higher order ARIMA (p,0,0) processes are extremely rare and, in any event, are more parsimoniously represented by the ARIMA (0,0,q) models which we will now introduce.

2.6 Moving Average Processes of Order q: ARIMA (0,0,q) Models

A white noise process is conveniently thought of as the "driving force" of all ARIMA (p,d,q) models. We have shown that integrated processes are realized as the sum of all past shocks and, thus, are well represented by ARIMA (0,d,0) models. Autoregressive processes are realized as an exponentially weighted sum of all past shocks and, thus, are well represented by ARIMA (p,0,0) models. The unifying factor between integrated and autoregressive processes is the persistence of a random shock. Each shock persists indefinitely, although for autoregressive processes, the impact of a shock diminishes rapidly. Moving average processes, in contrast, are characterized by a *finite* persistence. A random shock enters the system and then persists for no more than q observations before it vanishes entirely.

Moving average processes are well represented by ARIMA (0,0,q) models. An ARIMA (0,0,1) model is written as

$$y_t = a_t - \Theta_1 a_{t-1}$$

or

$$y_t = (1 - \Theta_1 B)a_t.$$

As are autoregressive processes, moving average processes are stationary. Again, there is no essential difference between ARIMA (0,0,q) and ARIMA (0,d,q) models except, of course, that ARIMA (0,d,q) models reflect a nonstationary process. To simplify our discussion, we will deal only with stationary processes and ARIMA (0,0,q) models, noting that when a process is nonstationary, our argument applies to the differenced process.

The general principle of moving average processes is that a random shock persists for exactly q observations and then is gone. Using the black box input-output analogy again, we can think of a first-order moving average process as

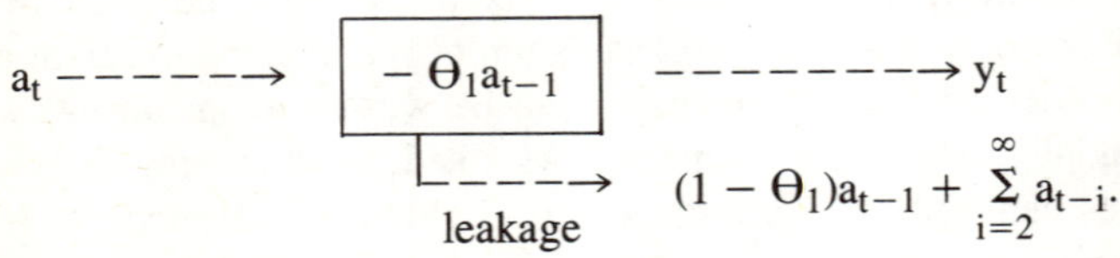

Here the random shock a_t enters the black box, is joined with a portion of the preceding random shock, a_{t-1}, and leaves the black box as the time series observation y_t. A portion of the preceding random shock, a_{t-1}, has already leaked out of the system along with all prior random shocks back into the distant past.

This basic principle of the moving average process can be generalized to higher order processes. For example, an ARIMA (0,0,2) process is written as

$$y_t = a_t - \Theta_1 a_{t-1} - \Theta_2 a_{t-2}$$

$$y_t = (1 - \Theta_1 B - \Theta_2 B^2)a_t.$$

Here the current time series observation consists of the current shock, a_t, and portions of the two preceding shocks, a_{t-1} and a_{t-2}. A shock persists for only two observations and then is gone. In the general case, an ARIMA (0,0,q) process is written as

$$y_t = (1 - \Theta_1 B - \Theta_2 B^2 - \ldots - \Theta_{q-1} B^{q-1} - \Theta_q B^q)a_t.$$

And in the general case, a shock persists for exactly q observations and then is gone. In practice, we have found that most moving average social processes are well represented by ARIMA (0,0,1) models. ARIMA (0,0,2) models are less common and higher order moving average models are extremely rare.

In Section 2.5, we noted that ARIMA (p,0,0) processes in which $p > 2$ will be more parsimoniously represented by lower order moving average models. This hints at an instructive relationship between ARIMA (p,0,0) and ARIMA (0,0,q) models. Writing an ARIMA (0,0,1) process at two points in time,

$$y_t = a_t - \Theta_1 a_{t-1}$$

and

$$y_{t-1} = a_{t-1} - \Theta_1 a_{t-2}$$

thus

$$a_{t-1} = y_{t-1} + \Theta_1 a_{t-2}.$$

If we substitute the expression for a_{t-1} into the expression for y_t,

$$\begin{aligned} y_t &= a_t - \Theta_1(y_{t-1} + \Theta_1 a_{t-2}) \\ &= a_t - \Theta_1 y_{t-1} - \Theta_1^2 a_{t-2}. \end{aligned}$$

Similarly, for a_{t-2},

$$\begin{aligned} y_t &= a_t - \Theta_1 y_{t-1} - \Theta_1^2(y_{t-2} + \Theta_1 a_{t-3}) \\ &= a_t - \Theta_1 y_{t-1} - \Theta_1^2 y_{t-2} - \Theta_1^3 a_{t-3}. \end{aligned}$$

And continuing this substitution back into time,

$$y_t = a_t - \sum_{i=1}^{\infty} \Theta_1^i y_{t-i}.$$

So an ARIMA (0,0,1) process can be expressed identically as the infinite sum of exponentially weighted past observations of the process. While we will not do so here, it can be demonstrated that any ARIMA (0,0,q) process can be expressed as an infinite series of exponentially weighted past observations.

Given this relationship, it is clear that we must constrain the values of moving average parameters. These constraints are identical to the constraints placed on autoregressive parameters. For an ARIMA (0,0,1) model,

$$-1 < \Theta_1 < +1$$

and for an ARIMA (0,0,2) model,

$$-1 < \Theta_2 < +1$$
$$\Theta_1 + \Theta_2 < +1$$
$$\Theta_2 - \Theta_1 < +1.$$

These are called the *bounds of invertibility for moving average parameters*. While the name is different, the bounds of invertibility play much the same role as the bounds of stationarity for autoregressive parameters. Suppose, for example, that $\Theta_1 = 1.5$ for some ARIMA (0,0,1) process. Then, writing this process as the infinite series,

$$y_t = a_t - 1.5y_{t-1} - 2.25y_{t-2} - 3.37y_{t-3} - \ldots$$

the weights associated with observations in the distant past become greater and greater.

In practice, when ϕ_p or Θ_q parameters exceed the bounds of stationarity or the bounds of invertability, the analyst may assume either that the series is nonstationary and must be differenced or that it was differenced too many times. Even if the parameters do not exceed their bounds, however, large values of ϕ_1 or Θ_1 may indicate that the model selected for the time series is inappropriate. We will return to these issues at a later point.

2.7 The General ARIMA (p,d,q) Model

Our development so far has treated p and q, the ARIMA autoregressive and moving average structural parameters, separately. Our development of autoregressive processes, for example, considered only ARIMA (p,0,0) models, models in which q = 0. Likewise, our development of moving average processes considered only ARIMA (0,0,q) models, models in which p = 0. This development reflects our experience with social science time series. If our experiences are typical, only a few social science time series in a thousand will have both p and q $\neq 0$.

While ARIMA (p,0,q) models are not logically *impossible,* the relationships between ARIMA (p,0,0) and ARIMA (0,0,q) models which we have discussed place some logical limits on ARIMA (p,0,q) models. These limitations are ordinarily referred to as the limitations of *parameter redundancy*. To illustrate parameter redundancy, write the ARIMA (1,0,1) process as

$$y_t = \phi_1 y_{t-1} + a_t - \Theta_1 a_{t-1}$$

or

$$(1 - \phi_1 B)y_t = (1 - \Theta_1 B)a_t$$

and solving for y_t

$$y_t = (1 - \phi_1 B)^{-1}(1 - \Theta_1 B)a_t.$$

In Section 2.4, we demonstrated that the inverse autoregressive operator, $(1 - \phi_1 B)^{-1}$, is identical to an infinite series:

$$(1 - \phi_1 B)^{-1} = 1 + \phi_1 B + \phi_1^2 B^2 + \ldots + \phi_1^n B^n + \ldots$$

so $$y_t = (1 + \phi_1 B + \phi_1^2 B^2 + \ldots + \phi_1^n B^n + \ldots)(1 - \Theta_1 B)a_t.$$

We can use this infinite series form of the ARIMA (1,0,1) model to examine the problem of parameter redundancy under certain conditions.

When $\phi_1 = \Theta_1$, both parameters are completely redundant. Substituting ϕ_1 for Θ_1 in the infinite series form of the ARIMA (1,0,1) process,

$$\begin{aligned}
y_t &= (1 + \phi_1 B + \phi_1^2 B^2 + \ldots + \phi_1^n B^n + \ldots)(1 - \phi_1 B)a_t \\
&= [(1 + \phi_1 B + \phi_1^2 B^2 + \ldots + \phi_1^n B^n + \ldots) \\
&\qquad - \phi_1 B(1 + \phi_1 B + \phi_1^2 B^2 + \ldots + \phi_1^n B^n + \ldots)]a_t \\
&= (1 - \phi_1 B + \phi_1 B - \phi_1^2 B^2 + \phi_1^2 B^2 - \ldots - \phi_1^n B^n + \phi_1^n B^n - \ldots)a_t \\
&= a_t.
\end{aligned}$$

So when $\phi_1 = \Theta_1$, an ARIMA (1,0,1) model reduces to an ARIMA (0,0,0) model.

Now suppose that $\phi_1 \neq \Theta_1$ but that the parameters are *close* to each other in value. If $\Theta_1 = \phi_1 + c$, where c is a very small number, the infinite series form of the ARIMA (1,0,1) model is:

$$\begin{aligned}
y_t &= (1 + \phi_1 B + \phi_1^2 B^2 + \ldots + \phi_1^n B^n + \ldots)(1 - \phi_1 B - cB)a_t \\
&= [(1 + \phi_1 B + \phi_1^2 B^2 + \ldots + \phi_1^n B^n + \ldots) \\
&\qquad - \phi_1 B(1 + \phi_1 B + \phi_1^2 B^2 + \ldots + \phi_1^n B^n + \ldots) \\
&\qquad - cB(1 + \phi_1 B + \phi_1^2 B^2 + \ldots + \phi_1^n B^n + \ldots)]a_t \\
&= (1 - cB - c\phi_1 B^2 - c\phi_1^2 B^3 - \ldots - c\phi_1^n B^{n+1} + \ldots)a_t.
\end{aligned}$$

Now when both c and ϕ_1 are small, $c\phi_1$ will be approximately zero and the ARIMA (1,0,1) model reduces approximately to an ARIMA (0,0,1) model:

$$y_t = (1 - \Theta_1 B)a_t,$$

where $c = \Theta_1$. And if $c\phi_1$ is not approximately zero, then the ARIMA

(1,0,1) model reduces approximately to an ARIMA (0,0,2) model:

$$yt = (1 - \Theta_1 B - \Theta_2 B^2)a_t$$

where $c\phi_1 = \Theta_2$. And of course, whenever $c \approx \phi_1$, the ARIMA (1,0,1) model reduces approximately to an ARIMA (1,0,0) model. In practice, the analyst must always remain skeptical of ARIMA (p,0,q) models. ARIMA (0,0,0), ARIMA (p,0,0), and ARIMA (0,0,q) models should always be ruled out before ARIMA (p,0,q) models are entertained.

2.8 The Autocorrelation Function

We have discussed the general ARIMA model so far without paying any attention to the task of model building. For a given time series, that is, how can the analyst select an appropriate ARIMA (p,d,q) model? *Identification,* the procedure whereby the values of p, d, and q are determined for a given time series, relies on a statistic called the autocorrelation function (ACF). For a time series process, Y_t, the ACF is defined as

$$ACF(k) = COV(Y_t\, Y_{t+k}) \,/\, VAR(Y_t).$$

Given a realization of the Y_t process, a finite time series of N observations, the ACF is estimated from the formula

$$ACF(k) = \frac{\sum_{t=1}^{N-k} (Y_t - \bar{Y})(Y_{t+k} - \bar{Y})}{\sum_{t=1}^{N} (Y_t - \bar{Y})^2} \left[\frac{N}{N-k}\right].$$

The ACF(k) is a measure of correlation between Y_t and Y_{t+k}. For a given lag k, however, variance (the denominator of the formula) is estimated over all N observations while covariance (the numerator of the formula) is estimated over only $N-k$ pairs of observations. Strictly speaking, then, ACF(k) is not the familiar Pearson product-moment correlation coefficient between time series observation k units apart.

To illustrate the formula for estimating ACF(k), lag the Y_t series forward in time.

lag-0	Y_1	Y_2	$Y_3 \ldots Y_N$			
lag-1		Y_1	$Y_2 \ldots Y_{N-1}$	Y_N		
lag-2			$Y_1 \ldots Y_{N-2}$	Y_{N-1}	Y_N	

and so forth. ACF(1) is the correlation coefficient estimated between the time series (lag-0) and its first lag (lag-1); ACF(2) is the correlation coefficient estimated between the time series (lag-0) and its second lag (lag-2); and, in general, ACF(k) is the correlation coefficient estimated between the time series (lag-0) and its kth lag (lag-k).

There are three points to be noted about the ACF. First, by definition, ACF(0) = 1; a time series is always perfectly correlated with itself. Second, also by definition, ACF(k) = ACF(−k); in other words, the ACF(k) is the same whether the series is lagged forward or backward. Because the ACF is symmetrical about lag-0, only the positive half of the ACF need be examined. Third, each time the series is lagged, one pair of observations is lost from the estimate of ACF(k). ACF(1) is estimated from N−1 pairs of observations; ACF(2) is estimated from N−2 pairs of observations; and so forth. As the value of k increases, confidence in the estimate of ACF(k) diminishes.

In theory, each time series process has a unique ACF. The Y_t process is fully determined by its ACF, and if two processes have the same ACF, they are identical. In Section 2.11, we will describe a procedure whereby the ACF estimated from a realization of the process (from a finite time series, that is) is used to *identify* the ARIMA structure of the underlying process. In this section, we will derive the theoretical or expected ACFs for a variety of ARIMA (p,d,q) processes.

First, an ARIMA (0,0,0) or white noise process written as

$$Y_t = a_t + \Theta_0$$

is expected to have a uniformly zero ACF. This follows from the definition of white noise. For all k,

$$COV(a_t\ a_{t+k}) = 0.$$

Second, an ARIMA (0,1,0) process written as

$$(1 - B)Y_t = a_t + \Theta_0$$

is expected to have an ACF that is positive and dies out slowly from lag to lag, that is,

$$ACF(1) \simeq ACF(2) \simeq \ldots \simeq ACF(k).$$

A trending process, for example, has the expected value

$$EY_t = \Theta_0 t.$$

A realization of this process will have a mean, $\bar{Y}$, approximately equal to some middle observation of the series. When $\bar{Y}$ is subtracted from each observation of the series, the resulting deviate series has the expected form

$$\ldots, -2\Theta_0, -1\Theta_0, 0, +1\Theta_0, +2\Theta_0, \ldots.$$

The first half of the deviate series will be negative numbers and the second half will be positive numbers. Thus, for a series of N observations, the estimate of ACF(1) will be based on N−2 pairs of observations with the same sign; the estimate of ACF(2) will be based on N−3 pairs of observations with the same sign; and so forth.

Figure 2.8(a) shows the expected ACFs for several ARIMA processes. The ACF of a nonstationary process is expected to have a relatively high positive value for ACF(1) and successive lags of the ACF are expected to die out slowly to zero. In particular, ACF(k) is expected to be approximately

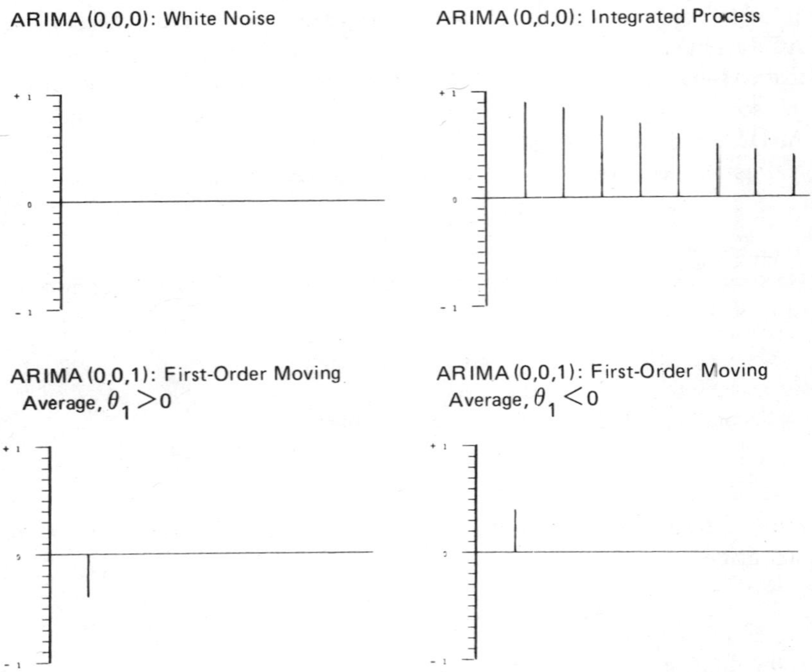

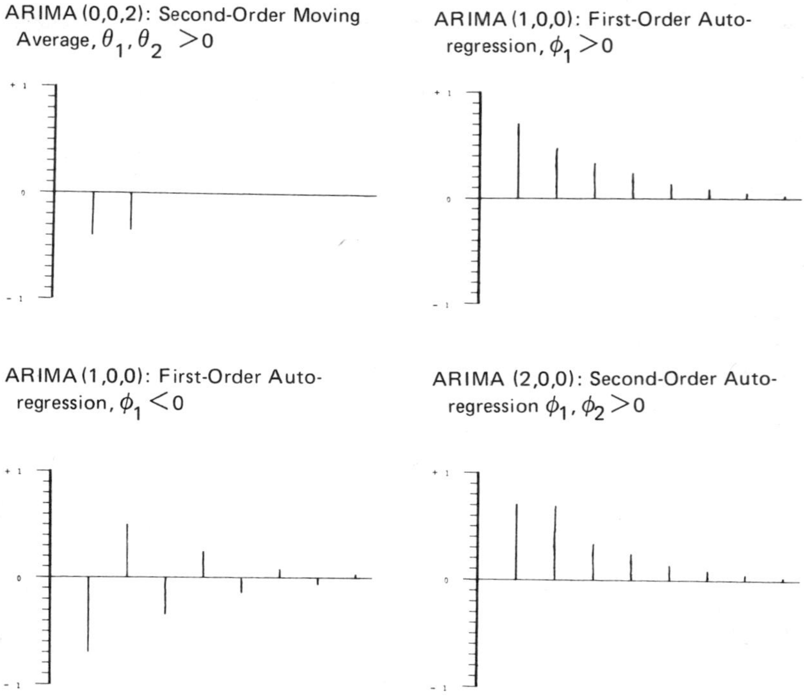

FIGURE 2.8(a) Expected ACFs for Several ARIMA Processes

equal to ACF(k+1) for all lags. In general, the ACF of any nonstationary process is expected to have the form indicated in Figure 2.8(a).

Third, an ARIMA (0,0,1) process written as

$$y_t = (1 - \Theta_1 B)a_t$$

is expected to have a nonzero ACF(1). All other lags of the ACF are expected to be zero. This is easily demonstrated by noting that COV $(y_t\, y_{t+1})$ is:

$$\begin{aligned} E(y_t y_{t+1}) &= E[(a_t - \Theta_1 a_{t-1})(a_{t+1} - \Theta_1 a_t)] \\ &= E(a_t a_{t+1} - \Theta_1 a_t^2 - \Theta_1 a_{t-1} a_{t+1} + \Theta_1^2 a_{t-1} a_t) \\ &= E a_t a_{t+1} - \Theta_1 E a_t^2 - \Theta_1 E a_{t-1} a_{t+1} + \Theta_1^2 E a_{t-1} a_t. \end{aligned}$$

As the property of white noise is that each random shock is independent of every other random shock, $Ea_t a_{t+k} = 0$ and all terms except one are zero:

$$E(y_t y_{t+1}) = -\Theta_1 E a_t^2 = -\Theta_1 \sigma_a^2.$$

To obtain the expected value of ACF (1), this result must be divided by the y_t process variance. VAR (y_t) is:

$$\begin{aligned} E\,(y_t^2) &= E\,[(a_t - \Theta_1 a_{t-1})^2] \\ &= E\,(a_t^2 - 2\Theta_1 a_t a_{t-1} + \Theta_1^2 a_{t-1}^2) \\ &= Ea_t^2 - 2\Theta_1 E a_t a_{t-1} + \Theta_1^2 E a_{t-1}^2 \\ &= \sigma_a^2(1 + \Theta_1^2). \end{aligned}$$

From these two results,

$$E\,[\text{ACF}\,(1)] = \frac{-\Theta_1 \sigma_a^2}{\sigma_a^2(1 + \Theta_1^2)} = \frac{-\Theta_1}{1 + \Theta_1^2}.$$

Through the same procedure,

$$\begin{aligned} E\,(y_t y_{t+2}) &= E\,[(a_t - \Theta_1 a_{t-1})(a_{t+2} - \Theta_1 a_{t+1})] \\ &= E\,(a_t a_{t+2} - \Theta_1 a_t a_{t+1} - \Theta_1 a_{t-1} a_{t+2} + \Theta_1^2 a_{t-1} a_{t+1}) \\ &= E a_t a_{t+2} - \Theta_1 E a_t a_{t+1} - \Theta_1 E a_{t-1} a_{t+2} + \Theta_1^2 E a_{t-1} a_{t+1} = 0. \end{aligned}$$

From this,

$$E\,[\text{ACF}\,(2)] = \frac{0}{\sigma_a^2(1 + \Theta_1^2)} = 0.$$

Through this same procedure, it can be demonstrated that ACF (3), ACF (4), . . . , ACF (k) are all expected to be zero.

It is important to note that an ARIMA (0,d,1) process is nonstationary and is thus expected to have an ACF typical of all nonstationary processes. If the process is differenced, however,

$$z_t = (1 - B)^d Y_t = a_t - \Theta_1 a_{t-1}.$$

An ACF for the z_t process will be that expected of an ARIMA (0,0,1) process.

Fourth, an ARIMA (0,0,2) process written as

$$y_t = (1 - \Theta_1 B - \Theta_2 B^2)a_t$$

is expected to have nonzero values for ACF (1) and ACF (2). The values of ACF (3) and all successive ACF (k) are expected to be zero. To demonstrate this, note that COV $(y_t\, y_{t+1})$ is:

$$\begin{aligned}
E\,(y_t y_{t+1}) &= E\,[(a_t - \Theta_1 a_{t-1} - \Theta_2 a_{t-2})\,(a_{t+1} - \Theta_1 a_t - \Theta_2 a_{t-1})] \\
&= E\,(a_t a_{t+1} - \Theta_1 a_t^2 - \Theta_2 a_t a_{t-1} - \Theta_1 a_{t-1} a_{t+1} + \Theta_1^2 a_{t-1} a_t \\
&\qquad + \Theta_1 \Theta_2 a_{t-1}^2 - \Theta_2 a_{t-2} a_{t+1} + \Theta_2 \Theta_1 a_{t-2} a_t \\
&\qquad + \Theta_2^2 a_{t-2} a_{t-1}) \\
&= -\,\Theta_1 E a_t^2 + \Theta_1 \Theta_2 E a_{t-1}^2 \\
&= \sigma_a^2 \Theta_1 (\Theta_2 - 1).
\end{aligned}$$

The process variance is:

$$\begin{aligned}
E\,(y_t^2) &= E\,[(a_t - \Theta_1 a_{t-1} - \Theta_2 a_{t-2})^2] \\
&= E\,(a_t^2 - 2\Theta_1 a_t a_{t-1} - \Theta_2 a_t a_{t-2} + \Theta_1^2 a_{t-1}^2 + 2\Theta_1 \Theta_2 a_{t-1} a_{t-2} \\
&\quad - \Theta_2 a_t a_{t-2} + \Theta_2^2 a_{t-2}^2) \\
&= E a_t^2 + \Theta_1^2 E a_{t-1}^2 + \Theta_2^2 E a_{t-2}^2 \\
&= \sigma_a^2 (1 + \Theta_1^2 + \Theta_2^2).
\end{aligned}$$

From these two results,

$$E\,[\text{ACF}\,(1)] = \frac{\sigma_a^2 \Theta_1 (\Theta_2 - 1)}{\sigma_a^2 (1 + \Theta_1^2 + \Theta_2^2} = \frac{\Theta_1 (\Theta_2 - 1)}{1 + \Theta_1^2 + \Theta_2^2}.$$

For the second lag of the ACF,

$$\begin{aligned}
E\,(y_t\, y_{t+2}) &= E\,(a_t - \Theta_1 a_{t-1} - \Theta_2 a_{t-2})\,(a_{t+2} - \Theta_1 a_{t+1} - \Theta_2 a_t) \\
&= -\,\Theta_2 E a_t^2 \\
&= -\,\Theta_2 \sigma_a^s.
\end{aligned}$$

This gives the result

$$E\,[\text{ACF}\,(2)] = \frac{-\,\Theta_2 \sigma_a^2}{\sigma_a^2 (1 + \Theta_1^2 + \Theta_2^2} = \frac{-\,\Theta_2}{1 + \Theta_1^2 + \Theta_2^2}.$$

For the third lag of the ACF,

$$\begin{aligned} E(y_t y_{t+3}) &= E[(a_t - \Theta_1 a_{t-1} - \Theta_2 a_{t-2})(a_{t+3} - \Theta_1 a_{t+2} - \Theta_2 a_{t+1})] \\ &= 0 \end{aligned}$$

so

$$E[ACF(3)] = 0.$$

The values of ACF (4), ACF (5), . . . , ACF (k) are all expected to be zero for the same reason.

Continuing this procedure, it can be demonstrated that an ARIMA (0,0,q) process is expected to have nonzero values for ACF (1), . . . , ACF (q) while ACF (q+1) and all successive lags are expected to be zero.

Fifth, an ARIMA (1,0,0) process written as

$$(1 - \phi_1 B)y_t = a_t$$

is expected to have an ACF that decays exponentially beginning with the first lag. To demonstrate this, note that COV ($y_t\ y_{t+1}$) is:

$$\begin{aligned} E(y_t y_{t+1}) &= E[(y_t)(y_{t+1})] = E[(y_t)(\phi_1 y_t + a_{t+1})] \\ &= E(\phi_1 y_t^2 + y_t a_{t+1}). \end{aligned}$$

As y_t and a_{t+1} are independent, $Ey_t a_{t+1} = 0$, so

$$E(y_t y_{t+1}) = \phi_1 E y_t^2 = \phi_1 \sigma_y^2.$$

The variance of the process, of course, is σ_y^2. Dividing covariance by variance,

$$E[ACF(1)] = \frac{\phi_1 \sigma_y^2}{\sigma_y^2} = \phi_1.$$

For lag-2 of the ACF,

$$\begin{aligned} E(y_t y_{t+2}) &= E[(y_t)(\phi_1 y_{t+1} + a_{t+2})] \\ &= E[(y_t)(\phi_1(\phi_1 y_t + a_{t+1}) + a_{t+2})] \\ &= E(\phi_1^2 y_t^2 + \phi_1 y_t a_{t+1} + y_t a_{t+2}) \\ &= \phi_1^2 E y_t^2 = \phi_1^2 \sigma_y^2. \end{aligned}$$

Dividing covariance by variance again,

$$E[ACF(2)] = \phi_1^2 \sigma_y^2 / \sigma_y^2 = \phi_1^2 .$$

By the same procedure,

$$E\,[ACF\,(3)] = \phi_1^3$$

$$\vdots$$

$$E\,[ACF\,(k)] = \phi_1^k.$$

This gives the ACF of an ARIMA (1,0,0) process a distinctive pattern of exponential decay from lag to lag. If $\phi_1 = .5$, for example, then,

$$ACF\,(1) = .5$$

$$ACF\,(2) = (.5)^2 = .25$$

$$ACF\,(3) = (.5)^3 = .125$$

$$\vdots$$

$$ACF\,(k) = (.5)^k \simeq 0\,.$$

Should ϕ_1 be negative, however, successive lags of the ACF are alternately negative and positive. So if $\phi_1 = -\,.5$,

$$ACF\,(1) = -\,.5$$

$$ACF\,(2) = (-\,.5)^2 = .25$$

$$ACF\,(3) = (-\,.5)^3 = -\,.125$$

$$\vdots$$

$$ACF\,(k) = (-\,.5)^k \simeq 0\,.$$

For both positive and negative values of ϕ_1, the expected ACF (k) grows smaller and smaller from lag to lag until, after three or four lags, ACF (k) is approximately zero.

Sixth, an ARIMA (2,0,0) process written as

$$(1 - \phi_1 B - \phi_2 B^2) y_t = a_t$$

is also expected to have an ACF that decays exponentially beginning with the first lag. For higher order ARIMA (p,0,0) processes, COV ($y_t\, y_{t+k}$) is difficult to derive. It can be demonstrated, nevertheless, that the expected ACF for an ARIMA (2,0,0) process is given by[7]:

$$ACF\,(k) = \phi_1\, ACF\,(k-1) + \phi_2\, ACF\,(k-2).$$

Recalling that ACF (0) = 1 and that ACF (−k) = ACF (k), the ACF of an ARIMA (2,0,0) process is expected to be:

$$
\begin{aligned}
\text{ACF}(1) &= \phi_1 \text{ACF}(0) + \phi_2 \text{ACF}(-1) = \phi_1 + \phi_2 \text{ACF}(1) \\
&= \frac{\phi_1}{1 - \phi_2} \\
\text{ACF}(2) &= \phi_1 \text{ACF}(1) + \phi_2 \text{ACF}(0) \\
&= \frac{\phi_1^2}{1 - \phi_2} + \phi_2 \\
\text{ACF}(3) &= \phi_1 \text{ACF}(2) + \phi_2 \text{ACF}(1) \\
&= \frac{\phi_1(\phi_2 + \phi_1^2)}{1 - \phi_2} + \phi_1\phi_2 \\
\text{ACF}(4) &= \phi_1 \text{ACF}(3) + \phi_2 \text{ACF}(2) \\
&= \frac{\phi_1^2(2\phi_2 + \phi_1^2)}{1 - \phi_2} + \phi_2(\phi_1^2 + \phi_2) \\
&\vdots \\
\text{ACF}(k) &= \phi_1 \text{ACF}(k-1) + \phi_2 \text{ACF}(k-2) .
\end{aligned}
$$

To illustrate this expected ACF, let $\phi_1 = \phi_2 = .4$, then,

$$
\begin{aligned}
\text{ACF}(1) &= .677 \\
\text{ACF}(2) &= .677 \\
\text{ACF}(3) &= .533 \\
\text{ACF}(4) &= .479 \\
\text{ACF}(5) &= .343 \\
\text{ACF}(6) &= .329
\end{aligned}
$$

and so forth. Alternatively, if $\phi_1 = .4$ and $\phi_2 = -.4$,

$$
\begin{aligned}
\text{ACF}(1) &= .286 \\
\text{ACF}(2) &= -.286 \\
\text{ACF}(3) &= -.229 \\
\text{ACF}(4) &= .023
\end{aligned}
$$

and so forth. The pattern of decay in the expected ACF is always determined by the values of ϕ_1 and ϕ_2. In the general case, an ARIMA (p,0,0) process is expected to have an ACF that decays from lag to lag with the rate of decay determined by the values of $\phi_1, \phi_2, \ldots, \phi_p$.

Examining the expected ACFs in Figure 2.8(a), a method for *identifying* the ARIMA structure of a time series process can be seen. Given a realization of the process, a finite time series, the ACF can be estimated and used to infer the process structure. If the estimated ACF is zero for all lags, the analyst can infer that the time series was generated by an ARIMA (0,0,0) process. If the estimated ACF (1) is large and positive, say ACF (1) $\geq$ .7, and if the ACF dies out slowly from lag to lag, the analyst can infer that the process is nonstationary; the series must be differenced. If the estimated ACF (1) is nonzero but ACF (2) and all successive lags are zero, the analyst can infer that the time series was generated by an ARIMA (0,0,1) process. Finally, if the estimated ACF dies out exponentially from lag to lag, the analyst can infer that the time series was generated by an ARIMA (1,0,0) process.

But in practice, *identification* may not always be a simple task. The ACFs shown in Figure 2.8(a) are *expected* ACFs which presume either a knowledge of the process itself or else an infinitely long realization of the process. In practice, the process itself is always unknown and only a finite realization of the process (an N-observation time series) is available. To be sure, the estimated ACFs of white noise and nonstationary processes are so distinctive that the analyst cannot mistake them. Similarly, the estimated ACFs of higher order ARIMA (0,0,q) and ARIMA (p,0,0) processes are so different that the analyst will not ordinarily mistake one for the other. The estimated ACFs of ARIMA (0,0,q) and ARIMA (1,0,0) processes are quite similar, however, and in practice, it is nearly impossible to distinguish between these two processes on the basis of an estimated ACF alone.

Fortunately, another *identification* statistic, the partial autocorrelation function (PACF), can be used to distinguish a higher order ARIMA (0,0,q) process from an ARIMA (p,0,0) process. The PACF has an interpretation not unlike that of any other measure of partial correlation. The lag-k PACF, PACF (k), is a measure of correlation between time series observations k units apart *after the correlation at intermediate lags has been controlled or "partialed out."*

Unlike the ACF, the PACF cannot be estimated from a simple, straightforward formula: PACF (k) is estimated from a solution of the Yule-Walker equation system (See Box and Jenkins, 1976:64).[8] While we will not do so here, it can be demonstrated that the solution gives the values of

$$\text{PACF}(1) = \text{ACF}(1)$$

$$\text{PACF}(2) = \frac{\text{ACF}(2) - [\text{ACF}(1)]^2}{1 - [\text{ACF}(1)]^2}$$

$$\text{ACF}(3) + \text{ACF}(1)\,[\text{ACF}(2)]^2 + [\text{ACF}(1)]^3$$

$$\text{PACF}(3) = \frac{-2\text{ACF}(1)\,\text{ACF}(2) - [\text{ACF}(1)]^2\,\text{ACF}(3)}{1 + 2[\text{ACF}(1)]^2\,\text{ACF}(2) - [\text{ACF}(2)]^2 - [\text{ACF}(1)]^2}$$

and so forth. Expressing the expected PACF in this form, the role of the PACF as a measure of partial correlation is made explicit. The PACF in fact is a "partial" (or "partialed") ACF. Expressing the PACF in this form also makes explicit the tedious arithmetic involved in its estimation. Without the proper software, the estimated PACF would be of little use to the time series analyst.

As the expected PACF is a function of the expected ACF, and as the expected ACFs of several ARIMA processes have already been derived, the expected PACFs are as given.

First, an ARIMA (1,0,0) process whose ACF is expected to be

$$\text{ACF}(k) = \phi_1^k$$

is expected to have a nonzero PACF (1) while PACF (2) and all successive lags are expected to be zero:

$$\text{PACF}(1) = \phi_1$$

$$\text{PACF}(2) = \frac{\phi_1^2 - \phi_1^2}{1 - \phi_1^2} = 0$$

$$\text{PACF}(3) = \frac{\phi_1^3 + \phi_1\phi_1^4 + \phi_1^3 - 2\phi_1\phi_1^2 - \phi_1^2\phi_1^3}{1 + 2\phi_1^2\phi_1^2 - \phi_1^4 - 2\phi_1^2}$$

$$= \frac{2\phi_1^3 + \phi_1^5 - 2\phi_1^3 - \phi_1^5}{1 + \phi_1^4 - 2\phi_1^2} = 0.$$

Successive lags of PACF (k) are also expected to be zero.

Second, an ARIMA (2,0,0) process whose ACF is expected to be

$$\text{ACF}(1) = \frac{\phi_1}{1 - \phi_2}$$

$$\text{ACF}(2) = \frac{\phi_1^2}{1 - \phi_2} + \phi_2$$

$$\text{ACF}(3) = \frac{\phi_1(\phi_2 + \phi_1^2)}{1 - \phi_2} + \phi_1\phi_2$$

is expected to have nonzero values of PACF (1) and PACF (2) while PACF (3) and all successive lags are expected to be zero. Substituting the expected ACF (k) in the formula (too tedious a procedure to be presented here),

$$\text{PACF}(1) = \frac{\phi_1}{1 - \phi_2}$$

$$\text{PACF}(2) = \frac{\phi_2(\phi_2 - 1)^2 - \phi_1\phi_2}{(1 - \phi_2)^2 - \phi_1^2}$$

$$\text{PACF}(3) = 0.$$

Successive lags are all expected to be zero. In the general case, an ARIMA (p,0,0) process is expected to have nonzero values for PACF (1), . . . , PACF (p) while PACF (p+1) and all successive lags are expected to be zero.

Third, an ARIMA (0,0,1) process whose ACF is expected to be

$$\text{ACF}(1) = \frac{-\Theta_1}{1 + \Theta_1^2}$$
$$\text{ACF}(2) = \ldots = \text{ACF}(k) = 0$$

has a decaying PACF, that is, all PACF (k) are expected to be nonzero:

$$\text{PACF}(1) \doteq \frac{-\Theta_1}{1 + \Theta_1^2}$$

$$\text{PACF}(2) = \frac{-\Theta_1^2}{1 + \Theta_1^2 + \Theta_1^4}$$

$$\text{PACF}(3) = \frac{-\Theta_1^3}{1 + \Theta_1^2 + \Theta_1^4 + \Theta_1^6}.$$

Successive lags of the expected PACF grow smaller and smaller in absolute value. If $\Theta_1 = .7$, for example,

$$\text{PACF}(1) = -.469$$
$$\text{PACF}(2) = -.283$$
$$\text{PACF}(3) = -.186$$

and so forth. In the general case, the PACF of an ARIMA (0,0,q) process is expected to decay in this same manner but at a rate determined by the values of $\Theta_1, \ldots, \Theta_q$.

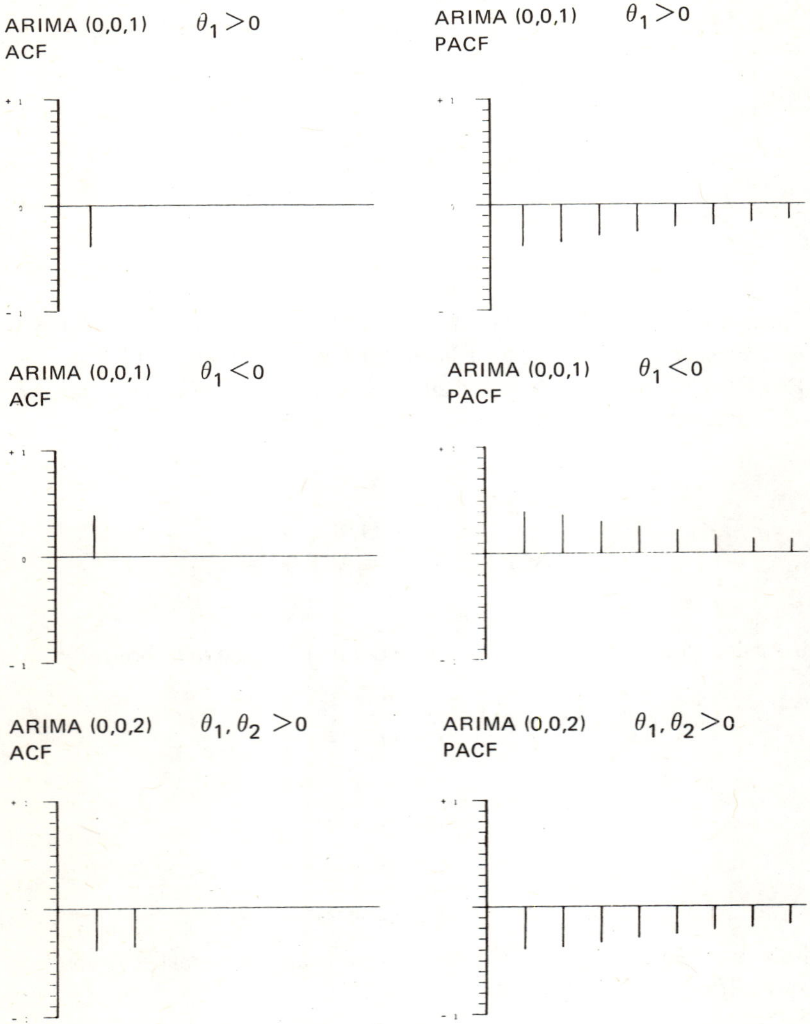

(Figure 2.8(b) continued on p. 79)

Figure 2.8(b) shows the expected ACFs and PACFs for several ARIMA (p,0,0) and ARIMA (0,0,q) processes. Autoregressive processes are characterized by decaying ACFs and spiking PACFs. An ARIMA (p,0,0) process is expected to have exactly p nonzero spikes in the first p lags of its PACF. All successive lags of the PACF are expected to be zero. Moving average

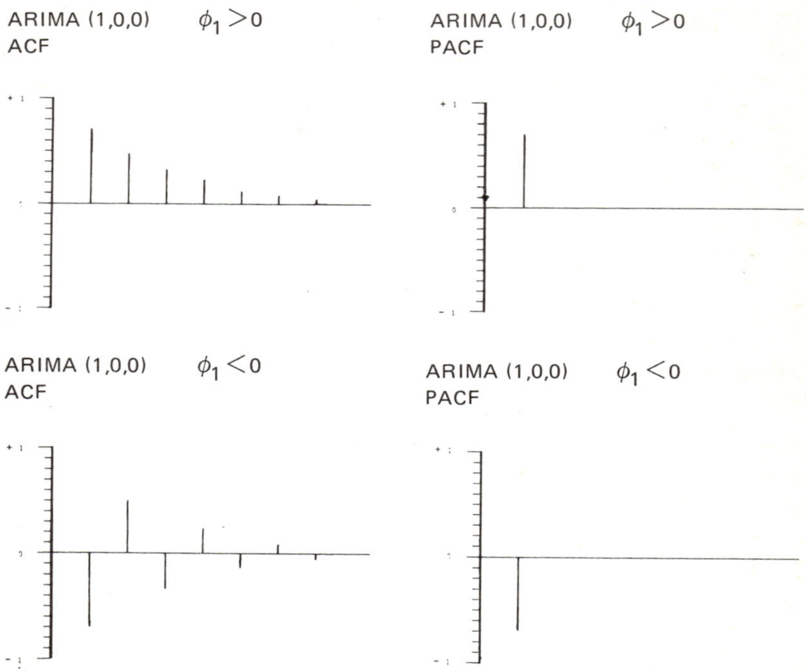

(Figure 2.8(b) continued on p. 80)

processes are characterized by spiking ACFs and decaying PACFs. An ARIMA (0,0,q) process is expected to have exactly q nonzero spikes in the first q lags of its ACF. All successive lags of the ACF are expected to be zero. Finally, an ARIMA (p,0,q) process is expected to have both decaying ACF and PACF.

There is one more theoretical issue to be covered before the practical issues of model identification can be considered. In Section 2.11, we will describe the procedures required to identify an appropriate ARIMA (p,d,q) model from estimated ACFs and PACFs. In Section 2.12, we will present four example analyses which illustrate in detail the model-building procedure.

2.9 Seasonality

If it were not for seasonality, time series analysis might become a rather simple, pleasant task. Most social science time series would be well repre-

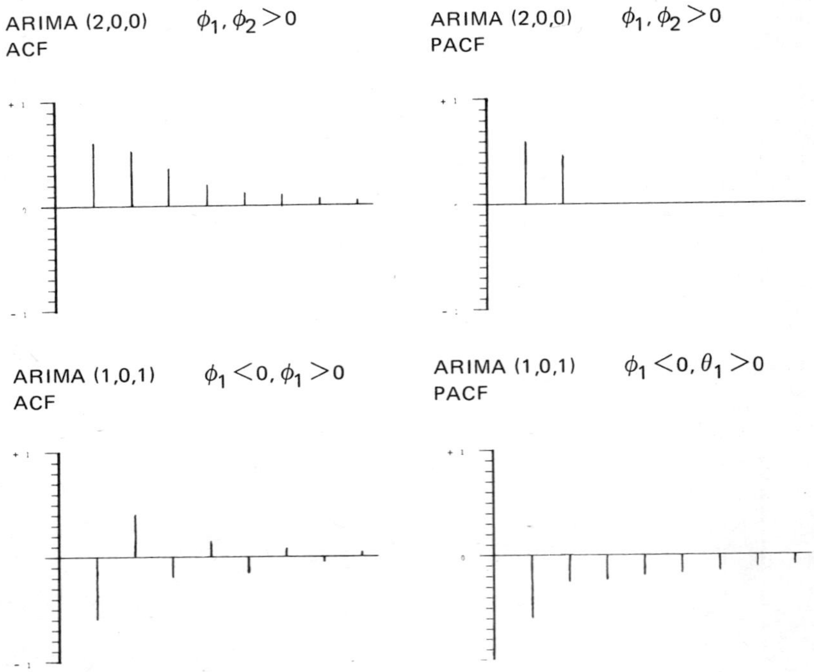

FIGURE 2.8(b) Expected ACFs and PACFs for Several ARIMA Processes

sented by lower order models such as ARIMA (1,0,0), ARIMA (0,0,1), and ARIMA (0,1,1) which are easily identified in most cases. But it is an unfortunte fact that almost all monthly or quarterly social science time series have strong seasonal components. This seasonality complicates the task of time series analysis generally and ARIMA modeling specifically.

We define seasonality as *any cyclical or periodic fluctuation in a time series that recurs or repeats itself at the same phase of the cycle or period.* Retail sales indicators, for example, normally peak in December when families shop for Christmas presents. If we knew nothing else about a retail sales indicator, then, we could guess that the series would reach a highpoint in each December.

To the time series analyst, seasonality is process variance which must be removed or controlled. One method of controlling for seasonal variance is to *deseasonalize* the time series prior to analysis. Johnston (1972: 186) describes a multiple regression deseasonalization method wherein dummy variables are used to estimate the seasonal variance of each month or quarter of the cycle. This estimated variance is then subtracted from the series. Makridakis and Wheelwright (1978: Chapter 16) describe similar (though more complicated) deseasonalization methods used by the U.S. Labor and Commerce Departments. All of these deseasonalization methods require that seasonal variation be "adjusted" or subtracted from the series prior to analysis. In *Design and Analysis of Time Series Experiments,* Glass et al. (1975) recommended deseasonalization, noting that it was the only practical means of handling seasonality. This recommendation was based on the limitations of time series software, however. Since that time, the state of the art in time series software has advanced to the point at which deseasonalization cannot generally be recommended. *Our comments in Section 2.1 about detrending a time series apply as well to deseasonalizing a time series: No adequate deseasonalization methods are available.*

The absolute "best" method of handling seasonality is to build a causal model of seasonal forces. On this point, we cite Nerlove:

> In one sense, the whole problem of seasonal adjustment of economic time series is a spurious one. Seasonal variations have causes (for example, variations in the weather), and insofar as these causes are measurable they should be used to explain changes that are normally regarded as seasonal . . . Ideally, one should formulate a complete econometric model in which the causes of seasonality are incorporated directly in the equations. . . . On the practical side the problems include the lack of availability of many relevant series, the non-measurability of key items, and the lack of appropriate statistical methodology . . . In addition, the precise structure of the model will very much affect the analysis of seasonal effects . . . On the conceptual side, the problem is basically one of continuing structural change, which is essentially the sort of thing which causes seasonality to show up [1964: 263].

In Chapter 5, where we develop bivariate ARIMA models, it will be clear that two time series

$$Y_t = N_t + S_t$$
$$X_t = N'_t + S'_t$$

may share the same set of seasonal causors. That is, for the seasonal components S_t and S'_t, there is some relationship

$$W_t \begin{array}{l} \nearrow S_t \\ \searrow S_t'. \end{array}$$

So in the bivariate model,

$$Y_t = f\,(X_{t-n}) + f\,(N_t, N_t')$$

the two seasonal components, S_t and S_t', will cancel each other out.

In the univariate situation, however, which is our only concern in this chapter, seasonality must be accounted for in the univariate ARIMA model. The seasonal ARIMA model, in contrast to deseasonalization methods, controls seasonal variance by incorporating seasonal correlations into the model. To illustrate this general principle, we may write out an N-month time series as

JAN	FEB	MAR	APR	MAY	JUN	JUL	AUG	SEP	OCT	NOV	DEC
Y_1	Y_2	Y_3	Y_4	Y_5	Y_6	Y_7	Y_8	Y_9	Y_{10}	Y_{11}	Y_{12}
Y_{13}	Y_{14}	Y_{15}	Y_{16}	Y_{17}	Y_{18}	Y_{19}	Y_{20}	Y_{21}	Y_{22}	Y_{23}	Y_{24}
$\vdots$											$\vdots$
$Y_{N-23} \cdots$											$\cdots Y_{N-12}$
$Y_{N-11} \cdots$											$\cdots Y_N$

Now an ARIMA (1,0,0) model uses the prior observation to predict the current observation, that is,

$$y_t = \phi_1 y_{t-1} + a_t.$$

The ARIMA (1,0,0) model defines a relationship between Y_1 and Y_2, between Y_2 and Y_3, and so forth. Similarly, an ARIMA (0,0,1) model

$$y_t = a_t - \Theta_1 a_{t-1}$$

defines a relationship between Y_1 and a_2, Y_2 and a_3, and so forth. If a time series is seasonal, however, it makes good sense to suspect that there will be similar relationships between Y_1, or January of the first year, and Y_{13}, January of the second year; between Y_2, February of the first year, and Y_{14}, February of the second year, and so forth.

Seasonal Nonstationarity

A process may drift or trend in annual steps or increments. Agricultural production series tend to exhibit nonstationarity of this sort due, we assume, to the prominence of crop seasons in the process. To account for seasonal drift or trend, difference the series seasonally. That is, for monthly data, subtract Y_1 from Y_{13}, Y_2 from Y_{14}, and so forth. We represent this procedure by the difference operator

$$(1 - B^{12})\, Y_t = \Theta_0$$

$$\text{or} \qquad Y_t = Y_{t-12} + \Theta_0.$$

Processes that are seasonally nonstationary drift or trend in annual steps such as

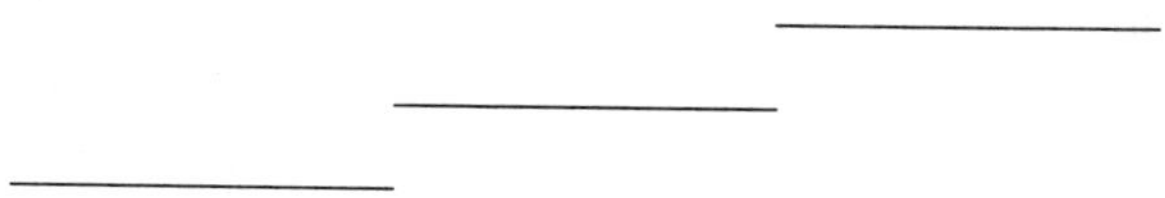

rather than as observation-to-observation steps.

Seasonal Autoregression

The current observation of a process may depend to some extent upon the corresponding observation from the preceding cycle or period. We express this relationship for monthly data as

$$y_t = \phi_{12} y_{t-12} + a_t$$

$$\text{or} \qquad (1 - \phi_{12} B^{12})\, y_t = a_t.$$

And, of course, the current observation may depend to some extent upon the corresponding observations from the two preceding cycles or periods:

$$(1 - \phi_{12} B^{12} - \phi_{24} B^{24}) y_t = a_t.$$

Of course, our comments on higher order autoregression apply here as well. For seasonal ARIMA structures, too, first-order structures are more common than second-order structures. Higher order structures are almost never encountered.

Seasonal Moving Averages

Finally, the current observation of a time series process may depend to some extent upon the random shocks from a year or two years earlier:

$$y_t = (1 - \Theta_{12}B^{12})a_t$$

or

$$y_t = (1 - \Theta_{12}B^{12} - \Theta_{24}B^{24})a_t.$$

And, of course, the seasonal ARIMA structure may be any combination of integrated, autoregressive, and moving average components.

The general ARIMA seasonal model is denoted by ARIMA (p,d,q) $(P,D,Q)_S$ where P, D, and Q are analogous to p, d, and q. The structural parameter S indicates the length of the naturally occurring period or cycle. Thus, for monthly data, S = 12. For quarterly data S = 4, and for weekly data S = 52.

Most time series with seasonal ARIMA behavior will exhibit regular ARIMA behavior as well. It might seem commonsensical to incorporate regular and seasonal structures additively into the ARIMA model. For example, a time series with both regular and seasonal autoregressive structures incorporated additively would be written as

$$(1 - \phi_1 B - \phi_{12}B^{12})y_t = a_t$$

or

$$y_t = \phi_1 y_{t-1} + \phi_{12} y_{t-12} + a_t.$$

However, regular and seasonal ARIMA structures are ordinarily incorporated *multiplicatively*. A time series with both regular and seasonal autoregressive structures would be written as

$$(1 - \phi_{12}B^{12})(1 - \phi_1 B)y_t = a_t.$$

The two autoregressive terms in this expression are called *factors*. The difference between additive and multiplicative seasonal models is made explicit by expanding the two-factor model:

$$(1 - \phi_{12}B^{12})(1 - \phi_1 B)y_t = a_t$$
$$(1 - \phi_1 B - \phi_{12}B^{12} + \phi_1\phi_{12}B^{13})y_t = a_t$$
$$y_t = \phi_1 y_{t-1} + \phi_{12} y_{t-12} - \phi_1\phi_{12} y_{t-13} + a_t.$$

The multiplicative model has a cross-product term, $\phi_1\phi_{12}B^{13}$, which an additive model lacks. Clearly, when both ϕ_1 and ϕ_{12} are small, their product, $\phi_1\phi_{12}$, is approximately zero and, as a result, there will be little difference between the additive and multiplicative models. When ϕ_1 and ϕ_{12} are not small, however, the two-factor multiplicative model reflects a much different process than the additive model. The two-factor model uses one extra piece of information (y_{t-13}) to predict the current observation, so it will ordinarily give a better fit to seasonal data than the additive model.

The model we have just demonstrated is an ARIMA (1,0,0) $(1,0,0)_{12}$ model. Other common seasonal models are (1) the ARIMA (0,0,1) $(0,0,1)_{12}$

$$\begin{aligned} y_t &= (1 - \Theta_1 B)(1 - \Theta_{12} B^{12}) a_t \\ &= (1 - \Theta_1 B - \Theta_{12} B^{12} + \Theta_1 \Theta_{12} B^{13}) a_t \\ &= a_t - \Theta_1 a_{t-1} - \Theta_{12} a_{t-12} + \Theta_1 \Theta_{12} a_{t-13} \end{aligned}$$

(2) the ARIMA (0,1,1) $(0,0,1)_{12}$

$$\begin{aligned} (1 - B) Y_t &= \Theta_0 + (1 - \Theta_1 B)(1 - \Theta_{12} B^{12}) a_t \\ &= \Theta_0 + (1 - \Theta_1 B - \Theta_{12} B^{12} + \Theta_1 \Theta_{12} B^{13}) a_t \\ &= \Theta_0 + a_t - \Theta_1 a_{t-1} - \Theta_{12} a_{t-12} + \Theta_1 \Theta_{12} a_{t-13} \end{aligned}$$

and (3) the ARIMA (0,1,1) $(0,1,1)_{12}$

$$\begin{aligned} (1 - B)(1 - B^{12}) Y_t &= \Theta_0 + (1 - \Theta_1 B)(1 - \Theta_{12} B^{12}) a_t \\ &= \Theta_0 + (1 - \Theta_1 B - \Theta_{12} B^{12} + \Theta_1 \Theta_{12} B^{13}) a_t \\ &= \Theta_0 + a_t - \Theta_1 a_{t-1} - \Theta_{12} a_{t-12} + \Theta_1 \Theta_{12} a_{t-13}. \end{aligned}$$

In the ARIMA (0,1,1) $(0,1,1)_{12}$ model, we describe a time series process that drifts or trends both regularly *and* seasonally. The order of differencing in practice (that is, regular differencing and then seasonal differencing or vice versa) does not matter because

$$\begin{aligned} (1 - B)(1 - B^{12}) &= (1 - B^{12})(1 - B) \\ &= (1 - B - B^{12} + B^{13}). \end{aligned}$$

So we difference the series regularly,

$$z_t = Y_t - Y_{t-1},$$

and then difference it seasonally,

$$z_t^* = z_t - z_{t-12},$$

or vice versa without changing the result. And, of course, we could accomplish both differences simultaneously with the operation

$$z_t^* = (1 - B)(1 - B^{12})\, Y_t$$
$$z_t^* = (1 - B - B^{12} + B^{13})\, Y_t$$
$$z_t^* = Y_t - Y_{t-1} - Y_{t-12} + Y_{t-13}.$$

Our comments about regular autoregression and moving averages are general to the seasonal cases as well. In particular, the ARIMA (p,d,q) $(P,D,Q)_S$ model must have parameters constrained to the bounds of stationarity which, for autoregressive models, are[9]

$$-1 < \phi_1, \phi_s < +1$$

and for the ARIMA (0,d,1) $(0,D,1)_S$ model, the bounds of invertibility are

$$-1 < \Theta_1, \Theta_S < +1.$$

Also, ARIMA (p,0,0) $(P,0,0)_S$ models may be written as an infinite series of exponentially weighted past shocks. Similarly, ARIMA (0,d,q) $(0,D,Q)_S$ models may be written as an infinite series of exponentially weighted past observations. The reader may demonstrate these truths by substitution. We will not do so here because it involves too much arithmetic. Nevertheless, it can be demonstrated that the seasonal autoregressive factor has as its inverse the infinite series

$$(1 - \phi_S B^S)^{-1} = 1 + \phi_S B^S + \phi_S^2 B^{2S} + \ldots + \phi_S^n B^{nS} + \ldots.$$

So to "solve" an ARIMA (1,0,0) $(1,0,0)_{12}$,

$$(1 - \phi_1 B)(1 - \phi_{12} B^{12}) y_t = a_t$$
$$y_t = (1 - \phi_1 B)^{-1} (1 - \phi_{12} B^{12})^{-1} a_t$$
$$= (1 + \phi_1 B + \phi_1^2 B^2 + \ldots + \phi_1^n B^n + \ldots)$$
$$(1 + \phi_{12} B^{12} + \phi_{12}^2 B^{24} + \ldots + \phi_{12}^n B^{12n} + \ldots) a_t.$$

As both infinite series converge, their product converges.

There should be some transfer of understanding here. The principles demonstrated for ARIMA (p,d,q) models generalize one-to-one to ARIMA $(P,D,Q)_S$ models. ARIMA(p,d,q)$(P,D,Q)_S$ models imply a stochastic behavior determined by the polynomial multiplication of ARIMA (p,d,q) and ARIMA $(P,D,Q)_S$ models. In the general case, the reader who understands the behavior of ARIMA (p,d,q) models and the rules of polynomial multiplication can deduce the behavior of ARIMA (p,d,q) $(P,D,Q)_S$ models. We will turn our attention to a more practical matter: identifying an appropriate seasonal model for a time series.

2.10 Identifying a Seasonal Model

As one might suspect, an ARIMA $(P,D,Q)_S$ model can be identified for a time series through an inspection of the *seasonal* lags of the ACF and PACF. For monthly data, the seasonal lags are -12, -24, and -36. For quarterly data, the seasonal lags are -4, -8, and -12. And in general, for the cycle S, the seasonal lags of the ACF and PACF are lags-S, -2S, and -3S. In Figure 2.10(a), we show the expected ACFs and PACFs for several ARIMA $(P,D,Q)_{12}$ processes. The patterns of spiking and decay for these ARIMA $(P,D,Q)_{12}$ processes are identical with the patterns expected of the analogous ARIMA (p,d,q) processes—except that the spiking and decay occur at seasonal lags.

First, seasonal nonstationarity is indicated by large and nearly equal values of the ACF at seasonal lags, that is,

$$\text{ACF (S)} \approx \text{ACF (2S)} \approx \ldots \approx \text{ACF (kS)}.$$

Seasonal differencing will make an ARIMA $(0,D,0)_S$ process stationary.

Second, ARIMA $(0,0,Q)_S$ processes are expected to have Q spikes at the first Q seasonal lags of the ACF. All successive lags are expected to be zero. The PACF of an ARIMA $(0,0,Q)_S$ process is expected to decay from seasonal lag to seasonal lag. The rate of decay is determined by the values of $\Theta_S, \Theta_{2S}, \ldots, \Theta_{QS}$.

Third, ARIMA (P,0,0) processes are expected to have a decaying ACF, the rate of decay determined by the values of $\phi_S, \phi_{2S}, \ldots, \phi_{PS}$. The PACF of an ARIMA (P,0,0) process is expected to have P spikes at the first P seasonal lags. All successive lags of the PACF are expected to be zero.

Of course, ARIMA $(P,D,Q)_S$ processes are rarely encountered in the social sciences. Most social science processes, if seasonal at all, are best represented by ARIMA (p,d,q) $(P,D,Q)_S$ models. The identification of an ARIMA (p,d,q) $(P,D,Q)_S$ model is complicated by interaction terms. Whereas an ARIMA (p,d,q) model can be identified on the basis of the first few lags of the ACF and PACF, and whereas an ARIMA $(P,D,Q)_S$ model can be identified on the basis of the first few seasonal lags of the ACF and PACF, an ARIMA (p,d,q) $(P,D,Q)_S$ model must be identified on the basis of the entire ACF and PACF.

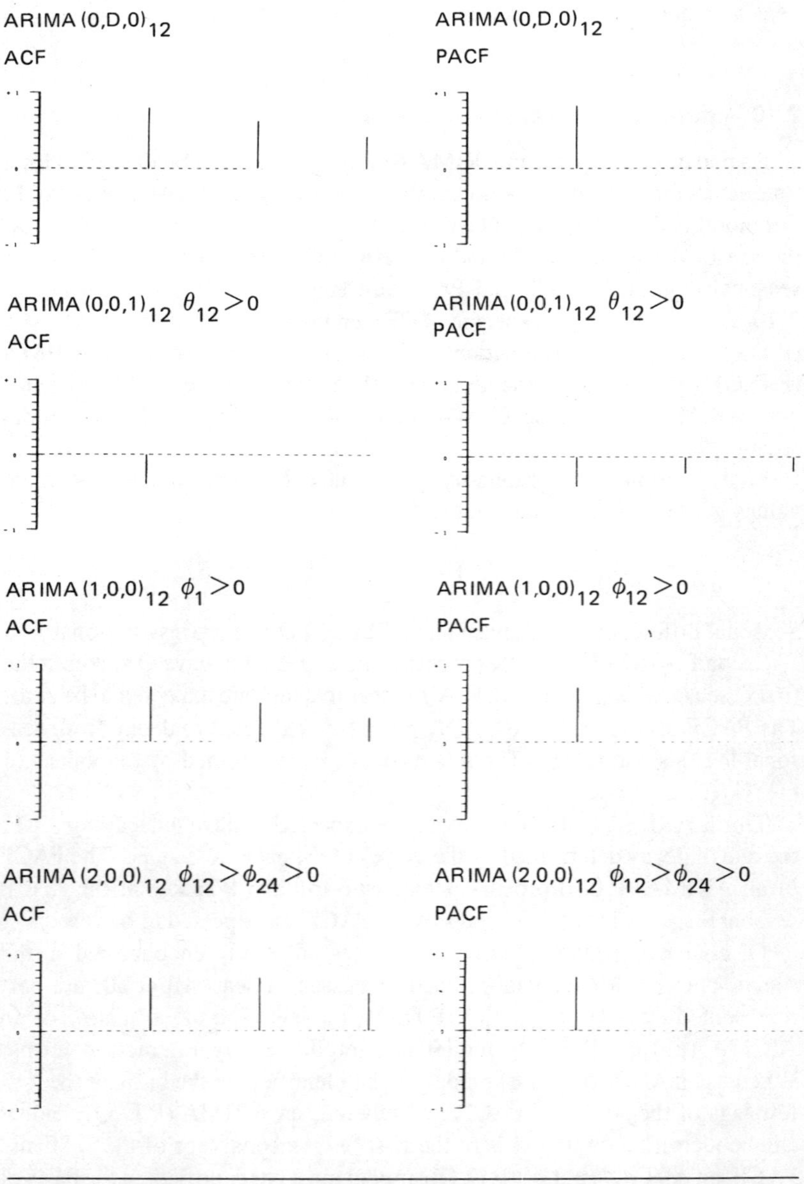

FIGURE 2.10(a) Expected ACFs for Several ARIMA $(P,D,Q)_{12}$ Processes

If it were not for the rather simple nature of social science time series processes, the identification of an ARIMA (p,d,q) $(P,D,Q)_S$ model would be an infinitely complicated task. There are a number of features which simplify the task, however.

For example, the analyst usually knows the value of S. For monthly data, S = 12. For quarterly data, S = 4, and so on. In other substantive areas, time series analysts may not know the value of S and this complicates model identification. Because we know the length of the seasonal cycle, we need only examine a few specific lags of the ACF and PACF to identify a model.

Similarly, social science processes typically have small integer values (almost always 0 or 1, sometimes 2) of p, q, P, and Q. If the ARIMA (p,d,q) $(P,D,Q)_S$ structural parameters took larger integer values, as seems to be the case in other substantive areas, the analyst would be forced to assess the statistical significance of many dozen lags of the ACF and PACF. This in turn would require much longer time series (say 300 observations or more) than are ordinarily available to the social scientist.

Finally, as a general rule, the regular and seasonal factors of an ARIMA (p,d,q) $(P,D,Q)_S$ model will be of the same type, that is, *either* autoregressive *or* moving average. If the regular factor is autoregressive, the analyst can usually rule out seasonal factors that are not autoregressive. As our discussion of parameter redundancy in section 2.7 suggests, ARIMA (p,0,0) $(0,0,Q)_S$ and ARIMA (0,0,q) $(P,0,0)_S$ models will often reduce to simpler ARIMA (p,0,0) $(P,0,0)_S$ and ARIMA (0,0,q) $(0,0,Q)_S$ models.

In Figure 2.10(b), we show the expected ACFs and PACFs of the most commonly encountered ARIMA (p,d,q) $(P,D,Q)_{12}$ processes. As shown, ARIMA (0,d,0) $(0,D,0)_{12}$ processes are characterized by persistently high values of the ACF at the regular *and* seasonal lags. The ACF shown indicates that the time series must be differenced both regularly and seasonally. Many social science time series processes are nonstationary only in the regular factor *or* the seasonal factor and, thus, should be differenced only regularly *or* seasonally. The ACFs of regularly, seasonally, and joint regularly/seasonally nonstationary processes are so distinctive that, in practice, the analyst will seldom mistake the type or number of differences required to make a process stationary.

An ARIMA (0,0,1) $(0,0,1)_{12}$ process is expected to have spikes at lags-11, -12, and -13 of the ACF. Higher lags of the ACF are expected to be zero. The PACF is expected to decay at both regular and seasonal lags but the key to identification is clearly the ACF. In the general case, an ARIMA (0,0,q) $(0,0,Q)_S$ process has a distinctive ACF with

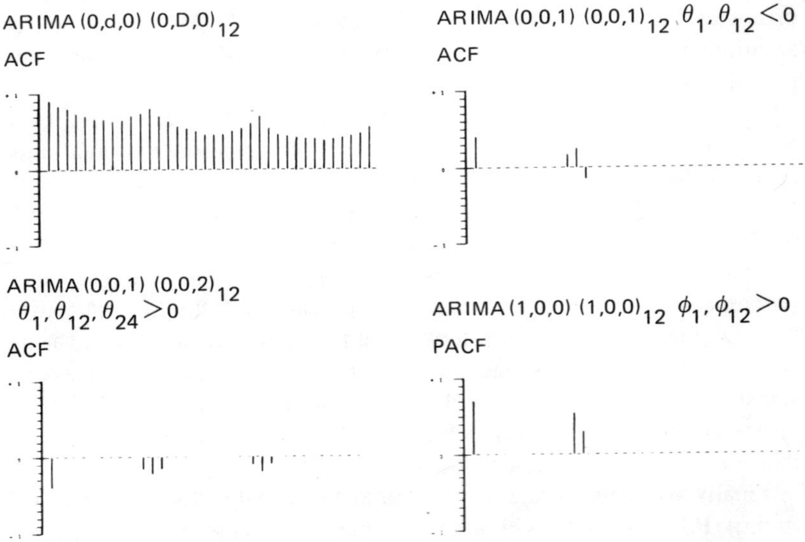

FIGURE 2.10(b) Expected ACF or PACF for Several ARIMA (p,d,q) $(P,D,Q)_{12}$ Processes

ACF (1), . . . , ACF (q) expected to be *non*zero
ACF (q+1) , . . . , ACF (S−q) expected to be zero
ACF (S−q+1) , . . . , ACF (S+q) expected to be *non*zero
ACF (S+q+1) , . . . , ACF (2S−q) expected to be zero
ACF (2S−q+1) , . . . , ACF (2S+q) expected to be *non*zero

and so forth. For Q>2, the process repeats itself. An ARIMA (0,0,1) $(0,0,2)_{12}$ process then is expected to have nonzero spikes at lags-11, -12, and -13 of the ACF and at lags-23, -24, and -25 of the ACF.

An ARIMA (1,0,0) $(1,0,0)_{12}$ process is expected to have an ACF which decays exponentially from seasonal lag to seasonal lag. The key to identification, however, is the PACF which will have a spike at lag-1 and at lags-12 and -13. In the general case, the PACF of an ARIMA (p,0,0) $(P,0,0)_{12}$ process is expected to have spikes at the first p lags, at lags S, . . . , SP, and at lags PS+p. For an ARIMA (2,0,0) $(2,0,0)_{12}$ process, then, we expect to see spikes at lags-1 and -2, at lags-12 and -13, and at lags-24 and -25 of the PACF. All other lags of the PACF are expected to be zero.

In theory, the identification of an ARIMA (p,d,q) $(P,D,Q)_S$ model for a time series reduces to a set of logical steps. First, use the estimated ACF to determine whether the series is stationary. If not, difference it appropriately. Second, use the ACF and PACF to determine the integer values of p and/or q. Third, use the ACF to determine the value of Q or the PACF to determine the value of P. Having thus identified an ARIMA (p,d,q) $(P,D,Q)_S$ model for the time series, we are ready to build the model.

2.11 Model Building

ARIMA modeling has been called an "art" by many authors. This seems to imply that one must be an artist (either by virtue of innate talent or lengthy training) to successfully create a model. We disagree. We prefer to think of ARIMA modeling as a *craft,* similar perhaps to carpentry. As a craft activity, basic ARIMA techniques are accessible to everyone after only a relatively short apprenticeship (usually an intensive workshop or a half-semester of coursework). And after acquiring a firm grasp of the essentials, one can develop journeyman skills by working with different types of data and more challenging substantive applications.

An ARIMA model is custom-built to fit a particular time series. Like a carpenter, the time series analyst uses tools (the statistical models we have developed), materials (the data), and plans (a model-building strategy) to create a model. In this section, we develop a detailed model-building strategy which, when followed by the analyst and supplemented by skills, will usually produce a sturdy, craftlike model. Our model-building strategy is essentially the one recommended by Box and Jenkins (1976) with the addition of several procedures we have found valuable in the analysis of social science data.

Our model-building strategy is generally conservative. We prefer to see a simple, robust model rather than a flimsy, flashy one. If a model does not fit the data well, we expect the craftsman to acknowledge that fact (rather than trying to force a fit by bending the model parts out of shape). Having built the model, the craftsman will critically evaluate its quality and, if found wanting, will make appropriate adjustments or will report its shortcomings.

ARIMA models are built through an iterative identification/estimation/diagnosis strategy which we have outlined as a flow chart in Figure 2.11(a). Before starting, the analyst should inspect a plot of the raw time series. Particular attention should be paid to sources of nonstationarity which may be visible in the series plot. While nonstationarity due only to systematic trend or drift is easily detected in an inspection of the series ACF, nonstationarity associated with other causes (variance nonstationarity, for exam-

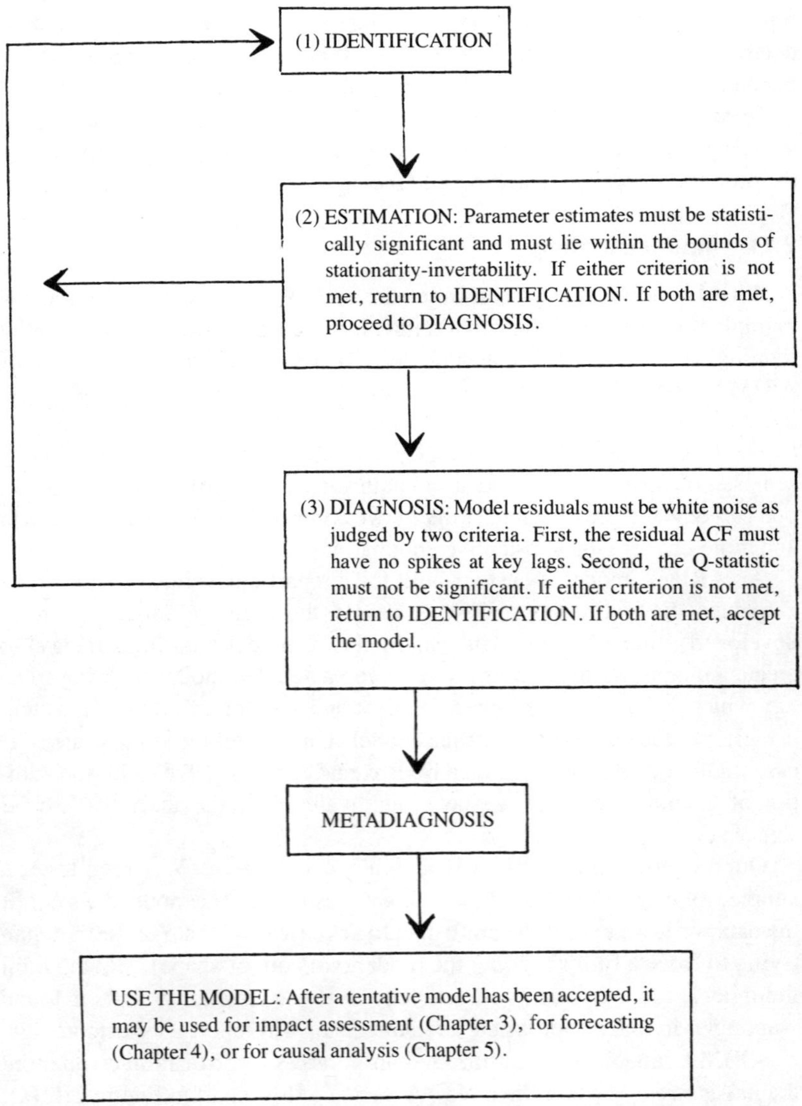

FIGURE 2.11(a) The ARIMA Model-Building Strategy

ple) can usually be detected only through an inspection of the time series plot.

The iterative identification/estimation/diagnosis strategy begins with the *identification* of a tentative ARIMA model for the series. Patterns of autocorrelation observed in the data are compared with the patterns expected of various ARIMA models. If nonstationarity is indicated—by an ACF which fails to "die out"; see Figure 2.8(a)—it will be necessary to difference and/or transform the series prior to identifying a tentative model.

Next, the parameters of the tentative model are *estimated*. All parameter estimates must lie within the bounds of stationarity-invertibility and must be statistically significant. If the parameter estimates do not satisfy these criteria, a new model must be identified and its parameters estimated.

After a tentative model has been identified and its parameters satisfactorily estimated, it must be *diagnosed*. To pass diagnosis, the residuals of the tentative model must be white noise. If this criterion is not satisfied, the tentative model is inadequate and must be rejected; the model-building procedure begins anew. Another model is identified, its parameters estimated, and its residuals diagnosed. The iterative identification/estimation/diagnosis procedure continues until an adequate model has been created for the time series.

The strategy we recommend is a model-*building* strategy. It leads to a model that is statistically adequate and yet parsimonious for a given time series. Alternative strategies might be better described as model-*fitting* strategies. The analyst might begin, for example, by fitting a general ARIMA (p,d,q) $(P,D,Q)_S$ model to the time series and then deleting unnecessary terms from the model. This alternative strategy will generally lead to a model that is statistically adequate (that "fits" the data, that is) but that is not necessarily parsimonious. A model that is adequate but not parsimonious is arbitrary and may lead to a set of confused and confusing inferences.

Our treatment of ARIMA modeling so far has been in the abstract. We have been concerned primarily with "expected" statistics and "general" cases. We will now consider the more practical aspects of modeling, starting with a more detailed description of the general model-building strategy.

Identification

Identification is the key to model building. An ARIMA model must have some empirical basis. That is, put simply, there should be some reason for selecting one tentative model over another. The empirical basis will ordinarily be the patterns of autocorrelation found in the ACFs and PACFs estimated from the time series. If two competing models are both adequate, the model that best fits the ACF and PACF is the "better" of the two models.

In practice, estimated ACFs and PACFs will not be identical to the expected ACFs and PACFs shown in Figures 2.8(a), 2.8(b), 2.10(a), and 2.10(b). An ARIMA (1,0,0) process, for example, is expected to show "perfect" exponential decay in the ACF and to have a single spike at PACF(1). These expected patterns can be counted on only when the process realization (the time series, that is) is infinitely long, however. If a time series is not infinitely long, the estimated ACF and PACF will not "perfectly" match the expected ACF and PACF of the underlying process.

Ambiguity in identification sometimes amounts to differences of opinion or interpretation. One analyst may see two spikes in the estimated ACF whereas some other analyst may see only one spike. The first analyst will then conclude that an ARIMA (0,0,2) model adequately represents the series

$$y_t = (1 - \Theta_1 B - \Theta_2 B^2)a_t$$

while the second analyst will conclude that an ARIMA (0,0,1) model adequately represents the series

$$y_t = (1 - \Theta_1 B)a_t.$$

When we discuss parameter estimation, it will be apparent that differences of opinion such as this can be decided absolutely. For the time being, however, we note that ambiguity in estimated ACFs and PACFs can be lessened somewhat by placing confidence bands around the ACFs and PACFs. For the ACF, standard errors (SE) of the ACF(k) are estimated from the formula

$$\text{SE [ACF (k)]} = \sqrt{1/N(1 + 2 \sum_{i=1}^{k} [\text{ACF (i)}]^2}.$$

For the PACF, standard errors of the PACF (k) are estimated from the formula

$$\text{SE [PACF (k)]} = \sqrt{1/N}.$$

In the example analyses of Section 2.12, we will plot confidence intervals around the ACF and PACF at ± 2 SE. Values of ACF (k) and PACF (k) which lie inside this interval will be considered not significantly different than zero.

The difference between the pattern of autocorrelation in the theoretical ACF of an infinitely long series and the pattern generated by finite samples is vividly illustrated in Figures 2.11(b) and 2.11(c). Figure 2.11(b) presents lags 1 to 10 of the theoretical ACF of the ARIMA (1,0,0) model

$$y_t = .5y_{t-1} + a_t.$$

Using this same model, we have generated six realizations of the process, each 100 observations long. Each realization was generated with NID (0,1) random shocks. Figure 2.11(c) shows the first 10 lags of the ACF for these realizations.

Note that, even though these six realizations were generated by the same ARIMA (1,0,0) process, there is a wide range in their estimated ACFs. The distinctive exponential decay pattern of an ARIMA (1,0,0) ACF is clearly seen in some of the estimated ACFs but is largely obscured in others. The estimated PACFs for these six realizations (not presented) clearly indicate an ARIMA (1,0,0) process, however, with statistically significant estimates of PACF(1) and all other lags not statistically significant. It is also important to note that, as the length of realization increases, the pattern of decay in the estimated ACF converges to the pattern expected of an ARIMA (1,0,0) process. If these six realizations were 200 observations long, identification of an ARIMA (1,0,0) model from the ACF would be more certain.

These simulated identifications illustrate the value of using all available data. The analyst should also use both the ACF and the PACF to identify a tentative model, rather than relying solely on the ACF and its SE. In general, a conservative approach to identification is urged. The more parsimonious

```
-1   -.8  -.6  -.4  -.2   0.   .2   .4   .6   .8  +1
+----+----+----+----+----+----+----+----+----+----+
                    (    IXXXX)XXXXXXXX
                   (     IXXXXX*
                   (     IXXXX )
                   (     IXX   )
                   (     IX    )
                   (     I     )
                   (     I     )
                   (     I     )
                   (     I     )
                   (     I     )
```

FIGURE 2.11(b) Expected ACF for the Process $Y_t = .5y_{t-1} + a_t$; to aid comparison with figure 2.11(c), standard errors have been calculated with T=100 observations

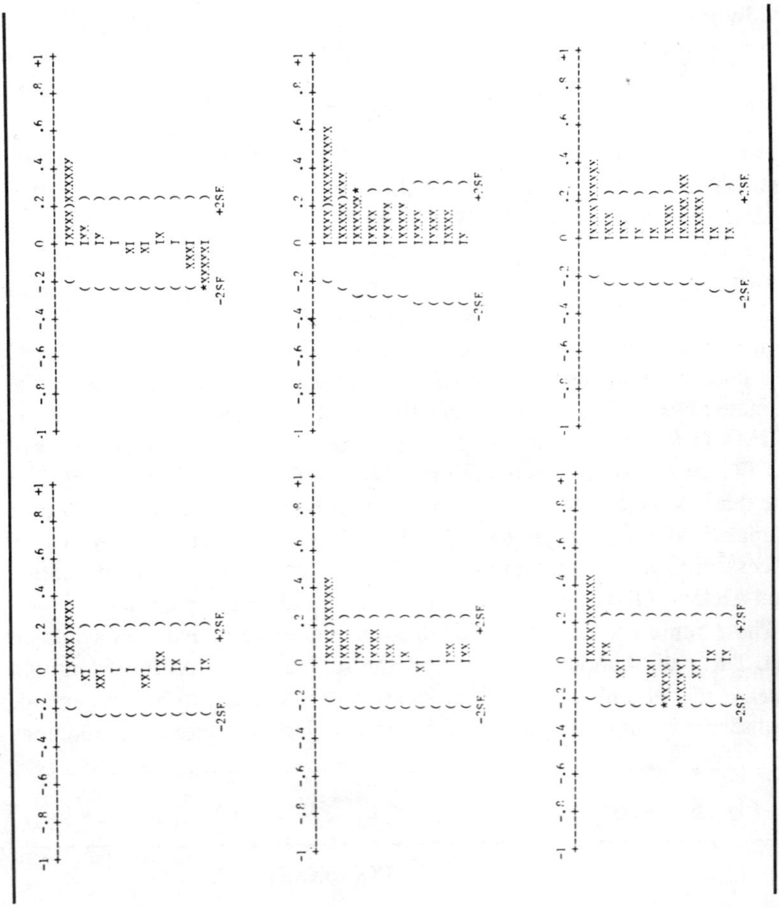

FIGURE 2.11(c) Estimated ACFs for the Process

models (p,P,q, and Q = 0 or 1) should be ruled out before more complicated models are entertained. Should a more parsimonious model prove inadequate, the inadequacy will become apparent at a later stage of the model-building strategy. An overly complicated model, on the other hand, may pass through all subsequent stages of the strategy without notice.

The conservative approach is especially urged when differencing is an issue. The impatient analyst may difference a time series generated by a stationary process and, as a result, may select an overly complicated, cumbersome model. The costs of *over*differencing a time series are easily demonstrated. A relatively simple ARIMA (0,0,1) process, for example,

$$y_t = (1 - \Theta_1 B)a_t,$$

when differenced becomes

$$\begin{aligned} y_t - y_{t-1} &= (1 - \Theta_1 B)a_t - (1 - \Theta_1 B)a_{t-1} \\ &= a_t - \Theta_1 a_{t-1} - a_{t-1} + \Theta_1 a_{t-2} \\ &= a_t - (\Theta_1 + 1)a_{t-1} + \Theta_1 a_{t-2} \end{aligned}$$

an ARIMA (2,0,0) process. Moreover, as the parameters of this process are likely to exceed the bounds of invertibility, the analyst may end up with a high-order ARIMA (p,0,0) model for the time series. Whenever the realization of a stationary process is differenced (when the time series is overdifferenced, that is), autoregressive or moving average structures must be incorporated in the model to remove the effects of differencing. This is a classic example of parameter redundancy. To avoid this problem, a time series should not be differenced unless its ACF clearly indicates a nonstationary process. And of course, a time series should never be differenced before its ACF is examined.

Estimation

Estimation follows identification. Having tentatively identified an ARIMA (p,d,q) $(P,D,Q)_S$ model for the time series, the ϕ_p, ϕ_P, Θ_q, and Θ_Q parameters of the tentative model must be estimated. *An ARIMA (p,d,q) $(P,D,Q)_S$ model is generally nonlinear in its parameters and this means that standard OLS regression software such as SPSS cannot be used for estimation.* All university computing centers will have either a general nonlinear regression program or a software package designed especially for ARIMA estimation.[10] The analyst will input the time series and an ARIMA (p,d,q) $(P,D,Q)_S$ model and will receive parameter estimates as output. The analyst will have two major concerns here.

First, the estimated autoregressive and moving average parameters should be statistically significant. Any parameter whose estimated value is not significantly different than zero should be dropped from the tentative model. In our discussion of identification, we cited a hypothetical situation in which ARIMA (0,0,1) and ARIMA (0,0,2) models were posited for the same time series, noting that this type of disagreement could be resolved absolutely. In fact, if the estimate of Θ_2 for the ARIMA (0,0,2) model is not statistically significant, the Θ_2 parameter should be dropped from the model.

Second, the estimated autoregressive and moving average parameters must lie within the bounds of stationarity and invertibility. If the estimated parameters of the tentative model do not satisfy the stationarity-invertibility conditions, then the tentative model must be rejected. If a stationary time series has been differenced incorrectly, or if a nonstationary time series has not be differenced, autoregressive and/or moving average parameter estimates will invariably exceed the bounds of stationarity or invertibility. Whatever the cause of the problem, however, the tentative model must be rejected if the estimated parameters do not satisfy the stationarity-invertibility conditions.

Diagnosis

Having identified a tentative ARIMA (p,d,q) $(P,D,Q)_S$ model, and having satisfactorily estimated its parameters, the model must be *diagnosed.* A statistically *adequate* model is defined as one whose residuals ($\hat{a}_t$) satisfy two diagnostic criteria.

First, the residuals of the tentative model must be independent at a first and second lag, that is,

$$E(\hat{a}_t \hat{a}_{t+1}) = E(\hat{a}_t \hat{a}_{t+2}) = 0.$$

To evaluate the statistical adequacy of the tentative model by this criterion, we estimate an ACF for the residuals. If the tentative model is statistically adequate by this criterion, then

$$\text{ACF}(1) = \text{ACF}(2) = 0.$$

The ACF for the model residuals must have no statistically significant values at the first two lags, that is, ACF (1) and ACF (2) must lie within the confidence intervals we have plotted around the ACF. For seasonal data, we will also require

$$\text{ACF}(S) = \text{ACF}(2S) = 0,$$

that is, the model residuals must be free of autocorrelation at the seasonal lags. Box and Jenkins (1976: 291) note that the standard error of the ACF may significantly underestimate the true standard error of the residual ACF depending on model form and parameter values. Therefore, in residual diagnosis, the ACF standard error should be considered an upper limit. If a low-order lag autocorrelation is slightly less than two standard errors in magnitude, then the prudent analyst may wish to consider it significant for diagnostic purposes.

If the tentative model is not statistically adequate by this criterion, the analyst may often remedy the situation by increasing the values of p, P, q, or Q. In the hypothetical situation of ARIMA (0,0,1) and ARIMA (0,0,2) models posited for the same time series, we see that the disagreement could be settled at the diagnosis stage also. If the analyst posits an ARIMA, (0,0,1) model but finds statistically significant autocorrelation at lag-2 of the residual ACF, then the ARIMA (0,0,1) model must be rejected. Similarly, if the analyst posits an ARIMA (p,d,q) model for a time series but finds statistically significant autocorrelation at seasonal lags of the residual ACF, then the ARIMA (p,d,q) model must be rejected in favor of an ARIMA (p,d,q) $(P,D,Q)_S$ model.

The second criterion of statistical adequacy is that the residuals of the tentative model must be distributed as white noise. The ACF of a white noise process is expected to be uniformly zero. For 20 or 30 lags of an ACF, however, given a .05 level of statistical significance, we anticipate that there will be two or three significant spikes by chance alone. To test whether the *entire* residual ACF is different from what would be expected of a white noise process, the analyst may use a Q statistic given by the formula[11]

$$Q = N \sum_{i=1}^{k} [\mathrm{ACF}(i)]^2$$

with df = k − p − q − P − Q. The Q statistic is distributed approximately chi-square with the degrees of freedom as indicated. A null hypothesis that the model residuals are white noise is:

$$H_0\text{: ACF}(1) = \ldots = \text{ACF}(k) = 0.$$

That is, the null hypothesis states that the residual ACF is not different than a white noise ACF. If the Q statistic for the residual ACF is significant, the null hypothesis must be rejected; the model residuals are not white noise, so the tentative model is not statistically adequate and must be rejected. If the Q statistic is not statistically significant, the null hypothesis is not rejected; the model residuals are not significantly different from white noise and the tentative model is accepted.

The Q statistic is sensitive to the value of k, that is, to the number of lags in the residual ACF. For a relatively long ACF, say 50 lags, the Q statistic is likely to understate the serial correlation in the model residuals. Even if the residuals are not white noise, the Q statistic for a relatively long ACF is not likely to be statistically significant. The problem here is that, using an ACF of 20 lags, the null hypothesis might be rejected. For the same set of residuals, using an ACF of 40 lags, the null hypothesis might not be rejected. To solve this problem, one should use an ACF that is neither "too long" nor "too short." In the example analyses in Section 2.12, we usually use ACFs of 25 lags. This length allows us to examine the seasonal lags of monthly data (lags 12 and 24 of the ACF and PACF) while still allowing a fair test for the Q statistic. In our experience, 25 lags is usually neither "too long" nor "too short" for most data but, of course, this length is somewhat arbitrary. Granger and Newbold (1977) do, however, recommend that a minimum of 20 lags always be used in the calculation of Q.

We note finally that each of the two criteria of statistical adequacy is *necessary* but not *sufficient* grounds for accepting a tentative model. Obviously, a set of model residuals can meet one criterion but not the other. To be accepted, however, the tentative model must be statistically adequate; and to be statistically adequate, the tentative model must meet both of these diagnostic criteria.

A variety of other residual checks may prove useful in diagnosing the estimated model. Box and Jenkins (1976) suggest methods for checking for Normality and for investigating possible seasonal dependencies in the residuals. The latter are often of doubtful utility given the limited length of many social science time series. We strongly recommend inspecting a plot of the residual series and a plot of the predicted values versus the observed values. Both of these are invaluable for assessing the fit and adequacy of the model, particularly with regard to potential sources of variance nonstationarity such as outliers and variance proportional to the series level.

Metadiagnosis

After the tentative model has been identified, its parameters estimated, and its residuals diagnosed, the analyst may accept the tentative model. However, the prudent analyst may wish to consider a set of factors over and above those implied by the identification/estimation/diagnosis strategy. We call this procedure *meta*diagnosis.

Having accepted the tentative model, the analyst can be sure only that the model is statistically adequate and parsimonious. These are *relative* qualities of the model which say very little about certain absolute concerns. In building an ARIMA model for a time series, the analyst plans to *use* the model for some purpose. We do not discuss the uses of ARIMA models in this chapter. But generally, the analyst plans to use the ARIMA model for impact assessment (which we discuss in Chapter 3), for forecasting (which we cover in Chapter 4), or for multivariate causal analyses (which we cover in Chapter 5). If the ARIMA model is used for any or all of these purposes, the analyst should consider the absolute qualities of the model in light of its projected use.

First, in absolute terms, how good is the ARIMA model? There are a number of criteria which could be used for this assessment. However, the most reasonable criterion would seem to be the R^2 statistic computed as

$$R^2 = 1 - \frac{\text{residual sum of squares}}{\text{total sum of squares}} = 1 - \sum_{t=1}^{N} \frac{\hat{a}_t^2}{Y_t^2}.$$

The R^2 statistic has the same interpretation here as in cross-sectional multiple regression analysis. It is the percent of variance in the time series that is "explained" by the model. The reader who is not familiar with time series methods may be surprised to learn that time series models routinely have R^2 statistics higher than .9. Of course, the greatest portion of this explained variance is due to the parameter Θ_0. A more realistic R^2 statistic, then, will be:

$$R^2 = 1 - \sum_{t=1}^{N} \frac{\hat{a}_t^2}{y_t^2}.$$

where $y_t = Y_t - \Theta_0$ for a stationary process and $y_t = z_t - \Theta_0$ for a nonstationary process. By subtracting the parameter Θ_0 from each time series observation, the analyst obtains an R^2 statistic which measures the percent variance explained only by the autoregressive and/or moving average parameters of the model. Naturally, the analyst will require a relatively high R^2 statistic for the model.

A statistic related to the R^2 is the residual mean square (RMS) statistic computed from the formula

$$\text{RMS} = 1/N \sqrt{\sum_{t=1}^{N} \hat{a}_t^2}$$

for a set of N residuals. Like the R^2, the RMS statistic gives a "goodness-of-fit" measure for the model. Unlike the R^2, however, the RMS statistic is not standardized. By tradition, the RMS statistic is more widely used in time series analysis than the R^2 and we will follow that tradition so far as possible.

Metadiagnosis ordinarily begins with *over*modeling. If the iterative identification/estimation/diagnosis procedure has lead to an ARIMA (0,1,1) model for a time series, the analyst should try to fit an ARIMA (0,1,2) model. If the ARIMA (0,1,1) model has been judged statistically adequate in diagnosis, the estimated Θ_2 parameter of the ARIMA (0,1,2) model should be statistically insignificant. In general, overmodeling amounts to increasing the values of p, P, q, and Q for the ARIMA (p,d,q) $(P,D,Q)_S$ model and, in general, if the accepted model has been judged statistically adequate, the estimated parameters of the higher order model should be statistically insignificant.

Another dimension of overmodeling might be called *under*modeling. In Sections 2.5 and 2.6, we demonstrated that an ARIMA (1,0,0) model could be expressed as an infinite order ARIMA (0,0,q) model, that is,

$$\begin{aligned} y_t &= \phi_1 y_{t-1} + a_t \\ &= a_t + \phi_1 a_{t-1} + \phi_1^2 a_{t-2} + \ldots + \phi_1^n a_{t-n} + \ldots, \end{aligned}$$

and that an ARIMA (0,0,1) model could be expressed as an infinite order ARIMA (p,0,0) model, that is,

$$\begin{aligned} y_t &= a_t - \Theta_1 a_{t-1} \\ &= a_t - \Theta_1 y_{t-1} - \Theta_1^2 y_{t-2} - \ldots - \Theta_1^n y_{t-n} - \ldots, \end{aligned}$$

and due to the conditions of stationarity-invertibility, the infinite series both converge absolutely to zero. More to the point, after two or three terms, for small values of ϕ_1 and Θ_1, the values of ϕ_1^n and Θ_1^n are approximately zero. Given this, it is possible that an ARIMA (0,0,2) model might be approximately identical with an ARIMA(1,0,0) model. In practice, then, if the accepted model is ARIMA(0,0,2),

$$y_t = a_t - \Theta_1 a_{t-1} - \Theta_2 a_{t-2}$$

and if $\Theta_2 \simeq \Theta_1^2$, the analyst can almost always find a statistically adequate ARIMA (1,0,0) model for the series

$$y_t = \phi_1 y_{t-1} + a_t$$

where $\phi_1 \simeq \Theta_1$ Now which of these two alternative models is the "better?"

In this case, the competition might be decided on the basis of parsimony alone. The ARIMA (1,0,0) model has only one parameter and this is "better" than the ARIMA (0,0,2) model. The analyst must remember, however, that the ARIMA (1,0,0) and ARIMA (0,0,2) models are only *approximately* identical. In fact, if the underlying process is ARIMA (0,0,2), an ARIMA (0,0,2) model will be the "better" under all circumstances. Parsimony not withstanding, the RMS statistics of the two models can be used as a measure of quality. If the ARIMA (1,0,0) model has a lower (or even approximately equal) RMS statistic, it must be deemed the "better" model because it is the more parsimonious of the two. But if the ARIMA (0,0,2) model has a significantly lower RMS statistic, the rule of parsimony may be waived.

The rough equivalence of autoregressive and moving average models becomes more of a problem when seasonal factors are considered. An ARIMA (0,0,1) $(0,0,2)_{12}$ model, for example,

$$y_t = (1 - \Theta_1 B)(1 - \Theta_{12}B^{12} - \Theta_{24}B^{24})a_t$$

will be approximately identical with an ARIMA (0,0,1) $(1,0,0)_{12}$ model

$$(1 - \phi_{12}B^{12})y_t = (1 - \Theta_1 B)a_t$$

whenever $\Theta_{24} \simeq \Theta_{12}^2$. The reader may demonstrate this simple fact by "solving" the ARIMA (0,0,1) $(1,0,0)_{12}$ model. Whenever P or Q is greater than one,then, the prudent analyst will undermodel the series, comparing alternatives.

Metadiagnosis is perhaps the most critical stage of the model-building strategy. In metadiagnosis, the analyst plays the role of devil's advocate, criticizing and arguing as best as possible that a "better" model can be found. Failing, the analyst should be convinced that the "best" possible model has been built for the time series. We will return to this topic in the next few chapters, discussing metadiagnostic techniques which pertain to the specific applications of ARIMA modeling: impact analysis, forecasting, and multivariate analysis.

2.12 Example Analyses

We suspect that, after digesting the material preceding this, the reader still has a number of unanswered questions. In this section, we hope to answer many of these questions by presenting a few in-depth example analyses. The series we will analyze here are listed in an appendix to this volume and the reader is invited to replicate our analyses, checking (or challenging) our results. If our experiences in teaching time series analysis are typical, many questions the reader may have can be answered only through personal experience. The surest way to learn time series analysis is through informed practice.

We have selected these series for analysis because each illustrates one or more practical problems that the analyst is likely to encounter. Collectively, we have analyzed hundreds of social science time series in the last few years and the series we analyze here are "typical." We include stationary series and nonstationary series and one time series that is nonstationary in both level and variance. One of the series is "not a time series" in the strictest sense and another is distorted by a deviant value or outlier.

The ACFs and PACFs for these analyses are the printed output of SCRUNCH (Hay, 1979), an interactive software package for Box-Jenkins time series analysis. The output of SCRUNCH is similar to most Box-Jenkins time series packages. Parentheses about the ACFs and PACFs indicate confidence intervals of ± 2 standard errors. Hence, any estimated ACF(k) or PACF(k) within the parentheses are not significantly different than zero. The values of each ACF (k) or PACF (k) along with their respective standard errors are listed alongside the correlogram plot.

2.12.1 Sutter County Workforce

The time series plotted in Figure 2.12.1(a) are monthly workforce statistics (the total number of people employed in the workforce) for Sutter County, California. The first observation of this series is January, 1946. The 252nd observation is December 1966. From an eyeball inspection of the plotted series, it seems obvious that this series is nonstationary and seasonal. In fact, as the series level appears to increase in annual steps, we suspect that this series may be seasonally nonstationary. We will reserve judgment on this issue until after we have inspected the ACFs and PACFs, however. *And of course, we will follow the model-building strategy we have outlined in Figure 2.11(a).*

Identification. The ACF of the raw time series shown in Figure 2.12.1(b) indicates nonstationarity as we had suspected. There is no evidence of decay and the high-order lags remain significant. The series must be differenced. The ACF of the regularly differenced series shows seasonal nonstationarity as well. The basis for this identification is seen at lags−12 and −24 of the ACF. Both lags are large and nearly equal. After differencing this series both regularly and seasonally, the ACF and PACF suggests an ARIMA (0,1,1) $(0,1,1)_{12}$ model. We arrive at this identification by noting the ACF spikes at lag−1 and lag−12 while the PACF exhibits rough patterns of decay beginning at lag−1 and lag−12. We write this tentative model as

$$(1 - B)(1 - B^{12})Y_t = \Theta_0 + (1 - \Theta_1 B)(1 - \Theta_{12} B^{12})a_t$$

or

$$Y_t = \frac{\Theta_0 + (1 - \Theta_1 B)(1 - \Theta_{12} B^{12})}{(1 - B)(1 - B^{12})} a_t.$$

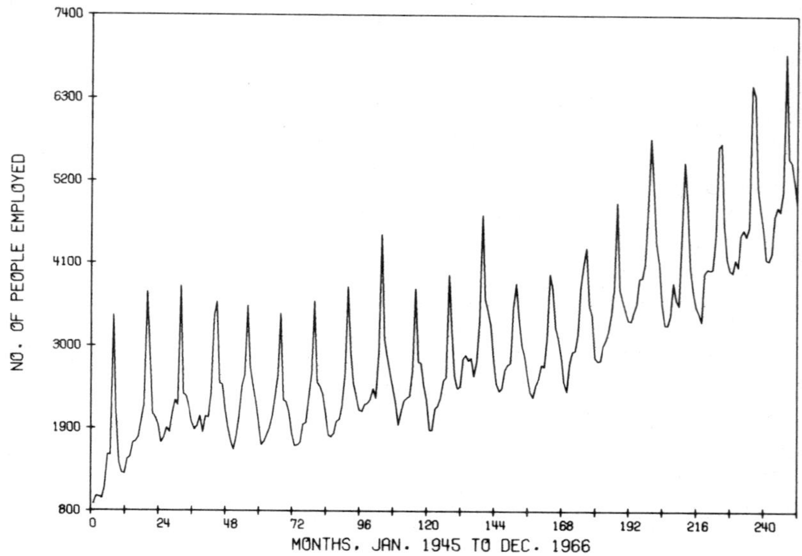

FIGURE 2.12.1(a) Sutter County Workforce

The model has three parameters, Θ_0, Θ_1, and Θ_{12}, which must now be estimated.

Estimation. Parameter estimates for the tentative model are[12]:

$$\hat{\Theta}_0 = .52 \text{ with t statistic } = .22$$
$$\hat{\Theta}_1 = .60 \text{ with t statistic } = 11.38$$
$$\hat{\Theta}_{12} = .68 \text{ with t statistic } = 13.33.$$

First, we note that both $\hat{\Theta}_1$ and $\hat{\Theta}_{12}$ lie within the bounds of invertibility for moving average parameters, so both parameters are acceptable by that criterion. Using a .05 level of significance, however, we require a t statistic of ± 1.96 and, at this level, the estimated value of Θ_0 is not statistically different than zero. What this means is that the upward motion of this time series is not significantly different than *drift* and, as a result, Θ_0 is dropped from our tentative model. Both $\hat{\Theta}_1$ and $\hat{\Theta}_{12}$ are statistically significant, so our tentative model is:

$$(1 - B)(1 - B^{12})Y_t = (1 - .60B)(1 - .68B^{12})a_t.$$

(text continued on p. 110)

```
SERIES.. EMPLOY   (NOBS= 252)  SUTTER COUNTY EMPLOYMENT, 1/45 TO 12/66
NO. OF VALID OBSERVATIONS  =   252.

AUTOCORRELATIONS OF LAGS 1 - 30.
Q( 30, 252) =  2277.8       SIG =     0.000

  LAG   CORR    SE  -1    -.8  -.6  -.4  -.2   0    .2   .4   .6   .8   +1
                     +----+----+----+----+----+----+----+----+----+----+
   1   .874  .063                       (    IXX)XXXXXXXXXXXXXXXXXXX
   2   .757  .100                      (     IXXXX)XXXXXXXXXXXXXX
   3   .664  .121                     (      IXXXXX)XXXXXXXXXXX
   4   .585  .134                    (       IXXXXXX)XXXXXXXX
   5   .514  .144                    (       IXXXXXX)XXXXXX
   6   .484  .151                   (        IXXXXXXX)XXXX
   7   .490  .157                   (        IXXXXXXX)XXXX
   8   .534  .163                   (        IXXXXXXX)XXXXX
   9   .580  .170                  (         IXXXXXXXX)XXXXX
  10   .633  .178                  (         IXXXXXXXX)XXXXXXX
  11   .722  .186                  (         IXXXXXXXX)XXXXXXXXX
  12   .792  .197                 (          IXXXXXXXXX)XXXXXXXXXX
  13   .700  .209                 (          IXXXXXXXXX)XXXXXXX
  14   .595  .219                (           IXXXXXXXXXX)XXXX
  15   .512  .225                (           IXXXXXXXXXX)XX
  16   .432  .229                (           IXXXXXXXXXXX*
  17   .368  .233               (            IXXXXXXXXX   )
  18   .341  .235               (            IXXXXXXXXX   )
  19   .347  .237               (            IXXXXXXXXX   )
  20   .392  .239               (            IXXXXXXXXXX )
  21   .444  .241               (            IXXXXXXXXXXX)
  22   .499  .245               (            IXXXXXXXXXXXX*
  23   .578  .249               (            IXXXXXXXXXXXX)XX
  24   .653  .254              (             IXXXXXXXXXXXXX)XXX
  25   .578  .261              (             IXXXXXXXXXXXXX)X
  26   .486  .266              (             IXXXXXXXXXXXXX)
  27   .405  .269              (             IXXXXXXXXXX   )
  28   .333  .272             (              IXXXXXXXX      )
  29   .278  .273             (              IXXXXXXX       )
  30   .249  .274             (              IXXXXXX        )
                            -2SE                          +2SE

PARTIAL AUTOCORRELATIONS OF LAGS 1 - 30.

  LAG   CORR    SE  -1    -.8  -.6  -.4  -.2   0    .2   .4   .6   .8   +1
                     +----+----+----+----+----+----+----+----+----+----+
   1   .874  .063                         (  IXX)XXXXXXXXXXXXXXXXXXX
   2  -.027  .063                         ( XI  )
   3   .038  .063                         (  IX )
   4   .007  .063                         (  I  )
   5  -.008  .063                         (  I  )
   6   .142  .063                         (  IXX)X
   7   .149  .063                         (  IXX)X
   8   .217  .063                         (  IXX)XX
   9   .114  .063                         (  IXX*
  10   .165  .063                         (  IXX)X
  11   .336  .063                         (  IXX)XXXXX
  12   .224  .063                         (  IXX)XXX
  13  -.512  .063               XXXXXXXXXX(XXI  )
  14  -.163  .063                        X(XXI  )
  15  -.005  .063                         (  I  )
  16  -.047  .063                         ( XI  )
  17   .024  .063                         (  IX )
  18   .003  .063                         (  I  )
  19  -.004  .063                         (  I  )
  20   .062  .063                         (  IXX)
  21   .106  .063                         (  IXX*
  22   .052  .063                         (  IX )
  23  -.009  .063                         (  I  )
  24   .200  .063                         (  IXX)XX
  25  -.182  .063                       XX(XXI  )
  26  -.072  .063                         (XXI  )
  27  -.018  .063                         (  I  )
  28   .012  .063                         (  I  )
  29   .022  .063                         (  IX )
  30  -.050  .063                         ( XI  )
                                       -2SE  +2SE
```

FIGURE 2.12.1(b) ACF and PACF for the Raw Series

```
SERIES.. EMPLOY   (NOBS= 252)  SUTTER COUNTY EMPLOYMENT, 1/45 TO 12/66
DIFFERENCED  1 TIME(S) OF ORDER  1.
NO. OF VALID OBSERVATIONS  =   251.

AUTOCORRELATIONS OF LAGS 1 - 30.
Q( 30, 251) =  296.02      SIG =    0.000

LAG   CORR    SE  -1  -.8  -.6  -.4  -.2    0   .2   .4   .6   .8   +1
                  +----+----+----+----+----+----+----+----+----+----+
  1  -.048  .063                        ( XI  )
  2  -.116  .063                        *XXI  )
  3  -.062  .064                        (XXI  )
  4  -.066  .064                        (XXI  )
  5  -.124  .065                        *XXI  )
  6  -.151  .066                       X(XXI  )
  7  -.116  .067                        *XXI  )
  8  -.041  .068                        ( XI  )
  9  -.043  .068                        ( XI  )
 10  -.157  .068                       X(XXI  )
 11   .080  .069                        (  IXX)
 12   .713  .070                        (  IXX)XXXXXXXXXXXXXXX
 13   .045  .094                      (    IX   )
 14  -.096  .095                      (  XXI    )
 15  -.051  .095                      (   XI    )
 16  -.074  .095                      (  XXI    )
 17  -.101  .095                      ( XXXI    )
 18  -.144  .096                      (XXXXI    )
 19  -.127  .097                      ( XXXI    )
 20  -.049  .097                      (   XI    )
 21  -.032  .097                      (   XI    )
 22  -.111  .097                      ( XXXI    )
 23   .021  .098                      (    IX   )
 24   .651  .098                      (    IXXXX)XXXXXXXXXXX
 25   .074  .114                     (     IXX   )
 26  -.057  .114                     (    XI     )
 27  -.072  .114                     (   XXI     )
 28  -.069  .114                     (   XXI     )
 29  -.077  .114                     (   XXI     )
 30  -.171  .115                     ( XXXXI     )
                                   -2SE        +2SE

PARTIAL AUTOCORRELATIONS OF LAGS 1 - 30.

LAG   CORR    SE  -1  -.8  -.6  -.4  -.2    0   .2   .4   .6   .8   +1
                  +----+----+----+----+----+----+----+----+----+----+
  1  -.048  .063                        ( XI  )
  2  -.119  .063                        *XXI  )
  3  -.075  .063                        (XXI  )
  4  -.090  .063                        (XXI  )
  5  -.156  .063                       X(XXI  )
  6  -.208  .063                      XX(XXI  )
  7  -.221  .063                     XXX(XXI  )
  8  -.199  .063                      XX(XXI  )
  9  -.249  .063                     XXX(XXI  )
 10  -.472  .063               XXXXXXXXX(XXI  )
 11  -.496  .063               XXXXXXXXX(XXI  )
 12   .406  .063                        (  IXX)XXXXXXX
 13   .183  .063                        (  IXX)XX
 14   .028  .063                        (  IX )
 15   .019  .063                        (  I  )
 16   .008  .063                        (  I  )
 17   .060  .063                        (  IX )
 18   .060  .063                        (  IX )
 19  -.013  .063                        (  I  )
 20  -.093  .063                        (XXI  )
 21  -.055  .063                        ( XI  )
 22   .055  .063                        (  IX )
 23  -.257  .063                     XXX(XXI  )
 24   .066  .063                        (  IXX)
 25   .033  .063                        (  IX )
 26   .029  .063                        (  IX )
 27  -.035  .063                        ( XI  )
 28  -.026  .063                        ( XI  )
 29   .020  .063                        (  IX )
 30  -.075  .063                        (XXI  )
                                      -2SE  +2SE
```

FIGURE 2.12.1(c) ACF and PACF for the Regularly Differenced Series

```
SERIES.. EMPLOY    (NOBS= 252)  SUTTER COUNTY EMPLOYMENT, 1/45 TO 12/66
DIFFERENCED  1 TIME(S) OF ORDER  1.
DIFFERENCED  1 TIME(S) OF ORDER 12.
NO. OF VALID OBSERVATIONS  =   239.

AUTOCORRELATIONS OF LAGS 1 - 30.
Q( 30, 239) =  151.93       SIG =      .000


   LAG    CORR     SE  -1    -.8   -.6   -.4   -.2    0     .2    .4    .6    .8   +1
                        +----+----+----+----+----+----+----+----+----+----+
     1  -.430  .065                       XXXXXXXX(XXI   )
     2   .065  .076                               (    IXX )
     3  -.076  .076                               ( XXI    )
     4   .002  .076                               (    I   )
     5  -.056  .076                               (  XI    )
     6   .014  .076                               (    I   )
     7   .049  .076                               (    IX  )
     8   .039  .077                               (    IX  )
     9   .040  .077                               (    IX  )
    10  -.144  .077                               *XXXI    )
    11   .306  .078                               (    IXXX)XXXX
    12  -.439  .083                       XXXXXXX(XXXI     )
    13   .120  .092                              (     IXXX )
    14  -.037  .093                              (   XI     )
    15   .066  .093                              (     IXX  )
    16  -.045  .093                              (   XI     )
    17   .025  .093                              (     IX   )
    18   .058  .093                              (     IX   )
    19  -.105  .093                              ( XXXI     )
    20   .030  .094                              (     IX   )
    21  -.021  .094                              (   XI     )
    22   .088  .094                              (     IXX  )
    23  -.150  .094                              (XXXXI     )
    24   .049  .095                              (     IX   )
    25   .001  .095                              (     I    )
    26   .131  .095                              (     IXXX )
    27  -.132  .096                              ( XXXI     )
    28   .030  .097                              (     IX   )
    29   .060  .097                              (     IXX  )
    30  -.123  .097                              ( XXXI     )
                                             -2SE       +2SE

PARTIAL AUTOCORRELATIONS OF LAGS 1 - 30.

   LAG    CORR     SE  -1    -.8   -.6   -.4   -.2    0     .2    .4    .6    .8   +1
                        +----+----+----+----+----+----+----+----+----+----+
     1  -.430  .065                       XXXXXXXX(XXI   )
     2  -.147  .065                              X(XXI   )
     3  -.135  .065                               *XXI   )
     4  -.103  .065                               *XXI   )
     5  -.136  .065                               *XXI   )
     6  -.104  .065                               *XXI   )
     7  -.007  .065                               (  I   )
     8   .061  .065                               (  IXX)
     9   .115  .065                               (  IXX*
    10  -.081  .065                               (XXI   )
    11   .300  .065                               (  IXX)XXXXX
    12  -.230  .065                            XXX(XXI   )
    13  -.191  .065                             XX(XXI   )
    14  -.134  .065                               *XXI   )
    15  -.090  .065                               (XXI   )
    16  -.119  .065                               *XXI   )
    17  -.129  .065                               *XXI   )
    18   .008  .065                               (  I   )
    19  -.103  .065                               *XXI   )
    20  -.011  .065                               (  I   )
    21   .082  .065                               (  IXX)
    22   .029  .065                               (  IX )
    23   .062  .065                               (  IXX)
    24  -.162  .065                              X(XXI   )
    25  -.141  .065                              X(XXI   )
    26   .040  .065                               (  IX  )
    27  -.093  .065                               (XXI   )
    28  -.129  .065                               *XXI   )
    29  -.041  .065                               ( XI   )
    30  -.074  .065                               (XXI   )
                                              -2SE  +2SE
```

FIGURE 2.12.1(d) ACF and PACF for the Regularly *and* Seasonally Differenced Series

```
SERIES.. RESIDUAL (NOBS= 239)  SUTTER COUNTY EMPLOYMENT RESIDUALS
NO. OF VALID OBSERVATIONS  =   239.

AUTOCORRELATIONS OF LAGS 1 - 30.
Q( 28, 239) =  28.304       SIG =      .448

  LAG   CORR    SE  -1   -.8  -.6  -.4  -.2   0    .2   .4   .6   .8  +1
                     +----+----+----+----+----+----+----+----+----+----+
   1    .050  .065                          (  IX )
   2   -.016  .065                          (  I  )
   3   -.112  .065                          *XXI  )
   4   -.054  .066                          ( XI  )
   5   -.024  .066                          ( XI  )
   6    .077  .066                          (  IXX)
   7    .080  .066                          (  IXX)
   8    .087  .067                          (  IXX)
   9   -.025  .067                          ( XI  )
  10   -.127  .067                          *XXI  )
  11    .037  .068                          (  IX )
  12   -.009  .068                          (  I  )
  13   -.034  .068                          ( XI  )
  14   -.031  .068                          ( XI  )
  15   -.065  .068                          (XXI  )
  16   -.061  .069                          (XXI  )
  17    .025  .069                          (  IX )
  18    .034  .069                          (  IX )
  19   -.033  .069                          ( XI  )
  20    .008  .069                          (  I  )
  21   -.020  .069                          ( XI  )
  22   -.034  .069                          ( XI  )
  23   -.113  .069                          *XXI  )
  24    .017  .070                          (  I  )
  25    .123  .070                          (  IXX*
  26    .098  .071                         (    IXX )
  27   -.072  .071                         (  XXI   )
  28   -.024  .072                         (   XI   )
  29    .037  .072                         (    IX  )
  30   -.043  .072                         (   XI   )
                                         -2SE     +2SE

PARTIAL AUTOCORRELATIONS OF LAGS 1 - 30.

  LAG   CORR    SE  -1   -.8  -.6  -.4  -.2   0    .2   .4   .6   .8  +1
                     +----+----+----+----+----+----+----+----+----+----+
   1    .050  .065                          (  IX )
   2   -.018  .065                          (  I  )
   3   -.111  .065                          *XXI  )
   4   -.044  .065                          ( XI  )
   5   -.023  .065                          ( XI  )
   6    .067  .065                          (  IXX)
   7    .064  .065                          (  IXX)
   8    .078  .065                          (  IXX)
   9   -.017  .065                          (  I  )
  10   -.106  .065                          *XXI  )
  11    .074  .065                          (  IXX)
  12   -.016  .065                          (  I  )
  13   -.065  .065                          (XXI  )
  14   -.044  .065                          ( XI  )
  15   -.078  .065                          (XXI  )
  16   -.056  .065                          ( XI  )
  17    .029  .065                          (  IX )
  18    .029  .065                          (  IX )
  19   -.062  .065                          (XXI  )
  20    .013  .065                          (  I  )
  21    .021  .065                          (  IX )
  22   -.025  .065                          ( XI  )
  23   -.113  .065                          *XXI  )
  24    .021  .065                          (  IX )
  25    .104  .065                          (  IXX*
  26    .060  .065                          (  IXX)
  27   -.074  .065                          (XXI  )
  28   -.007  .065                          (  I  )
  29    .064  .065                          (  IXX)
  30   -.032  .065                          ( XI  )
                                           -2SE  +2SE
```

FIGURE 2.12.1(e) *Diagnosis:* ACF and PACF for the Model Residuals

If the residuals from this tentative model fail to meet our diagnostic criteria, we will return to the identification stage.

Diagnosis. The ACF and PACF—Figure 2.12.1(d)—for the residuals of our tentative model show no spikes at lag-1 or at the seasonal lags. The Q statistic for the ACF is not statistically significant. With 28 degrees of freedom, the value of Q = 28.3 is associated with a .448 level of significance. The residuals of the tentative model meet both diagnostic criteria, so the model is accepted.

This time series, as well as the others used as examples, are listed in Appendix A of this volume. We suggest that the reader reanalyze the series as an exercise. The Sutter County Workforce series is a textbook example of an ARIMA (0,1,1) $(0,1,1)_{12}$ process. The ACFs and PACFs give clear and unambiguous evidence for this model.

In December 1955, the 120th observation of this series, a major flood forced the evacuation of Sutter County. In the next chapter, we will use the ARIMA model we have identified here to assess the impact of this flood on the level of the Sutter County Workforce series.

2.12.2 Boston Armed Robberies

The monthly number of reported armed robberies in Boston, Massachusetts, is plotted in Figure 2.12.2(a). The first observation is January 1966 and the 118th observation is October 1975. Deutsch and Alt (1977: Deutsch, 1979) analyzed this series along with a large number of other Uniform Crime Report time series. In our opinion, the model proposed by Deutsch and Alt does not adequately represent the series. By following the iterative model-building strategy presented in Section 2.11, we will contrast the inadequacies of their model with the empirical characteristics of the data.

It should be noted that there are a number of ways to construct an ARIMA model other than the empirically based procedure we have outlined. We refer to these other methods as *arbitrary methods* and emphasize that they usually will not result in a parsimonious *and* statistically adequate model. For example, an analyst might have identified the same ARIMA model for a number of series all belonging to the same substantive class, such as crime series. The analyst might then be tempted to assume that all other similar substantive time series (e.g., *all* crime series) could best be fit by the same ARIMA model. Furthermore, the analyst may attempt to infer that identification of the same ARIMA model for a number of series provides evidence that the same social process was generating all of the series.

These are fallacies, of course. A univariate ARIMA model is a stochastic or probabilistic description of the outcome of a process operating through time. It provides no information about the inputs generating that process. As

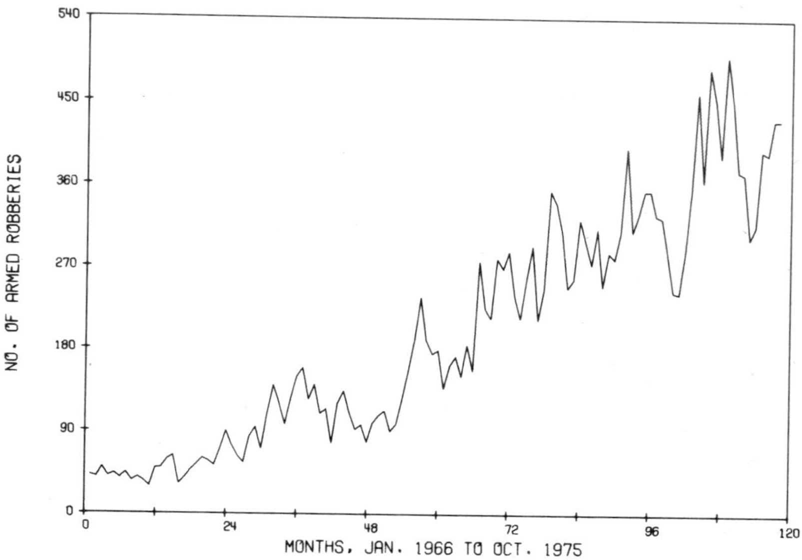

FIGURE 2.12.2(a) Boston Armed Robberies

Hibbs (1977) succinctly points out, "[Univariate] Box-Tiao or Box-Jenkins models are essentially models of ignorance that are not based in theory and, in this sense, are devoid of explanatory power." As in other areas of the social sciences, inference of a causal relationship in time series analysis can only be made through assessment of covariation betweeen one or more explanatory variables and a dependent variable—a crime rate in this case. We develop the methodology for this type of analysis in Chapter 5.

A careful reading of Deutsch and Alt does not clearly reveal their model selection procedure. They do not report identification statistics such as the ACF and PACF and this makes it difficult to assess the adequacy of their models. We recommend that such statistics be routinely reported in time series research so that the social science community may make informed appraisals of the quality of ARIMA models. Although we cannot second guess the Deutsch-Alt model selection procedure, their results are not incongruous with the arbitrary procedure described above. We will now contrast these results with those produced by use of the model-building procedure presented in Section 2.11. The reader is referred to Hay and McCleary (1979) for a more detailed discussion of these issues.

(text continued on p. 115)

```
SERIES.. BAR       (NOBS= 118)  BOSTON / MONTHLY ARMED ROBBERY
NO. OF VALID OBSERVATIONS  =   118.

AUTOCORRELATIONS OF LAGS 1 - 25.
Q( 25, 118) =  1221.4      SIG =    0.000

LAG    CORR    SE  -1   -.8  -.6  -.4  -.2  0    .2   .4   .6   .8  +1
                   +----+----+----+----+----+----+----+----+----+----+
  1    .928  .092                      (    IXXXX)XXXXXXXXXXXXXXXXXX
  2    .881  .152                   (       IXXXXXXX)XXXXXXXXXXXXX
  3    .853  .190                 (         IXXXXXXXXX)XXXXXXXXXXX
  4    .808  .220                (          IXXXXXXXXXX)XXXXXXXXX
  5    .797  .244               (           IXXXXXXXXXXX)XXXXXXXX
  6    .774  .265              (            IXXXXXXXXXXXX)XXXXXX
  7    .749  .284             (             IXXXXXXXXXXXXX)XXXXX
  8    .744  .300            (              IXXXXXXXXXXXXXX)XXXX
  9    .712  .315           (               IXXXXXXXXXXXXXXX)XX
 10    .681  .329           (               IXXXXXXXXXXXXXXX)X
 11    .684  .340          (                IXXXXXXXXXXXXXXXX*
 12    .668  .352         (                 IXXXXXXXXXXXXXXXXX)
 13    .628  .362         (                 IXXXXXXXXXXXXXXXX )
 14    .599  .372        (                  IXXXXXXXXXXXXXXX   )
 15    .550  .380        (                  IXXXXXXXXXXXXXX    )
 16    .519  .386        (                  IXXXXXXXXXXXXX     )
 17    .505  .392       (                   IXXXXXXXXXXXXX      )
 18    .483  .398       (                   IXXXXXXXXXXXX       )
 19    .478  .403       (                   IXXXXXXXXXXXX       )
 20    .465  .407       (                   IXXXXXXXXXXXX       )
 21    .434  .412      (                    IXXXXXXXXXXX         )
 22    .428  .416      (                    IXXXXXXXXXXX         )
 23    .412  .420      (                    IXXXXXXXXXX          )
 24    .394  .423      (                    IXXXXXXXXXX          )
 25    .377  .426      (                    IXXXXXXXXX           )
                     -2SE                                      +2SE


PARTIAL AUTOCORRELATIONS OF LAGS 1 - 25.

LAG    CORR    SE  -1   -.8  -.6  -.4  -.2  0    .2   .4   .6   .8  +1
                   +----+----+----+----+----+----+----+----+----+----+
  1    .928  .092                      (    IXXXX)XXXXXXXXXXXXXXXXXX
  2    .147  .092                      (    IXXXX)
  3    .145  .092                      (    IXXXX)
  4   -.092  .092                      (  XXI    )
  5    .217  .092                      (    IXXXX*
  6   -.034  .092                      (   XI    )
  7    .032  .092                      (    IX   )
  8    .092  .092                      (    IXX  )
  9   -.116  .092                      ( XXXI    )
 10   -.032  .092                      (   XI    )
 11    .188  .092                      (    IXXXX*
 12   -.007  .092                      (    I    )
 13   -.245  .092                     X(XXXXI    )
 14   -.025  .092                      (   XI    )
 15   -.124  .092                      ( XXXI    )
 16    .031  .092                      (    IX   )
 17    .065  .092                      (    IXX  )
 18    .065  .092                      (    IXX  )
 19   -.005  .092                      (    I    )
 20   -.017  .092                      (    I    )
 21   -.002  .092                      (    I    )
 22    .100  .092                      (    IXXX )
 23   -.060  .092                      (   XI    )
 24    .010  .092                      (    I    )
 25   -.078  .092                      (  XXI    )
                                     -2SE      +2SE
```

FIGURE 2.12.2(b) ACF and PACF for the Raw Series

```
SERIES.. BAR       (NOBS= 118)  BOSTON / MONTHLY ARMED ROBBERY
DIFFERENCED  1 TIME(S) OF ORDER  1.
NO. OF VALID OBSERVATIONS  =   117.

AUTOCORRELATIONS OF LAGS 1 - 25.
Q( 25, 117) =  63.074       SIG =      .000

LAG    CORR     SE     -1   -.8  -.6  -.4  -.2   0    .2   .4   .6   .8  +1
                        +----+----+----+----+----+----+----+----+----+----+
  1   -.259   .092                         X(XXXXI    )
  2   -.137   .098                          ( XXXI    )
  3    .156   .100                          (    IXXXX)
  4   -.238   .102                         X(XXXXI    )
  5    .113   .107                          (    IXXX )
  6   -.055   .108                          (   XI    )
  7   -.215   .108                          *XXXXI    )
  8    .177   .112                         (     IXXXX )
  9   -.092   .114                         (   XXI     )
 10   -.205   .115                         (XXXXXI     )
 11    .113   .118                         (     IXXX  )
 12    .190   .119                         (     IXXXXX)
 13    .040   .121                         (     IX    )
 14    .076   .121                         (     IXX   )
 15   -.082   .122                         (   XXI     )
 16   -.082   .122                         (   XXI     )
 17    .135   .123                         (     IXXX  )
 18   -.163   .124                         ( XXXXI     )
 19    .023   .126                         (     IX    )
 20    .114   .126                         (     IXXX  )
 21   -.251   .127                         *XXXXXI     )
 22    .079   .131                        (      IXX     )
 23    .024   .131                        (      IX      )
 24    .000   .131                        (      I       )
 25    .174   .131                        (      IXXXX   )
                                        -2SE          +2SE



PARTIAL AUTOCORRELATIONS OF LAGS 1 - 25.

LAG    CORR     SE     -1   -.8  -.6  -.4  -.2   0    .2   .4   .6   .8  +1
                        +----+----+----+----+----+----+----+----+----+----+
  1   -.259   .092                         X(XXXXI    )
  2   -.219   .092                          *XXXXI    )
  3    .062   .092                          (    IXX  )
  4   -.229   .092                         X(XXXXI    )
  5    .026   .092                          (    IX   )
  6   -.125   .092                          ( XXXI    )
  7   -.235   .092                         X(XXXXI    )
  8   -.042   .092                          (   XI    )
  9   -.131   .092                          ( XXXI    )
 10   -.330   .092                       XXX(XXXXI    )
 11   -.253   .092                         X(XXXXI    )
 12    .098   .092                          (    IXX  )
 13    .058   .092                          (    IX   )
 14    .083   .092                          (    IXX  )
 15   -.001   .092                          (    I    )
 16   -.130   .092                          ( XXXI    )
 17   -.011   .092                          (    I    )
 18   -.094   .092                          (  XXI    )
 19    .024   .092                          (    IX   )
 20    .051   .092                          (    IX   )
 21   -.132   .092                          ( XXXI    )
 22    .008   .092                          (    I    )
 23    .036   .092                          (    IX   )
 24    .078   .092                          (    IXX  )
 25    .020   .092                          (    I    )
                                          -2SE     +2SE
```

FIGURE 2.12.2(c) ACF and PACF for the Regularly Differenced Series

```
SERIES.. RESIDUAL (NOBS= 105)  BOSTON ARMED ROBBERY RESIDUALS
NO. OF VALID OBSERVATIONS  =  105.

AUTOCORRELATIONS OF LAGS 1 - 25.
Q( 23, 105) =  41.871      SIG =      .009

LAG   CORR   SE -1  -.8  -.6  -.4  -.2   0   .2   .4   .6   .8  +1
                +----+----+----+----+----+----+----+----+----+----+
  1   .010 .098                     (    I    )
  2  -.033 .098                     (   XI    )
  3   .158 .098                     (    IXXXX)
  4  -.215 .100                     *XXXXI    )
  5   .009 .104                     (    I    )
  6  -.135 .104                     ( XXXI    )
  7  -.307 .106                 XXX(XXXXI    )
  8   .094 .114                    (     IXX    )
  9  -.088 .115                    (   XXI     )
 10  -.193 .116                    (XXXXXI     )
 11   .070 .119                    (     IXX   )
 12   .025 .119                    (     IX    )
 13   .114 .119                    (     IXXX  )
 14   .190 .120                    (     IXXXXX)
 15  -.043 .123                    (    XI     )
 16  -.100 .123                    (   XXI     )
 17   .082 .124                    (     IXX   )
 18  -.105 .124                    (  XXXI     )
 19  -.018 .125                    (     I     )
 20   .046 .125                    (     IX    )
 21  -.137 .125                    (  XXXI     )
 22   .122 .127                    (     IXXX  )
 23   .006 .128                    (     I     )
 24  -.137 .128                    (  XXXI     )
 25   .146 .129                    (     IXXXX )
                                 -2SE        +2SE
```

```
PARTIAL AUTOCORRELATIONS OF LAGS 1 - 25.

LAG   CORR   SE -1  -.8  -.6  -.4  -.2   0   .2   .4   .6   .8  +1
                +----+----+----+----+----+----+----+----+----+----+
  1   .010 .098                     (    I    )
  2  -.033 .098                     (   XI    )
  3   .159 .098                     (    IXXXX)
  4  -.226 .098                    X(XXXXI    )
  5   .037 .098                     (    IX   )
  6  -.194 .098                     *XXXXI    )
  7  -.241 .098                    X(XXXXI    )
  8   .047 .098                     (    IX   )
  9  -.083 .098                     (   XXI    )
 10  -.194 .098                     *XXXXI    )
 11  -.072 .098                     (   XXI    )
 12   .029 .098                     (    IX   )
 13   .042 .098                     (    IX   )
 14   .081 .098                     (    IXX  )
 15  -.038 .098                     (   XI    )
 16  -.244 .098                    X(XXXXI    )
 17  -.002 .098                     (    I    )
 18  -.054 .098                     (   XI    )
 19   .024 .098                     (    IX   )
 20   .016 .098                     (    I    )
 21  -.095 .098                     (   XXI    )
 22   .046 .098                     (    IX   )
 23  -.029 .098                     (   XI    )
 24  -.037 .098                     (   XI    )
 25   .013 .098                     (    I    )
                                  -2SE      +2SE
```

FIGURE 2.12.2(d) *Diagnosis:* ACF and PACF for the Model Residuals

Identification. The ACFs and PACFs for this series are shown in Figures 2.12.2(b), 2.12.2(c), and 2.12.2(d). Deutsch and Alt picked an ARIMA (0,1,1) $(0,1,1)_{12}$ model for this series. This is the same model we identified for the Sutter County Workforce time series, so the reader can compare those ACFs and PACFs with these.

The ACF for the raw armed robbery time series indicates nonstationarity, so the series must be differenced. The ACF and PACF of the differenced series do not indicate *seasonal* nonstationarity, however. The ACF and PACF of the regularly and seasonally differenced series do not unambiguously suggest the ARIMA (0,1,1) $(0,1,1)_{12}$ model used by Deutsch and Alt. To be perfectly frank, these ACFs and PACFs baffle us. We see no clear patterns of spiking and/or decay which would lead us to accept any tentative model.

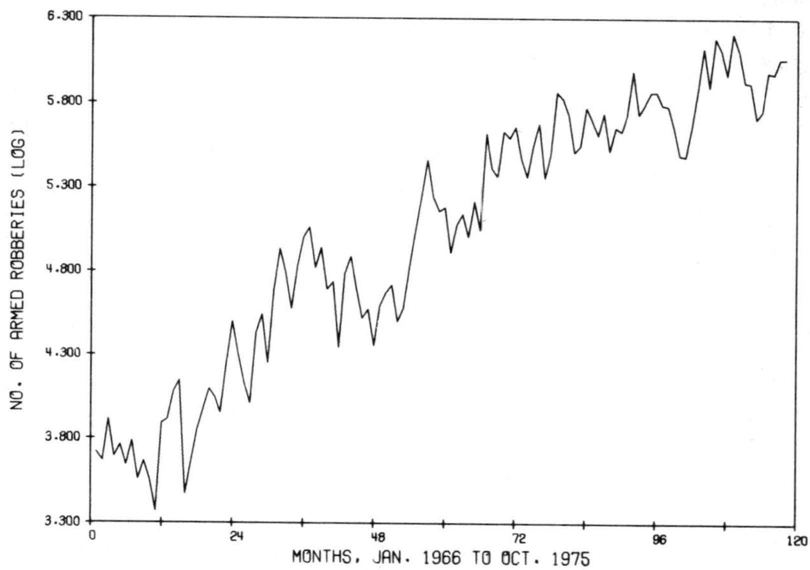

FIGURE 2.12.2(e) Boston Armed Robberies (Logged)

Estimation. For the ARIMA (0,1,1) $(0,1,1)_{12}$ model proposed by Deutsch and Alt, we obtain parameter estimates of

$$\hat{\Theta}_0 = .31 \text{ with t statistic} = .47$$
$$\hat{\Theta}_1 = .52 \text{ with t statistic} = 6.08$$
$$\hat{\Theta}_{12} = .73 \text{ with t statistic} = 8.29.$$

As $\hat{\Theta}_0$ is not statistically significant, we drop it from the tentative model. $\hat{\Theta}_1$ and $\hat{\Theta}_{12}$ are both statistically significant and both lie within the bounds of invertibility for moving average parameters. The tentative model is, then,

$$Y_t = \frac{(1 - .52B)(1 - .73B^{12})}{(1 - B)(1 - B^{12})} a_t.$$

Diagnosis. The ACF and PACF for the residuals of this model do not inspire confidence. There are statistically significant spikes at lags-4, -7, and -10 of the ACF and -16 of the PACF as well as a handful of "marginally significant" spikes at other lags. The Q statistic for this ACF is also quite large. With 23 degrees of freedom, Q = 41.87, a value of Q associated with a .009 significance level. As the Q statistic is significant, we must conclude that these residuals are not white noise. The tentative model fails our diagnostic criteria and we thus reject it.

Identification. Because we have rejected the tentative model, we return to the identification stage. The ACFs and PACFs of the raw and differenced series are not much help here. As noted, we see no evidence for a parsimonious ARIMA model in these statistics. One alternative at this point would be to include extra moving average parameters in the model, that is, to increase the values of q and/or Q. We see no evidence of higher order moving averages in these ACFs and PACFs, however; and, moreover, the model is already rather complicated and cumbersome.

Another alternative is to explore a transformation of the time series. In Section 2.4, we discussed variance stationarity, noting that a time series process must be made stationary in variance as well as in level. Examining the plotted armed robbery time series in Figure 2.12.2(a), we can see that the series variance is roughly proportional to the the series level. In the first half of the series, when the process is at its lowest level, month-to-month fluctuations are small. In the second half of the series, when the process is at its highest level, month-to-month fluctuations are relatively large. In Figure

(text continued on p. 120)

```
SERIES.. VAR3      (NOBS= 118)  VAR3      = LOG (BAR      )
NO. OF VALID OBSERVATIONS  =   118.

AUTOCORRELATIONS OF LAGS 1 - 25.
Q( 25, 118) =  1249.4       SIG =     0.000

 LAG    CORR     SE  -1    -.8  -.6  -.4  -.2   0    .2   .4   .6   .8  +1
                      +----+----+----+----+----+----+----+----+----+----+
   1   .946  .092                         (     IXXXX)XXXXXXXXXXXXXXXXXXX
   2   .906  .154                      (        IXXXXXXX)XXXXXXXXXXXXXXX
   3   .883  .194                    (          IXXXXXXXXX)XXXXXXXXXXXX
   4   .855  .225                   (           IXXXXXXXXXX)XXXXXXXXXX
   5   .836  .251                 (             IXXXXXXXXXXXX)XXXXXXXX
   6   .808  .274                (              IXXXXXXXXXXXXX)XXXXXX
   7   .782  .293               (               IXXXXXXXXXXXXXX)XXXXX
   8   .761  .311              (                IXXXXXXXXXXXXXXX)XXX
   9   .726  .326              (                IXXXXXXXXXXXXXXX)XX
  10   .695  .339             (                 IXXXXXXXXXXXXXXXX*
  11   .672  .351            (                  IXXXXXXXXXXXXXXXXX)
  12   .650  .362            (                  IXXXXXXXXXXXXXXXX )
  13   .616  .372           (                   IXXXXXXXXXXXXXXX    )
  14   .591  .380           (                   IXXXXXXXXXXXXXXX    )
  15   .560  .388           (                   IXXXXXXXXXXXXXX     )
  16   .528  .395          (                    IXXXXXXXXXXXXX       )
  17   .501  .401          (                    IXXXXXXXXXXXXX       )
  18   .479  .406          (                    IXXXXXXXXXXXX        )
  19   .463  .411         (                     IXXXXXXXXXXXX         )
  20   .442  .415         (                     IXXXXXXXXXXX          )
  21   .417  .419         (                     IXXXXXXXXXX           )
  22   .397  .423         (                     IXXXXXXXXXX           )
  23   .377  .426         (                     IXXXXXXXXX            )
  24   .363  .429         (                     IXXXXXXXXX            )
  25   .341  .431        (                      IXXXXXXXXX             )
                     -2SE                                            +2SE



PARTIAL AUTOCORRELATIONS OF LAGS 1 - 25.

 LAG    CORR     SE  -1    -.8  -.6  -.4  -.2   0    .2   .4   .6   .8  +1
                      +----+----+----+----+----+----+----+----+----+----+
   1   .946  .092                          (     IXXXX)XXXXXXXXXXXXXXXXXXX
   2   .107  .092                          (     IXXXX )
   3   .153  .092                          (     IXXXXX)
   4  -.011  .092                          (     I     )
   5   .099  .092                          (     IXX   )
   6  -.079  .092                          (   XXI     )
   7   .014  .092                          (     I     )
   8   .004  .092                          (     I     )
   9  -.111  .092                          (  XXXI     )
  10  -.021  .092                          (    XI     )
  11   .033  .092                          (     IX    )
  12   .023  .092                          (     IX    )
  13  -.124  .092                          (  XXXI     )
  14   .065  .092                          (     IXX   )
  15  -.094  .092                          (   XXI     )
  16  -.014  .092                          (     I     )
  17  -.011  .092                          (     I     )
  18   .074  .092                          (     IXX   )
  19   .032  .092                          (     IX    )
  20  -.012  .092                          (     I     )
  21  -.020  .092                          (    XI     )
  22   .009  .092                          (     I     )
  23  -.006  .092                          (     I     )
  24   .050  .092                          (     IX    )
  25  -.064  .092                          (   XXI     )
                                         -2SE        +2SE
```

FIGURE 2.12.2(f) ACF and PACF for the Raw (Logged) Series

```
SERIES.. VAR3      (NOBS= 118)  VAR3      = LOG (BAR      )
DIFFERENCED  1 TIME(S) OF ORDER  1.
NO. OF VALID OBSERVATIONS  =   117.

AUTOCORRELATIONS OF LAGS 1 - 25.
Q( 25, 117) =  28.966       SIG =      .265

  LAG   CORR    SE  -1    -.8   -.6   -.4   -.2   0     .2    .4    .6    .8   +1
                     +----+----+----+----+----+----+----+----+----+----+
    1  -.229  .092                           X(XXXXI    )
    2  -.162  .097                             (XXXXI    )
    3  -.004  .099                             (    I    )
    4  -.078  .099                             (  XXI    )
    5   .102  .100                             (    IXXX )
    6  -.040  .101                             (   XI    )
    7  -.134  .101                             ( XXXI    )
    8   .162  .103                             (    IXXXX)
    9  -.112  .105                             ( XXXI    )
   10  -.120  .106                             ( XXXI    )
   11   .083  .107                             (    IXX  )
   12   .144  .107                             (    IXXXX)
   13  -.029  .109                             (   XI    )
   14   .054  .109                             (    IX   )
   15  -.136  .109                             ( XXXI    )
   16   .023  .111                            (     IX    )
   17   .008  .111                            (     I     )
   18  -.064  .111                            (   XXI     )
   19   .051  .111                            (     IX    )
   20   .025  .111                            (     IX    )
   21  -.116  .111                            (  XXXI     )
   22   .055  .112                            (     IX    )
   23  -.027  .113                            (    XI     )
   24   .060  .113                            (     IX    )
   25   .032  .113                            (     IX    )
                                           -2SE        +2SE



PARTIAL AUTOCORRELATIONS OF LAGS 1 - 25.

  LAG   CORR    SE  -1    -.8   -.6   -.4   -.2   0     .2    .4    .6    .8   +1
                     +----+----+----+----+----+----+----+----+----+----+
    1  -.229  .092                           X(XXXXI    )
    2  -.227  .092                           X(XXXXI    )
    3  -.113  .092                            ( XXXI    )
    4  -.166  .092                            (XXXXI    )
    5   .012  .092                            (    I    )
    6  -.063  .092                            (  .XXI    )
    7  -.169  .092                            (XXXXI    )
    8   .062  .092                            (    IXX  )
    9  -.122  .092                            ( XXXI    )
   10  -.203  .092                            *XXXXI    )
   11  -.083  .092                            (  XXI    )
   12   .114  .092                            (    IXXX )
   13  -.023  .092                            (   XI    )
   14   .100  .092                            (    IXXX )
   15  -.044  .092                            (   XI    )
   16  -.037  .092                            (   XI    )
   17  -.070  .092                            (  XXI    )
   18  -.061  .092                            (  XXI    )
   19  -.027  .092                            (   XI    )
   20  -.021  .092                            (   XI    )
   21  -.086  .092                            (  XXI    )
   22  -.017  .092                            (    I    )
   23  -.052  .092                            (   XI    )
   24  -.018  .092                            (    I    )
   25  -.028  .092                            (   XI    )
                                           -2SE      +2SE
```

FIGURE 2.12.2(g) ACF and PACF for the Regularly Differenced (Logged) Series

```
SERIES.. RESDLOG   (NOBS= 117)  BOSTON ARMED ROBBERY (LOG) RESIDUALS
NO. OF VALID OBSERVATIONS   =   117.

AUTOCORRELATIONS OF LAGS 1 - 25.
Q( 23, 117) =   16.315       SIG =      .841

   LAG    CORR    SE    -1   -.8  -.6  -.4  -.2  0    .2   .4   .6   .8  +1
                        +----+----+----+----+----+----+----+----+----+----+
     1    .041  .092                        (    IX   )
     2   -.152  .093                        (XXXXI    )
     3   -.053  .095                        (   XI    )
     4   -.109  .095                        ( XXXI    )
     5    .077  .096                        (    IXX  )
     6   -.046  .097                        (   XI    )
     7   -.149  .097                        (XXXXI    )
     8    .094  .099                        (    IXX  )
     9   -.099  .099                        (  XXI    )
    10   -.099  .100                        (  XXI    )
    11    .097  .101                        (    IXX  )
    12   -.012  .102                        (    I    )
    13    .021  .102                        (    IX   )
    14    .040  .102                        (    IX   )
    15   -.098  .102                        (  XXI    )
    16   -.001  .103                        (    I    )
    17   -.031  .103                        (   XI    )
    18   -.040  .103                        (   XI    )
    19    .043  .103                        (    IX   )
    20   -.029  .103                        (   XI    )
    21   -.066  .103                        (  XXI    )
    22    .032  .104                        (    IX   )
    23   -.017  .104                        (    I    )
    24    .060  .104                        (    IXX  )
    25    .073  .104                        (    IXX  )
                                           -2SE      +2SE
```

```
PARTIAL AUTOCORRELATIONS OF LAGS 1 - 25.

   LAG    CORR    SE    -1   -.8  -.6  -.4  -.2  0    .2   .4   .6   .8  +1
                        +----+----+----+----+----+----+----+----+----+----+
     1    .041  .092                        (    IX   )
     2   -.154  .092                        (XXXXI    )
     3   -.040  .092                        (   XI    )
     4   -.132  .092                        ( XXXI    )
     5    .076  .092                        (    IXX  )
     6   -.098  .092                        (  XXI    )
     7   -.134  .092                        ( XXXI    )
     8    .080  .092                        (    IXX  )
     9   -.154  .092                        (XXXXI    )
    10   -.102  .092                        ( XXXI    )
    11    .053  .092                        (    IX   )
    12   -.036  .092                        (   XI    )
    13   -.029  .092                        (   XI    )
    14    .026  .092                        (    IX   )
    15   -.075  .092                        (  XXI    )
    16   -.062  .092                        (  XXI    )
    17   -.053  .092                        (   XI    )
    18   -.034  .092                        (   XI    )
    19   -.045  .092                        (   XI    )
    20   -.036  .092                        (   XI    )
    21   -.079  .092                        (  XXI    )
    22   -.028  .092                        (   XI    )
    23   -.043  .092                        (   XI    )
    24    .016  .092                        (    I    )
    25    .016  .092                        (    I    )
                                           -2SE      +2SE
```

FIGURE 2.12.2(h) *Diagnosis:* ACF and PACF for the Model Residuals (Logged)

2.12.2(e), we show the log-transformed series. In the natural log metric, month-to-month fluctuations are more or less the same size throughout the length of the series. We can now try to identify a simple ARIMA model for the log series.

The ACF and PACF for the raw log-transformed series—Figure 2.12.2(f)—indicate nonstationarity. The ACF and PACF for the first-differenced log series—(Figure 2.12.2(g)—indicate that one regular difference is sufficient. The absence of spikes at the seasonal lags indicates that seasonal differencing is not required. There is a rough pattern of decay beginning at lag-1 of the PACF and a single spike at lag-1 of the ACF which suggest a ARIMA (0,1,1) model for this series. Spiking and decay at lag-12 of the ACF and PACF lie within the confidence bands and, thus, are not statistically significant. Nevertheless, we see seasonal variation in the plotted series. To account for this seasonality, we propose an ARIMA (0,1,1) $(0,0,1)_{12}$ model. If $\hat{\Theta}_{12}$ proves not to be statistically significant, we can drop it from our model at the estimation stage.

Estimation. Our tentative model is ARIMA (0,1,1) $(0,0,1)_{12}$ for the log-transformed series. We write this as

$$\mathrm{Ln}(Y_t) = \frac{\Theta_0 + (1 - \Theta_1 B)(1 - \Theta_{12} B^{12})}{1 - B} a_t.$$

The parameters of this model are estimated as

$$\hat{\Theta}_0 = .0195 \text{ with t statistic} = 1.57$$
$$\hat{\Theta}_1 = .4321 \text{ with t statistic} = 4.99$$
$$\hat{\Theta}_{12} = .1884 \text{ with t statistic} = -1.97.$$

As the estimate of Θ_0 is not statistically significant, this parameter is dropped from the tentative model. The estimates of Θ_1 and Θ_{12} are both statistically significant (though $\hat{\Theta}_{12}$, just barely) and both lie within the bounds of invertibility for moving average parameters.

Diagnosis. The ACF and PACF for the model residuals—Figure 2.12.2(h)—have no spikes at lag-1 or at the seasonal lags. With 23 degrees of freedom, Q = 16.315, a value of the Q statistic associated with a .84 significance level. As the residuals from our tentative model appear to be white noise, we can accept the model.

Our decision to include a Θ_{12} parameter in the tentative model despite the lack of evidence for seasonality in the ACFs and PACFs might be criticized.

Our decision was based primarily on the seasonal appearance of the time series. In this case, however, had we started with a ARIMA (0,1,1) model, metadiagnosis would have lead to the ARIMA (0,1,1) $(0,0,1)_{12}$ model. Metadiagnosis, particularly overfitting the tentative model, is crucial when a seasonal effect is possible. While including a seasonal component in the ARIMA model when no seasonality is present is an error, failing to include a seasonal component when one is present is a more dangerous error.

The value of inspecting model residuals as a metadiagnostic step is apparent in this example analysis. It would be desirable to have a simple statistical test for variance nonstationarity. Unfortunately, as Granger and Newbold note, "No completely satisfactory techniques are available for testing whether or not a series contains a trend in mean and/or variance. A number of sensible procedures can be suggested, but a decision based on the plot of the data is likely to be a reasonable one." (1977: 37).

To test for a variance nonstationary process in this time series, Hay and McCleary (1979) devided the series into equal interval segments and noted that the mean and standard deviation of each segment showed a nearly monotonic increase over time. Two tests for variance homogeneity, Cochran's C and the Bartlett-Box F tests, also were applied to the segments. Yet none of these statistical tests is as compelling as the visual evidence. In Figure 2.12.2(i), we show a plot of the estimated residuals of the ARIMA (0,1,1) $(0,1,1)_{12}$ model for this series. The variance of the residuals increases as a function of time, clearly indicating variance nonstationarity.

Our point is this example is that, unless an ARIMA model is built through the iterative identification/estimation/diagnosis strategy that we recommend, an arbitrary ARIMA model may result. In the next chapter, we will return to this time series to demonstrate how serious this error can be. In April 1975, the Massachusetts legislature passed a strict gun control law which (presumably) would have an impact on the level of this time series. On the basis of an arbitrary and statistically inadequate ARIMA model, Deutsch and Alt concluded that the law had an abrupt and profound impact on this series. We will demonstrate in the next chapter that the evidence does not support this conclusion.

2.12.3 Swedish Harvest Index

When is a time series not a time series? In Figure 2.12.3(a), we show the annual Swedish Harvest Index for the period 1749–1850 as reported by Thomas (1940). These data come close to being a "time series that is not a time series." In each year, the Swedish grain harvest was rated on a nine-

(text continued on p. 124)

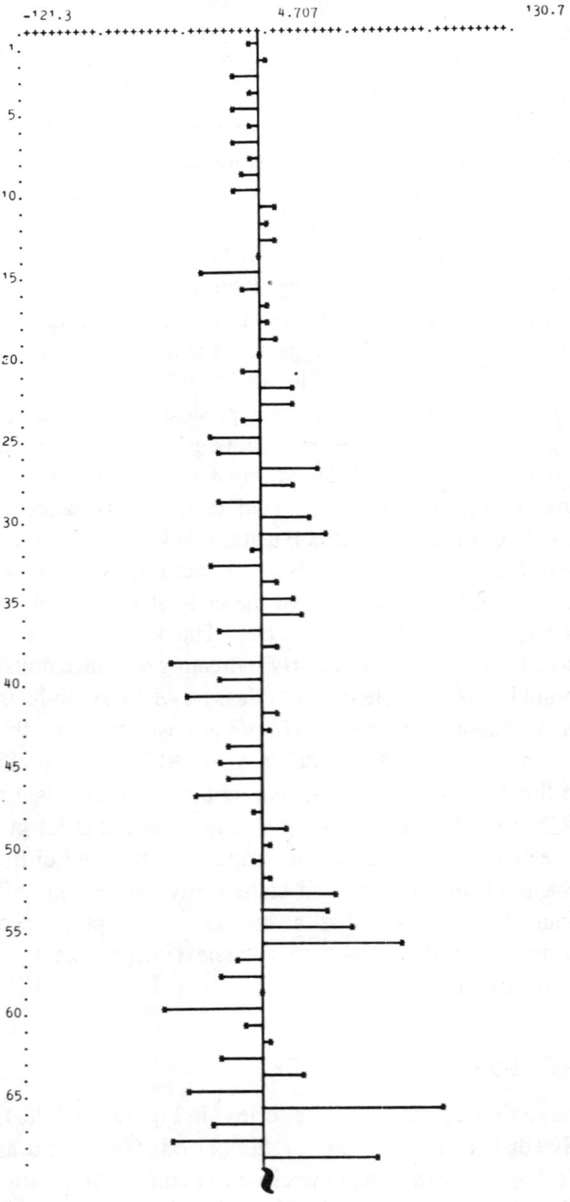
-121.3
4.707
130.7
1.
5.
10.
15.
20.
25.
30.
35.
40.
45.
50.
55.
60.
65.

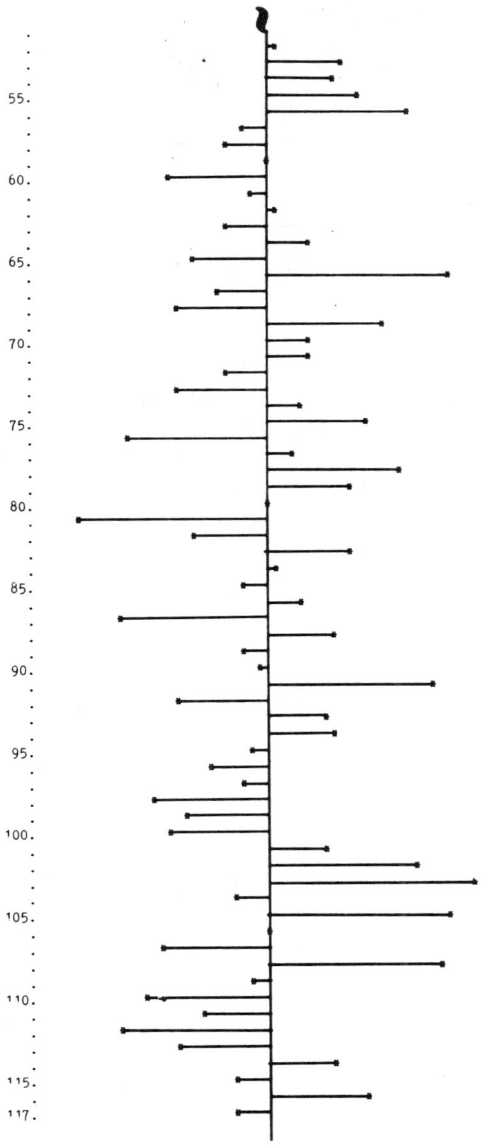

FIGURE 2.12.2(i) The Residuals of the Unlogged Model Indicate that the Generating Process is Nonstationary in Variance

point scale with a total crop failure scored as zero and a superabundant crop scored as nine. The problem with this time series is its level of measurement. While we require a time series process to be measured at the interval level, this one is measured at the ordinal level. In only 13 of the 102 years does the index take on a noninteger value. Because this series does not have a real interval level variance (and covariance), we are not optimistic about building a good ARIMA model for it.

Identification. The ACF of the raw time series indicates that the process is stationary. A single spike at lag-1 of the ACF and rough decay beginning at lag-1 of the PACF suggest an ARIMA (0,0,1) model. We write this tentative model as

$$Y_t = \Theta_0 + (1 - \Theta_1 B)a_t,$$

where the parameter Θ_0 is interpreted as the level of the stationary series.

Estimation. Parameter estimates for the tentative model are:

$$\hat{\Theta}_0 = \quad 5.21 \text{ with t statistic } = 15.19$$
$$\hat{\Theta}_{11} = - .39 \text{ with t statistic } = -4.20.$$

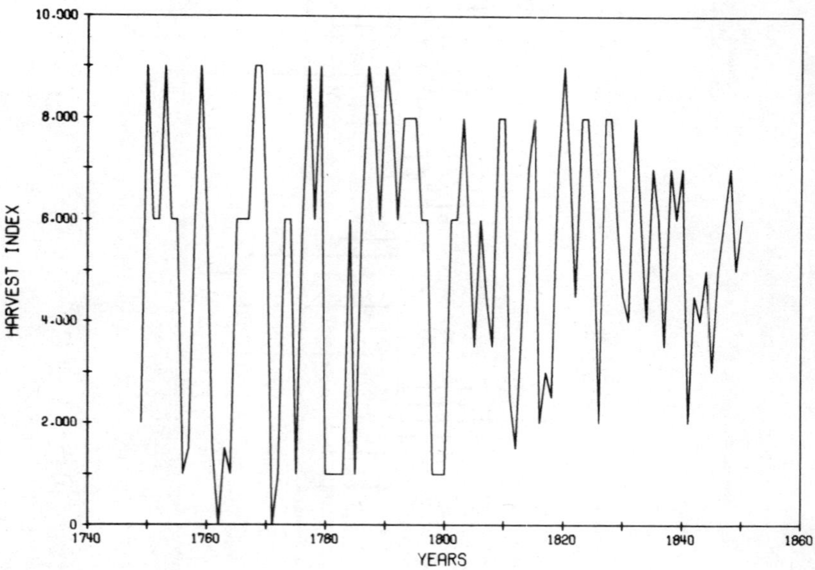

FIGURE 2.12.3(a) Swedish Harvest Index

```
SERIES.. HARVEST   (NOBS= 102)  SWEDISH HARVEST INDEX 1749-1860
NO. OF VALID OBSERVATIONS   =   102.

AUTOCORRELATIONS OF LAGS 1 - 25.
Q( 25, 102) =   36.048      SIG =      .071

   LAG    CORR    SE   -1    -.8   -.6   -.4   -.2    0    .2    .4    .6    .8   +1
                        +----+----+----+----+----+----+----+----+----+----+
     1    .340  .099                            (     IXXXX)XXX
     2   -.044  .110                            (    XI     )
     3   -.139  .110                            (  XXXI     )
     4   -.097  .112                            (   XXI     )
     5   -.065  .113                            (   XXI     )
     6   -.161  .113                            ( XXXXI     )
     7   -.095  .115                            (   XXI     )
     8    .029  .116                            (     IX    )
     9    .026  .116                            (     IX    )
    10    .018  .116                            (     I     )
    11   -.062  .116                            (   XXI     )
    12   -.170  .116                            ( XXXXI     )
    13   -.050  .119                            (    XI     )
    14    .042  .119                            (     IX    )
    15    .015  .119                            (     I     )
    16   -.062  .119                            (   XXI     )
    17    .040  .120                            (     IX    )
    18    .199  .120                            (     IXXXXX)
    19    .171  .123                            (     IXXXX )
    20   -.007  .125                            (     I     )
    21   -.167  .125                            ( XXXXI     )
    22   -.155  .127                            ( XXXXI     )
    23   -.047  .129                            (    XI     )
    24    .008  .129                            (     I     )
    25    .023  .129                            (     IX    )
                                              -2SE       +2SE


PARTIAL AUTOCORRELATIONS OF LAGS 1 - 25.

   LAG    CORR    SE   -1    -.8   -.6   -.4   -.2    0    .2    .4    .6    .8   +1
                        +----+----+----+----+----+----+----+----+----+----+
     1    .340  .099                             (    IXXXX)XXX
     2   -.181  .099                             *XXXXI     )
     3   -.070  .099                             (  XXI     )
     4   -.030  .099                             (   XI     )
     5   -.050  .099                             (   XI     )
     6   -.169  .099                             (XXXXI     )
     7    .000  .099                             (    I     )
     8    .030  .099                             (    IX    )
     9   -.057  .099                             (   XI     )
    10    .003  .099                             (    I     )
    11   -.089  .099                             (  XXI     )
    12   -.175  .099                             (XXXXI     )
    13    .046  .099                             (    IX    )
    14    .011  .099                             (    I     )
    15   -.075  .099                             (  XXI     )
    16   -.083  .099                             (  XXI     )
    17    .093  .099                             (    IXX   )
    18    .102  .099                             (    IXXX  )
    19    .041  .099                             (    IX    )
    20   -.037  .099                             (   XI     )
    21   -.133  .099                             ( XXXI     )
    22   -.067  .099                             (  XXI     )
    23    .019  .099                             (    I     )
    24    .002  .099                             (    I     )
    25    .014  .099                             (    I     )
                                               -2SE      +2SE
```

FIGURE 2.12.3(b) ACF and PACF for the Raw Series

```
SERIES.. RESIDUAL (NOBS= 102)  SWEDISH HARVEST INDEX RESIDUALS
NO. OF VALID OBSERVATIONS  =   102.

AUTOCORRELATIONS OF LAGS 1 - 25.
Q( 23, 102) =  15.058       SIG =      .893

  LAG   CORR    SE  -1  -.8  -.6  -.4  -.2    0   .2   .4   .6   .8   +1
                     +----+----+----+----+----+----+----+----+----+----+
    1   .010  .099                       (    I    )
    2  -.009  .099                       (    I    )
    3  -.117  .099                       ( XXXI    )
    4  -.058  .100                       (   XI    )
    5   .001  .101                       (    I    )
    6  -.141  .101                       (XXXXI    )
    7  -.065  .103                       (  XXI    )
    8   .050  .103                       (    IX   )
    9   .002  .103                       (    I    )
   10   .020  .103                       (    I    )
   11  -.013  .103                       (    I    )
   12  -.164  .103                       (XXXXI    )
   13  -.005  .106                       (    I    )
   14   .033  .106                       (    IX   )
   15   .030  .106                       (    IX   )
   16  -.076  .106                       (  XXI    )
   17   .013  .107                       (    I    )
   18   .154  .107                       (    IXXXX)
   19   .123  .109                       (    IXXX )
   20  -.005  .110                      (     I     )
   21  -.131  .110                      (  XXXI     )
   22  -.107  .112                      (  XXXI     )
   23  -.015  .113                      (     I     )
   24   .017  .113                      (     I     )
   25  -.007  .113                      (     I     )
                                      -2SE        +2SE


PARTIAL AUTOCORRELATIONS OF LAGS 1 - 25.

  LAG   CORR    SE  -1  -.8  -.6  -.4  -.2    0   .2   .4   .6   .8   +1
                     +----+----+----+----+----+----+----+----+----+----+
    1   .010  .099                       (    I    )
    2  -.009  .099                       (    I    )
    3  -.117  .099                       ( XXXI    )
    4  -.057  .099                       (   XI    )
    5  -.000  .099                       (    I    )
    6  -.159  .099                       (XXXXI    )
    7  -.081  .099                       (  XXI    )
    8   .044  .099                       (    IX   )
    9  -.039  .099                       (   XI    )
   10  -.017  .099                       (    I    )
   11  -.010  .099                       (    I    )
   12  -.196  .099                       *XXXXI    )
   13  -.035  .099                       (   XI    )
   14   .036  .099                       (    IX   )
   15  -.024  .099                       (   XI    )
   16  -.118  .099                       ( XXXI    )
   17   .017  .099                       (    I    )
   18   .108  .099                       (    IXXX )
   19   .077  .099                       (    IXX  )
   20   .016  .099                       (    I    )
   21  -.104  .099                       ( XXXI    )
   22  -.113  .099                       ( XXXI    )
   23  -.012  .099                       (    I    )
   24   .017  .099                       (    I    )
   25  -.014  .099                       (    I    )
                                       -2SE      +2SE
```

FIGURE 2.12.3(c) *Diagnosis:* ACF and PACF for the Model Residuals

Both parameter estimates are statistically significant. The estimate of Θ_1 is well within the bounds of invertibility for moving average parameters, so the tentative model poses no problems at this stage.

Diagnosis. The residual ACF has no spikes at early lags. Moreover, the Q statistic for this ACF is not significant. As these residuals satisfy our diagnostic criteria, we infer that they are white noise and accept the model.

The model for the Swedish Harvest Index time series is:

$$Y_t = 5.21 + a_t + .39a_{t-1}.$$

Due to the form of these data, however, we should not immediately conclude that this model is of the highest quality. While the model is clearly adequate in the sense that its residuals are white noise, the model may not necessarily be of much use to us. In general, the utility of a model depends upon its predictive ability. How well does it fit the data? To answer this question, we note that the R^2 statistic is:

$$R^2 = 1 - \frac{\text{residual sum of squares}}{\text{total sum of squares}} = 1 - \frac{604.47}{3489.25}$$

$$= .83$$

We interpret the R^2 statistic to mean that 83% of the variance in the time series is explained by the model. This is a respectable figure. However, the largest portion of the explained variance is due to the mean. We see that the total sum of squares can be divided into three parts:

residual sum of squares	604.47
sum of squares due to Θ_0	2758.89
sum of squares due to Θ_1	125.89
total sum of squares	3489.25

This partition of the total sum of squares is analogous to an analysis of variance partition. We see that the moving average accounts for less than 4% of the variance in the time series. Though this percentage is statistically significant and not at all trivial, it is small compared to the variance explained by the model mean.

Our point in this example analysis is that, sometimes, data that appear to be a time series do not really constitute a time series. This is nearly true of the Harvest Index time series. When data do not constitute a time series, of course, an ARIMA model fit to the data might lead to incorrect conclusions. We will return to this time series in Chapter 5.

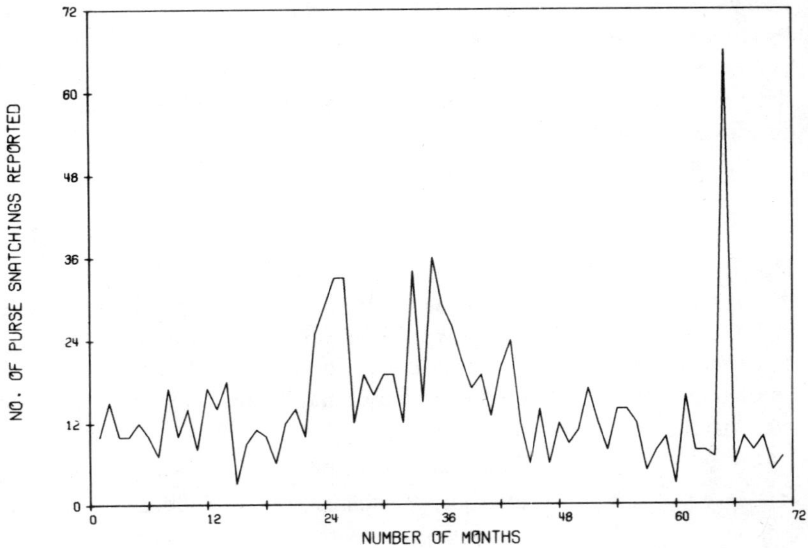

FIGURE 2.12.4(a) Hyde Park Purse Snatchings

2.12.4 Hyde Park Purse Snatchings: Outliers

In Figure 2.12.4(a), we show a time series of purse snatchings reported 13 times per year (every 28 days) in the Hyde Park neighborhood of Chicago from January 1969 to September 1973. These data were collected by Reed (1978) for an evaluation of Operation Whistlestop, a community crime prevention program. Our attention is immediately drawn to the 65th observation of the series which is approximately five times larger than adjacent observations. This is an outlier. In an analysis of these data, Reed arrived at an ambiguous conclusion and suspected that the problem was due to this outlier. Since there was no apparent explanation for this single extreme observation, Reed concluded that it was due to an error in recording the data. Returning to the primary data collection sheets, Reed was able to arrive at a more reasonable, correct number for the 65th observation.

Outliers are a somewhat obscure topic in the literature of time series statistical analysis. While it appears that extreme values may have a distorting effect on the identification and estimation of ARIMA models (we will demonstrate the size of this distortion shortly), there is little consensus on

```
SERIES.. HPPS        (NOBS=  71)  HYDE PARK PURSE SNATCHINGS (W/OUTLIER)
NO. OF VALID OBSERVATIONS  =    71.

AUTOCORRELATIONS OF LAGS 1 - [illegible].
Q( 15,   71) =  12.359        SI[illegible] =       .652

  LAG   CORR    SE  -1   -.8  -.6  -.4  -.2   0    .2   .4   .6   .8  +1
                     +----+----+----+----+----+----+----+----+----+----+
   1    .166  .119                      (     IXXXX )
   2    .230  .122                      (     IXXXXX*
   3    .111  .128                      (     IXXX  )
   4    .150  .129                      (     IXXXX )
   5   -.017  .132                     (      I      )
   6   -.004  .132                     (      I      )
   7    .089  .132                     (      IXX    )
   8    .028  .133                     (      IX     )
   9    .082  .133                     (      IXX    )
  10    .136  .133                     (      IXXX   )
  11    .112  .135                     (      IXXX   )
  12    .075  .137                     (      IXX    )
  13    .060  .137                     (      IX     )
  14   -.001  .138                     (      I      )
  15   -.053  .138                     (     XI      )
                                     -2SE         +2SE

PARTIAL AUTOCORRELATIONS OF LAGS 1 - 15.

  LAG   CORR    SE  -1   -.8  -.6  -.4  -.2   0    .2   .4   .6   .8  +1
                     +----+----+----+----+----+----+----+----+----+----+
   1    .166  .119                      (     IXXXX )
   2    .209  .119                      (     IXXXXX)
   3    .049  .119                      (     IX    )
   4    .088  .119                      (     IXX   )
   5   -.086  .119                      (   XXI     )
   6   -.049  .119                      (    XI     )
   7    .108  .119                      (     IXXX  )
   8    .012  .119                      (     I     )
   9    .064  .119                      (     IXX   )
  10    .114  .119                      (     IXXX  )
  11    .029  .119                      (     IX    )
  12    .008  .119                      (     I     )
  13   -.003  .119                      (     I     )
  14   -.068  .119                      (   XXI     )
  15   -.065  .119                      (   XXI     )
                                     -2SE        +2SE
```

FIGURE 2.12.4(b) ACF and PACF for the Raw Series (Outlier Included)

just what characteristics define an outlier. How (relatively) large must an observation be before it is called an outlier?

It is important to note here that outliers are *sample* phenomena, not *population* phenomena. In cross-sectional analyses, for example, it would be unusual to find a seven-foot-tall person in a sample of five people. This person would be considered an outlier. If the sample size was increased, however, there would come a point at which a seven-foot-tall person would no longer be considered an outlier. This principle holds for longitudinal analyses as well. The deviant 65th observation of the Hyde Park time series

```
SERIES.. HPPS2     (NOBS=  71)  HYDE PARK PURSE SNATCHINGS (WO/OUTLIER)
NO. OF VALID OBSERVATIONS  =     71.

AUTOCORRELATIONS OF LAGS 1 - 15.
Q( 15,   71) =  81.002       SIG =       .000

 LAG   CORR     SE  -1   -.8  -.6  -.4  -.2   0    .2   .4   .6   .8  +1
                     +----+----+----+----+----+----+----+----+----+----+
  1    .493  .119                       (     IXXXXX)XXXXXX
  2    .534  .145                      (      IXXXXXX)XXXXXX
  3    .363  .170                    (        IXXXXXXXX*
  4    .294  .181                    (        IXXXXXXX )
  5    .261  .187                    (        IXXXXXXX )
  6    .163  .192                   (         IXXXX     )
  7    .243  .194                   (         IXXXXXX   )
  8    .183  .199                   (         IXXXXX    )
  9    .179  .201                   (         IXXXX     )
 10    .243  .203                   (         IXXXXXX   )
 11    .204  .207                   (         IXXXXX    )
 12    .227  .210                  (          IXXXXXX    )
 13    .147  .214                  (          IXXXX      )
 14   -.022  .215                  (         XI          )
 15   -.023  .215                  (         XI          )
                                 -2SE                 +2SE


PARTIAL AUTOCORRELATIONS OF LAGS 1 - 15.

 LAG   CORR     SE  -1   -.8  -.6  -.4  -.2   0    .2   .4   .6   .8  +1
                     +----+----+----+----+----+----+----+----+----+----+
  1    .493  .119                       (     IXXXXX)XXXXXX
  2    .385  .119                       (     IXXXXX)XXXX
  3    .018  .119                       (     I     )
  4   -.038  .119                       (    XI     )
  5    .053  .119                       (     IX    )
  6   -.048  .119                       (    XI     )
  7    .141  .119                       (     IXXXX )
  8    .044  .119                       (     IX    )
  9   -.026  .119                       (    XI     )
 10    .136  .119                       (     IXXX  )
 11    .036  .119                       (     IX    )
 12    .001  .119                       (     I     )
 13   -.056  .119                       (    XI     )
 14   -.296  .119                      X(XXXXXI     )
 15   -.067  .119                       (   XXI     )
                                      -2SE        +2SE
```

FIGURE 2.12.4(c) ACF and PACF for the Raw Series (Outlier Excluded)

is considered an outlier in a time series of 71 observations. If the length of this series was increased, however, there would come a point at which the 65th observation would no longer be considered an outlier. The question, of course, is how long must the series be before an outlier can be accommodated by an ARIMA model?

We will answer this question indirectly. Although detection of the outlier is obvious in this example, it will be of some value to analyze this series both with and without the deviant 65th observation. By comparing the two analyses, we can make a statement about the effect that this outlier has on the identification and estimation of an ARIMA model.

Identification. Figure 2.12.4(b) shows the ACF and PACF for the series including the deviant 65th observation. There is no identifiable pattern in either the ACF or PACF, so we must conclude that this series is white noise. This judgment is confirmed by the Q statistic (Q = 12.36 with 15 degrees of freedom) which is significant only at the .65 level. In Figure 2.12.4(c), however, we show the ACF and the PACF for the series excluding the deviant 65th observation. An ARIMA (2,0,0) process is now indicated. The ACF decays and the PACF has spikes at the first two lags. The presence of only one outlier in a series of 71 observations has severely distorted the ACF, leading to an incorrect identification of an ARIMA (0,0,0) model for this series.

A single outlier can have such a profound effect because of the nature of the *estimated* ACF. In estimating an ACF (k), deviated time series observations are weighted by their absolute distance to the series mean. Outlying observations thus exert a profound effect on the estimated ACF (k). We can see this effect more clearly by comparing the components of the lag-1 ACF estimated with and without the outlier:

	Covariance	Variance	ACF (1)
Estimated with the correct observation	27.47	55.71	.49
Estimated with the outlier included	15.49	93.29	.17

Examining these figures, we note that removing the outlier results in a 44% *in*crease in covariancc and a 67% *de*crease in variance. Since the estimated value of ACF (1) is the ratio of covariance to variance, it is not surprising that a decrease in numerator and an increase in denominator seriously underestimates ACF (1). This will generally be true: A single outlier in a short time series will result in a biased underestimation of low lags of the ACF. Because this biased estimate of ACF (1) is not statistically significant, we incorrectly identified an ARIMA (0,0,0) model rather than the more appropriate ARIMA (2,0,0) model.

Estimation. For the ARIMA (2,0,0) model, our parameter estimates are:

$$\hat{\phi}_1 = .31 \text{ with t statistic} = 2.67$$

$$\hat{\phi}_2 = .40 \text{ with t statistic} = 3.45.$$

Both estimates are statistically significant and otherwise acceptable.

It is difficult in practice to untangle the distortions in model estimation introduced by outliers. In this time series, for example, the inappropriate ARIMA (0,0,0) model (with the outlier included) has RMS = 94.62. Fitting the more appropriate ARIMA (2,0,0) model to the series (with the outlier included) results in RMS = 92.82; and both autoregressive parameter estimates are statistically insignificant at a .05 level. The ARIMA (2,0,0) model fit to the series with the outlier excluded results in RMS = 37.75. Much of this difference in the RMS statistics is due only to the outlier.

Diagnosis. The residuals from the ARIMA (2,0,0) model without the outlier appear to be white noise. There are no statistically significant spikes at low lags of the ACF, and Q = 17.8 with 22 degrees of freedom is statistically significant only at the .71 level. We accept this tentative model, then, for the time series with the corrected value for the 65th observation.

We will have more to say about this time series and about outliers generally in the next chapter. Since univariate ARIMA models are always built with some purpose in mind (such as impact assessment, forecasting, or multivariate analyses), outliers can confound an analysis in more complicated ways than those we have discussed. In impact assessment, for example, the location of outliers relative to the point of intervention may directly bias the estimate of impact. This issue will be discussed in detail in the next chapter.

2.13 Conclusion

We conclude this chapter with an overview of ARIMA models and modeling. It is often instructive to see an ARIMA model as a series of linear filters as shown in Figure 2.13(a). In general, a linear filter expresses an output time series as a function of an input white noise process. For any ARIMA model, we start with a random shock input drawn from a Normal (Gaussian) distribution with a zero mean and constant variance. As indicated in this figure (and as explained in Sections 2.2, 2.5, and 2.6), a moving average filter, an autoregressive filter, and/or a nonstationary summation filter is applied to the input to produce an output. Autocorrelation and stochastic behavior in the output time series is determined by the ϕ and Θ parameter values of the filter.

ARIMA model-building procedures, of course, can be represented as inverse filtering. We start with an observed time series and, through the empirical model-building strategy, determine the likely filters and parameter values which will produce an output series of white noise. This process is illustrated in Figure 2.13(b). Note that in both of these diagrams, the input to

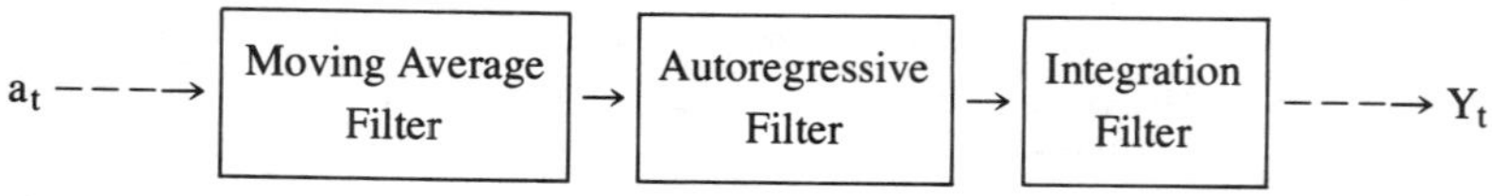

FIGURE 2.13(a) An Input-Output Representation of the ARIMA Model

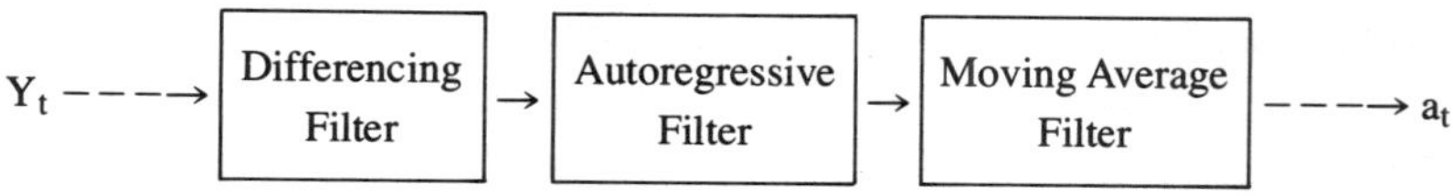

FIGURE 2.13(b) An Input-Output Representation of the ARIMA Model-Building Strategy

both the autoregressive and moving average filters is a stationary series. Application of the model to nonstationary time series is achieved through use of the summation and differencing filters. As we explained in Section 2.2, each of these filters is the inverse of the other. In Figure 2.13(a), the summation filter sums (or integrates) a stationary input series to produce a nonstationary output series. In Figure 2.13(b), the differencing filter differences a nonstationary input to produce a stationary output (which is then passed through autoregressive and/or moving average filters).

An attractive statistical property of ARIMA models is that they can be "run" in both directions. Given an observed time series, we "run the model forward" as in Figure 2.13(b) and reduce the Y_t input to white noise. This requires that statistically adequate parameters be estimated for an appropriate and parsimonious filter structure. Having determined these parameters, we then "run the model backward" as in Figure 2.13(a) to produce predicted values of the Y_t series.

At this point, we suggest that the reader pause to review the material developed in this chapter. We first discussed the statistical properties of ARIMA models: trend, drift, integrated processes, stationarity, variance, autoregressive processes, moving average processes, the expected ACF and PACF, and seasonality. Algebraic operators and formulae were presented to facilitate expression and manipulation of the models. We then developed an iterative model-building strategy wherein the analyst applies these concepts to the problem of constructing a model for an observed time series. Finally, the model-building strategy was applied to the analysis of four time series typical of those encountered in social science research.

We suspect that the reader may feel somewhat frustrated at this point. So far, we have expended a great deal of effort explaining how to build a time series model without explaining how to *use* the model. A univariate ARIMA model in and of itself is admittedly of little interest or utility. This necessary but unfortunate state of affairs will now be remedied. In Chapter 3, we will combine univariate ARIMA models with intervention components to build a variety of models for social impact assessment. In Chapter 4, we will employ univariate ARIMA models to forecast future values of a time series. And in Chapter 5, we will use several univariate ARIMA models to create multivariate time series models of social phenomena.

For Further Reading

There are several treatments of univariate ARIMA modeling which develop this same material at various levels of sophistication. McCain and McCleary (1979), for example, require only an introductory course in statistics while Box and Jenkins (1976: Chapters 6–8) require a solid mathematical background. Intermediate level treatments are given by Granger and Newbold (1977: Chapter 3), Pindyck and Rubinfeld (1976: Chapters 14–15), and Nelson (1973: Chapter 5).

NOTES TO CHAPTER 2

1. The symbol "~" means "is distributed as." The meanings of other symbols and conventions are given in a glossary appendix.

2. See Feller (1968: Chapter III) for an illuminating discussion of drift in the random walk. The reader who wishes to study stochastic processes generally is directed to this work and to Feller (1971). While Feller's development lacks nothing in mathematical rigor, he is clearly the most understandable and readable authority on stochastic processes.

3. We use this example only to illustrate the random walk process. In Chapter 5, we build a population growth model, but of course that model is much more complicated than this one. The reader who plans to do a time series analysis of population statistics would do well to first read Keyfitz (1977: Chapter 1).

4. The two equations

$$Y_t = Y_0 + \Theta_0 t$$

and

$$Y_t - Y_{t-1} = \Theta_0$$

are related in a rather straightforward manner. The first equation is the unique solution of the second. We require no background in difference equations for this volume. However, the interested reader is directed to Goldberg (1958) for an introduction to difference equations written especially for social scientists.

5. Our discussion of stationarity is necessarily a conceptual discussion. A process that is stationary in the *widest* sense is one in which both the process variance, $VAR(y_t)$, and the process covariance, $COV(y_t y_{t+k})$, are independent of t. Such a process is fully described by its variance and covariance. For a more precise definition of widest sense stationarity, see Dhrymes (1974: 385) or Malinvaud (1970: 418–419). Our discussion of transformations is similarly conceptual. See Box and Cox (1964) for a general discussion of transformations.

6. The bounds of stationarity for an ARIMA (p,0,0) process are determined by the roots of the characteristic equation

$$1 + \phi_1 B + \phi_2 B^2 + \ldots + \phi_p B^p = 0.$$

If the process is nonstationary, then all roots must be greater than unity in absolute value. Thus, for an ARIMA (1,0,0) model

$$1 + \phi_1 B = 0$$
$$\phi_1 B = -1$$
$$B = -(1/\phi_1).$$

This root will be greater than unity in absolute value only when ϕ_1 is less than unity in absolute value. Similarly, for an ARIMA (2,0,0) model

$$1 + \phi_1 B + \phi_2 B^2 = 0.$$

The roots of this characteristic equation are given by the formula

$$B = \frac{-b \pm \sqrt{b^2 - 4ac}}{2a} = \frac{-\phi_1 \pm \sqrt{\phi_1^2 - 4\phi_2}}{2\phi_2}$$

These roots will be greater than unity in absolute value only when the bounds of stationarity are satisfied.

7. In the general case, an ARIMA (p,0,0) process is:

$$y_t = \phi_1 y_{t-1} + \phi_2 y_{t-2} + \ldots + \phi_p y_{t-p} + a_t.$$

Multiplying the process by y_{t-k} gives us

$$y_t y_{t-k} = \phi_1 y_{t-1} y_{t-k} + \phi_2 y_{t-2} y_{t-k} + \ldots + \phi_p y_{t-p} y_{t-k} + a_t y_{t-k}.$$

Then taking the expectation of this process and dividing by σ_y^2, we obtain the expected value of ACF (k):

$$ACF(k) = \phi_1 ACF(k-1) + \phi_2 ACF(k-2) + \ldots + \phi_p ACF(k-p).$$

The reader may use this general expression to derive the expected ACF of higher order ARIMA (p,0,0) processes.

8. The YuleWalker equation system is:

$$\begin{bmatrix} ACF(0) & ACF(1) & \ldots & ACF(k-1) \\ ACF(1) & ACF(2) & \ldots & ACF(k-2) \\ \cdot & \cdot & \cdots & \\ \cdot & \cdot & \cdots & \\ \cdot & \cdot & \cdots & \\ ACF(k-2) & ACF(k-3) & \ldots & ACF(1) \\ ACF(k-1) & ACF(k-2) & \ldots & ACF(0) \end{bmatrix} \times \begin{bmatrix} PACF(1) \\ PACF(2) \\ \cdot \\ \cdot \\ \cdot \\ PACF(k-1) \\ PACF(k) \end{bmatrix} = \begin{bmatrix} ACF(1) \\ ACF(2) \\ \cdot \\ \cdot \\ \cdot \\ ACF(k-1) \\ ACF(k) \end{bmatrix}$$

Cramer's Rule can be applied to this k-equation system to obtain solutions for the k unknown values of the PACF. Of course, this assumes that the true values of ACF (1), ACF (2), . . . , ACF (k) are available. Box and Jenkins (1976: 82–84) present a recursive method for calculating PACF (k) which is attributed to Durbin (1960). In practice, a time series computer program routinely estimates the PACF. Our concern here is largely with the *interpretation* of an empirical PACF in the context of model building.

9. Cf. Note 6 above. If

$$(1 - \phi_1 B)(1 - \phi_S B^S) = 0,$$

$$\text{then either } (1 - \phi_1 B) = 0$$

$$\text{or } \quad (1 - \phi_S B^S) = 0.$$

This implies that the two factors will have identical bounds of stationarity-invertibility.

10. Nonlinear estimation is covered as a separate topic in Chapter 6. We have two motives for relegating this topic to the last chapter of the volume. First, many readers will have no interest in the topic. Indeed, the analyst can perform and competently interpret the results of an ARIMA analysis without ever knowing the mechanical details of parameter estimation. Second, nonlinear estimation requires a slightly higher degree of mathematical sophistication of the reader. A superficial knowledge of calculus is assumed, for example. The topic of nonlinear estimation (and Chapter 6) in this sense is not consistent with the more general topics of time series analysis (and the other chapters). In any event, the details of estimation will be better understood after the reader has absorbed the basics of ARIMA modeling.

11. Ljung and Box (1976) have proposed a modification of the Q statistic which increases its value slightly and thus makes it a slightly more conservative test of the hypothesis that the ACF is not different from white noise. We have used the original formula for the Q statistic (Box and Jenkins, 1976: Chapter 8.2) throughout this volume.

12. Model parameter estimates presented in this volume were obtained without backcasting initialization. See Chapter 4, Note 2 and Chapter 6, Notes 1 and 2 for a description of this method and the underlying issues.

Appendix to Chapter 2: Expected Values

The reader who has a working knowledge of calculus is directed to Feller (1966: Chapter 9) for a rigorous but readable discussion of expected values. The reader who lacks this background is directed instead to any introductory statistics text (e.g., Hays, 1973: 871) for an introduction to these concepts. This appendix will deal only with the algebra of expectations required for an understanding of Chapters 2, 3, 4, and 5.

If a random variable, x, is *discrete*, it takes on only a finite set of values. This is ordinarily written as

$$x = \{x_1, \ldots, x_n\}.$$

If each element of this set has an associated probability, $p(x_i)$, then the *expected value* of x is defined as

$$E(x) = \sum_{i=1}^{n} x_i \cdot p(x_i).$$

For example, if x is the number observed on the roll of a die, then x takes on only six values:

$$x = \{1, 2, 3, 4, 5, 6\}.$$

Each of these six values is equiprobable, so the expected value of x is:

$$E(x) = \sum_{i=1}^{6} x_i (1/6) = 3.5.$$

On the other hand, if the random variable is *continuous,* it may take on any value in the real line. This is ordinarily written as

$$-\infty < x < +\infty.$$

If a probability density function, f(x), is defined for the real line, then the expected value of x is:

$$E(x) = \int_{-\infty}^{+\infty} x \cdot f(x) \cdot dx .$$

A random shock, may take on any real value, so the expected value of a random shock is:

$$E(a)_t = \int_{-\infty}^{+\infty} a_t \cdot f(a_t) \cdot da_t = 0,$$

where $f(a_t)$ is the Normal probability density function.

The proof of any expected value theorem is done by summing (in the case of a discrete random variable) or integrating (in the case of a continuous random variable) the product of a random variable and its probability function. A random variable of particular interest to the time series analyst is white noise: the random shock. Each shock is distributed Normally and independently with zero-mean and constant variance, that is:

$$a_t \sim NID(0,\sigma_a^2).$$

The implication of this distribution is that

$$E(a_t) = 0$$
$$E(a_t^2) = \sigma_a^2$$
$$E(a_t a_{t+n}) = 0.$$

Each of these expected values can be derived by integrating the product of the term and its density function as indicated. We will take these expected values as givens.

The expectation operator, E, is applied to a random variable, or to a combination of random variables, to derive the expected value. The expectation operator is a *linear* operator, so the procedure of applying the operator follows the common rules of linear algebra. These rules consist of the five listed below.

First, the operator is applied to a function only after all other operations have been performed. For example, to take the expected value of the term

$$(a_t - \Theta_1 a_{t-1})^2,$$

the expectation operator is applied

$$E\,[(a_t - \Theta_1 a_{t-1})^2].$$

However, all operations indicated inside the brackets ([ ]) must be performed before the expected value is taken, that is,

$$E\,[(a_t - \Theta_1 a_{t-1})^2] = E\,[a_t^2 - 2\Theta_1^2 a_t a_{t-1} + a_{t-1}^2].$$

Taking the expected value before the bracketed operations have been performed will ordinarily not give the expected value of the function.

Second, the expected value of a constant is the constant, for example,

$$E\,(\Theta_0) = \Theta_0.$$

Third, the expected value of the product of a constant and a random variable is the product of the constant and the expected value of the random variable, for example,

$$E\,(\Theta_1 a_{t-1}) = \Theta_1\, E\,(a_{t-1})\,.$$

Fourth, the expected value of the sum of two random variables is the sum of the expected values, for example,

$$E\,(a_t - \Theta_1 a_{t-1}) = E\,(a_t) - \Theta_1 E\,(a_{t-1}).$$

This rule generalizes to any linear combination of random variables.

Fifth, the expected value of the product of two independent random variables is the product of their expected values, for example,

$$E\,(a_t a_{t-1}) = E\,(a_t)\, E\,(a_{t-1})$$

The key word here is *independence*. Random shocks are independent by definition, so the expected value of random shock products is equal to the product of their expected values. If two random variables are *not* independent, however, the expected value of their product is not generally the product of their expected values. For example, successive realizations of an ARIMA (0,0,1) process

$$y_t = a_t - \Theta_1 a_{t-1}$$

$$\text{and } y_{t+1} = a_{t+1} - \Theta_1 a_t$$

have the same zero expected values:

$$E(y_t) = E(y_{t+1}) = 0$$

but because these two random variables are not independent, the expected value of their product is *not* the product of their expected values:

$$E(y_t y_{t+1}) \neq E(y_t)E(y_{t+1}).$$

Instead, the expected value of their product is:

$$\begin{aligned} E(y_t\, y_{t+1}) &= E[(a_t - \Theta_1 a_{t-1})(a_{t+1} - \Theta_1 a_t)] \\ &= E[a_t a_{t+1} - \Theta_1 a_t^2 - \Theta_1 a_{t-1} a_{t+1} + \Theta_1^2 a_{t-1} a_t] \\ &= E(a_t a_{t+1}) - \Theta_1 E(a_t^2) - \Theta_1 E(a_{t-1} a_{t+1}) \\ &\qquad + \Theta_1^2 E(a_{t-1} a_t) \\ &= 0 - \Theta_1 \sigma_a^2 - 0 + 0 = -\Theta_1 \sigma_a^2. \end{aligned}$$

Our use of the expectation operator in Chapters 2, 3, 4, and 5 employs these five rules. Although a particular demonstration may appear formidable, the algebraic manipulations are all straightforward. The reader is urged to learn the rules for applying expectation operators, as developed in this appendix, and to replicate each derivation.

3 ARIMA Impact Assessment

The time series quasi-experiment was proposed originally by Campbell (1963; Campbell and Stanley, 1966) as a means of assessing the impact of a discrete social intervention (or an "event" as we shall soon call it) on behavioral processes. The reader must not assume that a time series quasi-experiment is always the best method of impact assessment for it requires a rather simple theory of impact. There appear nevertheless to be many situations in which simple theories of impact are justified and, in these situations, the time series quasi-experiment may be the most useful of all designs.

Time series quasi-experiments have been used to measure the impacts of new traffic laws (Campbell and Ross, 1968; Glass, 1968; Ross et al., 1970); the impact of decriminalization (Aaronson et al., 1978); the impact of gun control laws (Zimring, 1975); and the impact of air pollution control laws (Box and Tiao, 1975). The widest use of this design has clearly been in the area of legal impact assessment. Time series quasi-experiments have also been used by political scientists to measure the impacts of political realignments (Caporaso and Pelowski, 1971; Lewis-Beck, 1979; Smoker, 1969), however, and by experimental psychologists to measure the impacts of treatments on behavior (Gottman and McFall, 1972; Hall et al., 1971; Tyler and Brown, 1968). This list is representative (but by no means exhaustive) of the situations in which time series quasi-experiments have been used to assess social impacts.

Our major concern in this chapter is with the analysis of time series quasi-experiments: impact assessment as we call it. Time series analysis cannot be divorced from the design of the time series quasi-experiment, however. According to Cook and Campbell (1979), the design of any quasi-experiment must recognize the threats to four types of validity: internal, external, statistical conclusion, and construct validities. We know of no statistical methods for correcting or controlling flaws in design. Our development of statistical models thus assumes an adequately designed quasi-experiment. Design is the sine qua non. The reader who is unfamiliar with quasi-experimental design is directed to Cook and Campbell for an authoritative treatment.

We will use the term *impact assessment* to refer to the statistical analysis of an adequately designed time series quasi-experiment. More generally, we define this term as "a test of the null hypothesis that a postulated event caused a change in a social process measured as a time series." Acknowledging the faults and limitations of this definition, we must comment on its two key elements.

First, impact assessment is concerned with the effects of a "postulated event." An event for our purposes is a qualitative change in state or, in common terms, "something that happens." Events can be represented as binary variables which indicate the *absence* of the state prior to the event and the *presence* of the state during and (possibly) after the event. In the parlance of experimental psychology, for example, introduction of a treatment is the event associated with a change in state from "no treatment" to "treatment." In legal studies, enactment of a new law is the event associated with a change in state from "no regulation" to "regulation."

Qualitative changes in states (events) are often indistinguishable from quantitative changes in levels (processes). In studying national arms expenditures over time, for example, some social scientists prefer to think of "war" as an *event* which affects expenditures. Other social scientists prefer to think of "the propensity to war" as a continuous *process* which affects expenditures. We will develop multivariate ARIMA models for the case in which the causal agent is a continuous process (as measured by an independent variable time series) in Chapter 5. For now, however, it is important to remember that an impact assessment analysis is concerned with the effect of an *event* on some social or behavioral process.

Because the change agent is an event, it is represented in the impact assessment model as a "dummy" variable or step function such that

$$\begin{aligned} I_t &= 0 \text{ prior to the event} \\ &= 1 \text{ thereafter.} \end{aligned}$$

If the change process is not an event in the technical sense, however, the impact assessment may lead to invalid conclusions. As a general rule, the independent variable of an impact assessment model should give as accurate a representation of the change agent as possible. In an analysis of a Washington, D.C., gun control law, for example, Zimring (1975) had information on the actual level of enforcement. Using this information, Zimring defined the event in terms of I_t as

$$
\begin{aligned}
I_t &= 0 \text{ prior to enactment} \\
&= 1/6 \text{ in the first month after enactment} \\
&= 2/6 \text{ in the second month} \\
&\;\;\vdots \\
&= 6/6 = 1 \text{ in the sixth and subsequent months.}
\end{aligned}
$$

With this definition of I_t, the event corresponding to a change in state from "no regulation" to "regulation" is distributed across a six-month period. The fundamental principle illustrated here is that an impact assessment requires a theory of change. If the change agent is an event, then it can and must be represented by a simple step function. If the change agent is *not* an event in the strictest sense, however, the analysis may lead to invalid conclusions. A more valid analysis can be ensured by modifying the step function, as Zimring did, to accommodate known properties of the change agent.

A second element of the "impact assessment" definition is the a priori specification of the onset of an event. A null hypothesis that an event "caused" a change in some behavior can be tested only because the time of the event is known a priori. It would indeed be possible to search the length of a time series for statistically significant changes but it would be logically impossible to then associate each change with the infinite number of events which might be the causes. An impact analysis based on a blind search (see, e.g., Deutsch, 1978; also, Section 4.3 below) might generously be called "exploratory analysis." Its results are quite uninterpretable. An impact assessment based on an event whose onset is specified a priori, in contrast, is a "confirmatory analysis." It is used only to test theoretically generated hypotheses according to a rigorous set of validity criteria.

These two elements of the definition are so important that we reiterate them. First, impact assessment is concerned only with *events* and, second, impact assessment requires that the *onset* of an event be specified a priori. Lacking these two elements, the results of an impact analysis will be uninterpretable.

Impact assessment (or the time series analysis of impacts) begins with an

ARIMA model for the time series. Since this ARIMA model describes the stochastic behavior of the time series process, we refer to it as the "noise component" of the model. An intervention component is then added to the model. The full impact assessment model may be written as

$$Y_t = f(I_t) + N_t,$$

where N_t denotes the noise component, an ARIMA model, and where $f(I_t)$ denotes a "function of the variable I_t," the intervention component. The intervention component itself describes the deterministic relationship between an event (as represented by the variable I_t) and the time series. The noise component describes the stochastic behavior of the time series around the $Y_t = f(I_t)$ relationship.

The general principles of ARIMA modeling which we developed in Chapter 2 apply as well to impact assessment modeling. Analysis begins with construction of an ARIMA model for the Y_t time series. In some cases, the real impact in the time series may be so large that it overwhelms and distorts the ACF and PACF; this phenomenon is similar to distortions associated with outliers. To avoid problems in identification, the analyst may have to estimate ACFs and PACFs from the preintervention series only.

After an adequate ARIMA model has been identified, its parameters satisfactorily estimated, and its residuals diagnosed, an intervention component is added. The intervention component will ideally be selected on the basis of a theoretically generated null hypothesis. The parameters of the full impact assessment model (both noise and intervention components) are then estimated. If a parameter estimate is not statistically significant or is otherwise unacceptable (if the estimate of a noise parameter lies outside the bounds of stationarity-invertibility, for example), the tentative model must be respecified and its parameters reestimated.

Once a tentative model has been specified and significant, acceptable parameter estimates have been obtained, the impact assessment model must be diagnosed. As in the case of univariate ARIMA models, residuals must not be different than white noise. Although the noise component alone may have white noise residuals, it sometimes happens that the full impact assessment model (noise *and* intervention components) does not. When this happens, a new tentative model must be specified, its parameters estimated, and its residuals diagnosed. The model-building procedure continues iteratively until a parsimonious but statistically adequate impact assessment model is generated.

Impact parameters may then be tested for statistical significance and, more generally, *the model may be interpreted*. We will illustrate the general model-building strategy for impact assessments with several example analyses. These will not be as definitively instructive as the example analyses of the previous chapter, however. While ARIMA models per se (the noise component of the impact assessment model) are atheoretical and uninterpretable, an impact assessment model (noise *and* intervention components) is built for no reason other than interpretation. The analyst must draw conclusions from the impact assessment analysis and, in every case, these conclusions must be reconciled with the prevailing theory of a substantive area. An impact assessment model may then be the "best" possible model in a statistical sense but not in the substantive sense. Interpretability is everything, and for this reason, impact assessment modeling cannot be reduced to a set of objective, mechanical steps.

3.1 The Zero-Order Transfer Function

Denoting the full impact assessment model as

$$Y_t = f(I_t) + N_t,$$

we are now concerned with the "function of I_t," that is, with the intervention component of the full model. Some writers refer to the intervention component as a "transfer function," a term derived from engineering contexts. We will use both terms synonymously here, although our preference is for the more straightforward "intervention component."

The simplest possible intervention component is the zero-order transfer function

$$f(I_t) = \omega_0 I_t.$$

This is a *zero*-order transfer function because the highest power of B in the function is zero. Where the variable I_t is defined as a step function such that

$$\begin{aligned} I_t &= 0 \text{ prior to the event} \\ &= 1 \text{ thereafter,} \end{aligned}$$

the impact assessment model is:

$$Y_t = \omega_0 I_t + N_t.$$

Now because the impact assessment model is linear in its components, the noise component, N_t, may be subtracted from the time series:

$$\begin{aligned} Y_t^* &= Y_t - N_t \\ &= \omega_0 I_t. \end{aligned}$$

So long as the N_t component is statistically adequate, that is, so long as it has been built along the lines described in the previous chapter, subtracting it from the Y_t time series results in a deterministic intervention component. Working with the Y_t^* series (instead of the Y_t series), the deterministic effects of the transfer function may be examined.

Prior to the event, when $I_t = 0$, the level of the Y_t^* series is:

$$\begin{aligned} Y_t^* &= \omega_0(0) \\ &= 0. \end{aligned}$$

But with the onset of the event, when $I_t = 1$, the level of the Y_t^* series is:

$$\begin{aligned} Y_t^* &= \omega_0(1) \\ &= \omega_0. \end{aligned}$$

The zero-order transfer function thus determines an *abrupt, permanent* shift in process level from pre- to postintervention, a pattern of impact such as

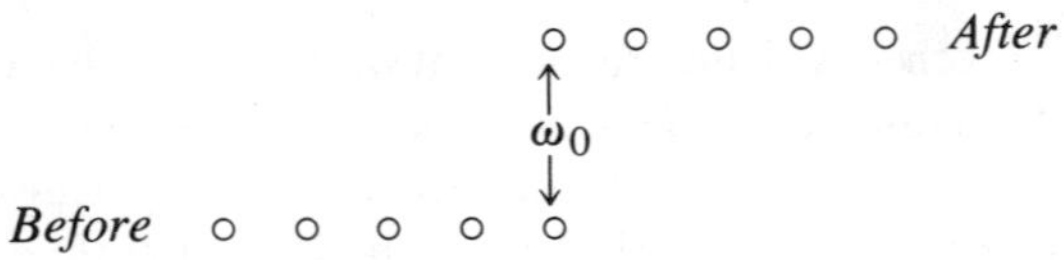

At the onset of the event, the level of the process increases by the quantity ω_0 (or decreases if ω_0 is negative).

A few comments on the concept of "level" may be helpful here. An impact assessment model describes a change in level and/or (sometimes) trend for the generating process of the time series. Some writers use the term *equilibrium* rather than "level," but whichever term is used, it is important to remember that a statistical concept (not a substantive concept) is implied. For a stationary time series process, the parameter ω_0 is an estimate of the difference between the pre- and postintervention process levels.

For a nonstationary series, an analogous interpretation is possible. Nonstationary time series generated by ARIMA(p,d,q) processes can be represented by stationary ARIMA(p,0,q) models after an appropriate differencing. As noted in Chapter 2, the inverse relationship between differencing and summation operators (or filters; see Figure 2.13) allows for a mapping between the stationary model and the nonstationary process. This is also true of impact assessment models. An *abrupt, permanent* pattern of impact, for example, as determined by the zero-order transfer function, in a trending

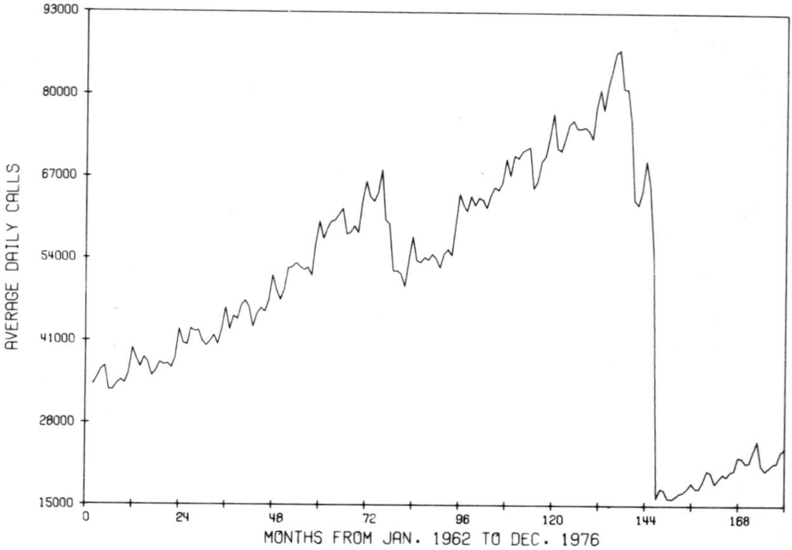

FIGURE 3.1(a) Directory Assistance, Monthly Average Calls per Day

series may appear as

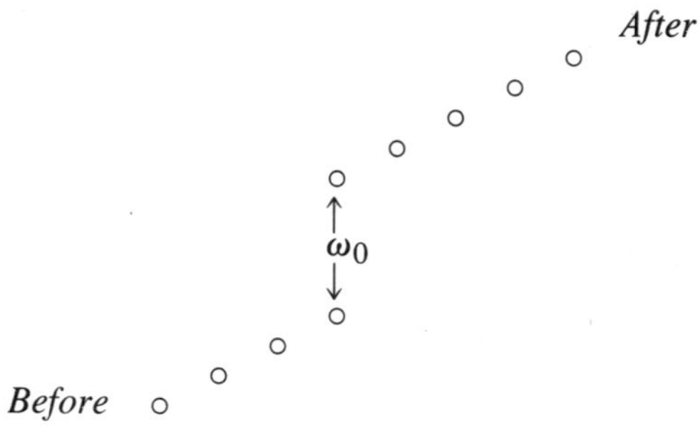

The interpretation of the parameter ω_0 is more or less the same, then, whether the N_t component is a stationary ARIMA(p,0,q) model or a nonstationary ARIMA(p,d,q) model.

Using the simple zero-order transfer function, we will now demonstrate the impact assessment model-building strategy. Figure 3.1(a) shows a monthly time series of calls to Directory Assistance in Cincinnati, Ohio, as reported by McSweeny (1978). The first observation of this series is Janu-

(text continued on p. 152)

```
SERIES.. TEL146    (NOBS= 146)  DIRECTORY ASST - PRE INTERVENTION
NO. OF VALID OBSERVATIONS   =   146.

AUTOCORRELATIONS OF LAGS 1 - 25.
Q( 25, 146) =  1727.3       SIG =     0.000

LAG    CORR     SE  -1   -.8  -.6  -.4  -.2   0    .2   .4   .6   .8  +1
                    +----+----+----+----+----+----+----+----+----+----+
  1    .971   .083                       (   IXXX)XXXXXXXXXXXXXXXXXXXX
  2    .939   .141                    (      IXXXXXX)XXXXXXXXXXXXXXXX
  3    .917   .178                  (        IXXXXXXXX)XXXXXXXXXXXXXX
  4    .894   .208                 (         IXXXXXXXXX)XXXXXXXXXXXX
  5    .869   .233               (           IXXXXXXXXXXX)XXXXXXXXXX
  6    .842   .254              (            IXXXXXXXXXXXX)XXXXXXXX
  7    .814   .273             (             IXXXXXXXXXXXXX)XXXXXX
  8    .790   .289             (             IXXXXXXXXXXXXX)XXXXXX
  9    .761   .303            (              IXXXXXXXXXXXXXX)XXXX
 10    .732   .316           (               IXXXXXXXXXXXXXXX)XX
 11    .713   .328           (               IXXXXXXXXXXXXXXX)XX
 12    .697   .338          (                IXXXXXXXXXXXXXXXX*
 13    .665   .348          (                IXXXXXXXXXXXXXXXX*
 14    .632   .356         (                 IXXXXXXXXXXXXXXXX )
 15    .606   .364         (                 IXXXXXXXXXXXXXXX  )
 16    .579   .371        (                  IXXXXXXXXXXXXXX    )
 17    .551   .377        (                  IXXXXXXXXXXXXXX    )
 18    .526   .382        (                  IXXXXXXXXXXXXX     )
 19    .502   .387        (                  IXXXXXXXXXXXXX     )
 20    .481   .392       (                   IXXXXXXXXXXXX       )
 21    .459   .396       (                   IXXXXXXXXXXX        )
 22    .435   .399       (                   IXXXXXXXXXXX        )
 23    .421   .403       (                   IXXXXXXXXXXX        )
 24    .412   .406       (                   IXXXXXXXXXX         )
 25    .387   .408       (                   IXXXXXXXXXX         )
                        -2SE                                   +2SE
```

```
PARTIAL AUTOCORRELATIONS OF LAGS 1 - 25.

LAG    CORR     SE  -1   -.8  -.6  -.4  -.2   0    .2   .4   .6   .8  +1
                    +----+----+----+----+----+----+----+----+----+----+
  1    .971   .083                       (   IXXX)XXXXXXXXXXXXXXXXXXXX
  2   -.047   .083                       (  XI   )
  3    .145   .083                       (   IXXX*
  4   -.052   .083                       (  XI   )
  5    .009   .083                       (   I   )
  6   -.082   .083                       ( XXI   )
  7   -.016   .083                       (   I   )
  8    .029   .083                       (   IX  )
  9   -.092   .083                       ( XXI   )
 10   -.015   .083                       (   I   )
 11    .156   .083                       (   IXXX*
 12    .039   .083                       (   IX  )
 13   -.277   .083                    XXX(XXXI   )
 14    .005   .083                       (   I   )
 15    .055   .083                       (   IX  )
 16   -.079   .083                       ( XXI   )
 17   -.012   .083                       (   I   )
 18    .089   .083                       (   IXX )
 19    .001   .083                       (   I   )
 20   -.024   .083                       (  XI   )
 21    .040   .083                       (   IX  )
 22   -.017   .083                       (   I   )
 23    .080   .083                       (   IXX )
 24    .010   .083                       (   I   )
 25   -.172   .083                       *XXXI   )
                                       -2SE    +2SE
```

FIGURE 3.1(b) ACF and PACF for the Raw Preintervention Series

```
SERIES.. TEL146    (NOBS= 146)  DIRECTORY ASST - PRE INTERVENTION
DIFFERENCED  1 TIME(S) OF ORDER  1.
NO. OF VALID OBSERVATIONS  =   145.

AUTOCORRELATIONS OF LAGS 1 - 25.
Q( 25, 145) =  63.683      SIG =     .000

  LAG   CORR    SE  -1  -.8  -.6  -.4  -.2    0   .2   .4   .6   .8  +1
                     +----+----+----+----+----+----+----+----+----+----+
    1   .032  .083                        (   IX  )
    2  -.195  .083                       X(XXXI   )
    3   .053  .086                        (   IX  )
    4   .012  .086                        (   I   )
    5   .006  .086                        (   I   )
    6   .022  .086                        (   IX  )
    7  -.069  .087                        ( XXI   )
    8   .104  .087                        (   IXXX)
    9   .021  .088                        (   IX  )
   10  -.235  .088                      XX(XXXI   )
   11  -.017  .092                        (   I    )
   12   .360  .092                        (   IXXXX)XXXX
   13  -.014  .101                       (    I    )
   14  -.150  .101                       (XXXXI    )
   15   .026  .103                       (    IX   )
   16  -.045  .103                       (   XI    )
   17  -.058  .103                       (   XI    )
   18  -.010  .103                       (    I    )
   19  -.072  .103                       (  XXI    )
   20   .023  .104                       (    IX   )
   21   .045  .104                       (    IX   )
   22  -.239  .104                      X(XXXXI    )
   23  -.017  .107                       (    I    )
   24   .319  .107                       (    IXXXX)XXX
   25  -.002  .114                      (     I     )
                                     -2SE        +2SE
```

```
PARTIAL AUTOCORRELATIONS OF LAGS 1 - 25.

  LAG   CORR    SE  -1  -.8  -.6  -.4  -.2    0   .2   .4   .6   .8  +1
                     +----+----+----+----+----+----+----+----+----+----+
    1   .032  .083                        (   IX  )
    2  -.196  .083                       X(XXXI   )
    3   .070  .083                        (   IXX )
    4  -.033  .083                        (  XI   )
    5   .033  .083                        (   IX  )
    6   .012  .083                        (   I   )
    7  -.065  .083                        ( XXI   )
    8   .123  .083                        (   IXXX)
    9  -.023  .083                        (  XI   )
   10  -.193  .083                       X(XXXI   )
   11  -.007  .083                        (   I   )
   12   .311  .083                        (   IXXX)XXXX
   13  -.038  .083                        (  XI   )
   14  -.069  .083                        ( XXI   )
   15   .020  .083                        (   I   )
   16  -.073  .083                        ( XXI   )
   17  -.089  .083                        ( XXI   )
   18  -.005  .083                        (   I   )
   19  -.033  .083                        (  XI   )
   20  -.082  .083                        ( XXI   )
   21   .026  .083                        (   IX  )
   22  -.116  .083                        (XXXI   )
   23   .003  .083                        (   I   )
   24   .179  .083                        (   IXXX*
   25   .016  .083                        (   I   )
                                       -2SE    +2SE
```

FIGURE 3.1(c) ACF and PACF for the Regularly Differenced Preintervention Series

```
SERIES.. TEL146    (NOBS= 146)  DIRECTORY ASST - PRE INTERVENTION
DIFFERENCED  1 TIME(S) OF ORDER  1.
DIFFERENCED  1 TIME(S) OF ORDER 12.
NO. OF VALID OBSERVATIONS  =   133.

AUTOCORRELATIONS OF LAGS 1 - 25.
Q( 25, 133) =  25.266      SIG =      .448

  LAG    CORR    SE  -1   -.8  -.6  -.4  -.2  0   .2   .4   .6   .8  +1
                     +----+----+----+----+----+----+----+----+----+----+
    1   -.036  .087                     (  XI   )
    2    .033  .087                     (   IX  )
    3    .083  .087                     (   IXX )
    4    .009  .088                     (   I   )
    5    .087  .088                     (   IXX )
    6    .073  .088                     (   IXX )
    7    .021  .089                     (   IX  )
    8    .110  .089                     (   IXXX)
    9   -.034  .090                     (  XI   )
   10   -.023  .090                     (  XI   )
   11    .070  .090                     (   IXX )
   12   -.294  .090                   XX(XXXXI    )
   13    .021  .097                     (    IX   )
   14   -.023  .097                     (   XI    )
   15   -.035  .097                     (   XI    )
   16   -.069  .097                     (  XXI    )
   17    .102  .098                     (    IXXX )
   18   -.121  .098                     ( XXXI    )
   19   -.061  .100                     (  XXI    )
   20   -.038  .100                     (   XI    )
   21    .116  .100                     (    IXXX )
   22   -.095  .101                     (  XXI    )
   23   -.038  .102                     (   XI    )
   24   -.015  .102                     (    I    )
   25    .032  .102                     (    IX   )
                                      -2SE      +2SE

PARTIAL AUTOCORRELATIONS OF LAGS 1 - 25.

  LAG    CORR    SE  -1   -.8  -.6  -.4  -.2  0   .2   .4   .6   .8  +1
                     +----+----+----+----+----+----+----+----+----+----+
    1   -.036  .087                      (  XI   )
    2    .031  .087                      (   IX  )
    3    .086  .087                      (   IXX )
    4    .014  .087                      (   I   )
    5    .083  .087                      (   IXX )
    6    .073  .087                      (   IXX )
    7    .021  .087                      (   IX  )
    8    .096  .087                      (   IXX )
    9   -.040  .087                      (  XI   )
   10   -.045  .087                      (  XI   )
   11    .042  .087                      (   IX  )
   12   -.305  .087                  XXXX(XXXI   )
   13   -.022  .087                      (  XI   )
   14   -.032  .087                      (  XI   )
   15    .004  .087                      (   I   )
   16   -.075  .087                      ( XXI   )
   17    .173  .087                      (   IXXX*
   18   -.069  .087                      ( XXI   )
   19   -.049  .087                      (  XI   )
   20    .025  .087                      (   IX  )
   21    .150  .087                      (   IXXX*
   22   -.133  .087                      (XXXI   )
   23    .002  .087                      (   I   )
   24   -.108  .087                      (XXXI   )
   25    .039  .087                      (   IX  )
                                       -2SE    +2SE
```

FIGURE 3.1(d) ACF and PACF for the Regularly *and* Seasonally Differenced
Preintervention Series

```
SERIES.. RESIDUAL (NOBS= 133)  DIRECTORY ASST - PRE IMPACT RESIDUALS
NO. OF VALID OBSERVATIONS  =   133.

AUTOCORRELATIONS OF LAGS 1 - 25.
Q( 24, 133) =  12.522       SIG =      .973

LAG  CORR    SE   -1   -.8  -.6  -.4  -.2   0    .2   .4   .6   .8  +1
                  +----+----+----+----+----+----+----+----+----+----+
  1  -.026  .087                       (  XI   )
  2   .009  .087                       (   I   )
  3   .089  .087                       (   IXX )
  4  -.032  .087                       (  XI   )
  5   .165  .088                       (   IXXX*
  6   .029  .090                       (   IX  )
  7  -.041  .090                       (  XI   )
  8   .111  .090                      (    IXXX )
  9   .023  .091                      (    IX   )
 10  -.061  .091                      (  XXI    )
 11  -.033  .091                      (   XI    )
 12   .012  .092                      (    I    )
 13  -.051  .092                      (   XI    )
 14  -.001  .092                      (    I    )
 15  -.053  .092                      (   XI    )
 16  -.088  .092                      (  XXI    )
 17   .055  .093                      (    IX   )
 18  -.060  .093                      (   XI    )
 19  -.019  .093                      (    I    )
 20  -.014  .093                      (    I    )
 21   .087  .093                      (    IXX  )
 22  -.061  .094                      (  XXI    )
 23  -.064  .094                      (  XXI    )
 24  -.018  .094                      (    I    )
 25  -.029  .094                      (   XI    )
                                  -2SE         +2SE

PARTIAL AUTOCORRELATIONS OF LAGS 1 - 25.

LAG  CORR    SE   -1   -.8  -.6  -.4  -.2   0    .2   .4   .6   .8  +1
                  +----+----+----+----+----+----+----+----+----+----+
  1  -.026  .087                       (  XI   )
  2   .008  .087                       (   I   )
  3   .089  .087                       (   IXX )
  4  -.028  .087                       (  XI   )
  5   .164  .087                       (   IXXX*
  6   .029  .087                       (   IX  )
  7  -.037  .087                       (  XI   )
  8   .083  .087                       (   IXX )
  9   .033  .087                       (   IX  )
 10  -.083  .087                       ( XXI   )
 11  -.066  .087                       ( XXI   )
 12   .024  .087                       (   IX  )
 13  -.072  .087                       ( XXI   )
 14  -.018  .087                       (   I   )
 15  -.030  .087                       (  XI   )
 16  -.072  .087                       ( XXI   )
 17   .038  .087                       (   IX  )
 18  -.023  .087                       (  XI   )
 19   .009  .087                       (   I   )
 20  -.020  .087                       (  XI   )
 21   .139  .087                       (   IXXX)
 22  -.073  .087                       ( XXI   )
 23  -.062  .087                       ( XXI   )
 24  -.015  .087                       (   I   )
 25  -.023  .087                       (  XI   )
                                   -2SE     +2SE
```

FIGURE 3.1(e) *Diagnosis:* ACF and PACF for the Model Residuals (Preintervention Series Only)

```
SERIES.. RESIDUAL (NOBS= 167)  ESTIMATED MODEL RESIDUALS
NO. OF VALID OBSERVATIONS   =  167.

AUTOCORRELATIONS OF LAGS 1 - 25.
Q( 24, 167) =  17.218       SIG =      .839

 LAG    CORR    SE   -1   -.8  -.6  -.4  -.2   0   .2   .4   .6   .8  +1
                      +----+----+----+----+----+----+----+----+----+----+
  1   -.030   .080                       (  XI    )
  2   -.030   .080                       (  XI    )
  3    .080   .080                       (    IXX )
  4   -.030   .080                       (  XI    )
  5    .110   .080                       (    IXXX)
  6    .040   .080                       (    IX  )
  7   -.060   .080                       ( XXI    )
  8    .070   .080                       (    IXX )
  9    .050   .080                       (    IX  )
 10   -.060   .080                       ( XXI    )
 11   -.090   .080                       ( XXI    )
 12    .040   .080                       (    IX  )
 13   -.080   .080                       ( XXI    )
 14    .010   .080                       (    I   )
 15    .020   .080                       (    IX  )
 16   -.100   .080                       (XXXI    )
 17   -.020   .080                       (  XI    )
 18   -.040   .080                       (  XI    )
 19   -.030   .080                       (  XI    )
 20   -.040   .080                       (  XI    )
 21    .090   .080                       (    IXX )
 22   -.090   .080                       ( XXI    )
 23   -.120   .080                       (XXXI    )
 24    .020   .080                       (    IX  )
 25   -.060   .080                       ( XXI    )
                                        -2SE    +2SE
```

FIGURE 3.1(f) *Diagnosis:* ACF for the Model Residuals

ary 1962 and the 180th and last observation is December 1976. In March 1974, the 147th month, Cincinnati Bell initiated a 20-cent charge for each call to Directory Assistance. Prior to this time, there was no charge for these calls. The impact of this event is visually striking. In the 147th month, the level of this time series drops abruptly and profoundly.

When an impact is as large as the one in this example, the change in process level complicates identification of the noise component. The change in level is a significant proportion of the series variance which tends to overwhelm the ACF and PACF. To avoid biased estimates of the ACFs and PACFs, only the first 146 observations of the series will be used. ACFs and PACFs used in the identification are shown in Figures 3.1(b) to 3.1(f).

Identification

The ACF and PACF estimated from the raw series indicate a nonstationary ARIMA process. The ACF and PACF estimated from the regularly differenced series, shown in Figure 3.1(c), indicate the ARIMA process is seasonally nonstationary as well. The ACF and PACF estimated from the regularly *and* seasonally differenced series, shown in Figure 3.1(d), suggest an ARIMA$(0,1,0)(0,1,1)_{12}$ model:

$$N_t = \frac{\Theta_0 + (1 - \Theta_{12}B^{12})}{(1 - B)(1 - B^{12})} a_t .$$

This is an interesting model and somewhat rare. The only autocorrelation is at the seasonal lags.

Estimation

Parameter estimates for the N_t model are:

$$\hat{\Theta}_0 = -.70 \text{ with t-statistic} = -\ 1.51$$
$$\hat{\Theta}_{12} = .85 \text{ with t-statistic} = 15.30.$$

The estimate of Θ_0 is not statistically significant, so it is dropped from the tentative model. The estimate of Θ_{12} is statistically significant and lies within the bounds of invertibility.

Diagnosis

The residual ACF and PACF, shown in Figure 3.1(e), indicate that the residuals of this model are not different than white noise. There is a significant spike at ACF(5) but nothing else: $Q = 12.52$ with 24 degrees of freedom is not statistically significant, so the tentative model is accepted. As an aside, we note that another analyst might be concerned about the spike at ACF(5) which might indicate the need for a more elaborate model. Also, the estimated value of Θ_0 is "marginally" significant and some other analyst might decide to keep that parameter in the model. The reader is invited to explore these possibilities.

Impact Assessment

The full impact assessment model is tentatively set as

$$Y_t = \omega_0 I_{147} + \frac{1 - \Theta_{12}B^{12}}{(1 - B)(1 - B^{12})} a_t$$

where $I_{147} = 0$ for the first 146 observations
$= 1$ for the 147th and subsequent observations.

Parameter estimates for the tentative model are:

$$\hat{\Theta}_{12} = .81 \text{ with t statistic} = 11.21$$
$$\hat{\omega}_0 = -39{,}931 \text{ with t statistic} = -17.41.$$

Both estimates are statistically significant and the estimate of Θ_{12} lies within the bounds of invertibility. A final diagnosis indicates that the estimated model is statistically adequate. The residual ACF, shown in Figure 3.1(f), has no significant spikes at all; the Q statistic for this ACF is not statistically significant.

Our interpretation of these findings is obvious. In the 147th month, the level of this series dropped by nearly 40,000 average daily calls to Directory Assistance.

The model-building strategy outlined in this example can be followed generally in all analyses. Each analysis will present a unique set of problems, however, which may require a slight adaptation of the strategy. We note finally that, in this example, a test of the null hypothesis was not at all in question. The impact was visually obvious. Impact assessment analysis nonetheless provided a precise estimate of the form and magnitude of the effect.

3.2 The First-Order Transfer Function

When impacts are as abrupt and dramatic as the one in the Directory Assistance example, the zero-order transfer function will adequately model the impact. Such abrupt, dramatic patterns of impact are rare, however. Most social impacts will be realized gradually, so the zero-order transfer function will not adequately reflect the expected impact. Returning to Zimring's reformulation of the step function in an impact assessment of a gun control law as

$$\begin{aligned} I_t &= 0 \text{ prior to enactment} \\ &= 1/6 \text{ in the first month after enactment} \\ &= 2/6 \text{ in the second month} \\ &\quad \vdots \\ &= 6/6 = 1 \text{ in the sixth and subsequent months,} \end{aligned}$$

the reformulation of I_t reflects a *gradual* impact of the new law. Zimring could have measured such an impact by using a first-order transfer function rather than a zero-order function with a reformulated I_t variable.

To date, most social science impact assessments have used the zero-order transfer function exclusively. There are two reasons for this. First, there has been little discussion in the methodological literature of any impact patterns other than the *abrupt, constant* pattern associated with the zero-order transfer function; and the computer software required for estimation of higher order transfer functions has not been widely available. Second, social science time series are often reported as annual statistics; an impact realized gradually over an eight-month period, for example, will appear as an abrupt impact if annual (rather than monthly) data are analyzed. If the data are aggregated in a way that obscures the form of an impact (abrupt *versus* gradual, for example), the zero-order transfer function will adequately model the impact. If the data are aggregated so that the form is not obscured, however, and if an impact is not abrupt, the zero-order transfer function will not adequately model the impact.

A *gradual, permanent* change in process level is implied by the first-order transfer function

$$f(I_t) = \frac{\omega_0}{1 - \delta_1 B} I_t$$

where the parameter δ_1 is constrained to the interval

$$-1 < \delta_1 < +1.$$

These constraints are called the *bounds of system stability*. If the value of δ_1 lies outside these bounds, the impact assessment model is unstable. It will be demonstrated later that system instability is identical with nonstationarity. When $\delta_1 \geqslant 1$, the postintervention time series is nonstationary and this is generally interpreted to mean that the event has affected a trend in the time series process.

Again, because the impact assessment model is linear in its two components, a Y_t^* time series can be defined as

$$Y_t^* = Y_t - N_t$$

$$= \frac{\omega_0}{1 - \delta_1 B} I_t.$$

So $$(1 - \delta_1 B) Y_t^* = \omega_0 I_t$$
$$Y_t^* = \delta_1 Y_{t-1}^* + \omega_0 I_t.$$

This formulation of Y_t^* can be used recursively to examine the behavior of this first-order transfer function. Prior to the event, when $I_t = 0$, the Y_t^* series has a zero level:

$$\begin{aligned} Y_t^* &= \delta_1 Y_{t-1}^* + \omega_0 I_t \\ &= \delta_1(0) + \omega_0(0) = 0. \end{aligned}$$

Now if the final preintervention observation of the series is Y_i^*, the event occurs at $t = i+1$ and $I_{i+1} = 1$. The value of Y_{i+1}^* is thus

$$\begin{aligned} Y_{i+1}^* &= \delta_1 Y_i^* + \omega_0 I_{i+1} \\ &= \delta_1(0) + \omega_0(1) = \omega_0. \end{aligned}$$

In the next postintervention observation, $I_{i+2} = 1$, and the value of Y_{i+2}^* is:

$$\begin{aligned} Y_{i+2}^* &= \delta_1 Y_{i+1}^* + \omega_0 I_{i+2} \\ &= \delta_1 (\omega_0) + \omega_0(1) = \delta_1\omega_0 + \omega_0. \end{aligned}$$

And in the next postintervention observation, $I_{i+3} = 1$, and the value of Y_{i+3}^* is:

$$\begin{aligned} Y_{i+3}^* &= \delta_1 Y_{i+2}^* + \omega_0 I_{i+3} \\ &= \delta_1 (\delta_1\omega_0 + \omega_0) + \omega_0(1) = \delta_1^2\omega_0 + \delta_1\omega_0 + \omega_0. \end{aligned}$$

Continuing this procedure, it can be shown that, in the nth postintervention observation, $I_{i+n} = 1$, and the value of Y_{i+n}^* is:

$$\begin{aligned} Y_{i+n}^* &= \delta_1 Y_{i+n-1}^* + \omega_0 I_{i+n} \\ &= \delta_1(\delta_1^{n-1}\omega_0 + \ldots + \delta_1\omega_0 + \omega_0) + \omega_0(1) \\ &= \sum_{k=0}^{n} \delta_1^k\omega_0. \end{aligned}$$

The importance of constraining the value of δ_1 to the bounds of system stability may now be obvious. So long as δ_1 is a fraction,

$$\left|\delta_1^{n-1}\omega_0\right| > \left|\delta_1^n\omega_0\right|.$$

Each successive term of the series $\sum_{k=0}^{n} \delta_1^k\omega_0$ is smaller than the previous term. As time passes then, the postintervention series level continues to change (increasing or decreasing, depending upon whether ω_0 is positive or negative) but by smaller and smaller increments (or decrements).

Figure 3.2(a) shows the change expected in the Y_t^* time series with each observation. At the moment of intervention, the series level changes from zero to ω_0, and in the next moment, from ω_0 to $(\omega_0 + \delta_1\omega_0)$. The change in level of Y_t^* from the n−1st to the nth postintervention moment is:

$$\left|\sum_{k=0}^{n} \delta_1^k\omega_0\right| - \left|\sum_{k=0}^{n-1} \delta_1^k\omega_0\right| = \left|\delta_1^n\omega_0\right|.$$

This will be a very small number, approaching zero as a limit.

The *asymptotic* or eventual change in the level of Y_t^* can be calculated by summing the infinite series

$$\sum_{k=0}^{\infty} \delta_1^k \omega_0 = \text{ asymptotic change in level.}$$

Because δ_1 is smaller than unity in absolute value, this infinite series can be evaluated as

$$\sum_{k=0}^{\infty} \delta_1^k \omega_0 = \frac{\omega_0}{1 - \delta_1} = \text{ asymptotic change in level.}$$

The asymptotic impact is generally realized at a rate determined by the value of δ_1. Figure 3.2(b) shows the expected patterns of impact for various values of this parameter. When δ_1 is small, near zero for example, the

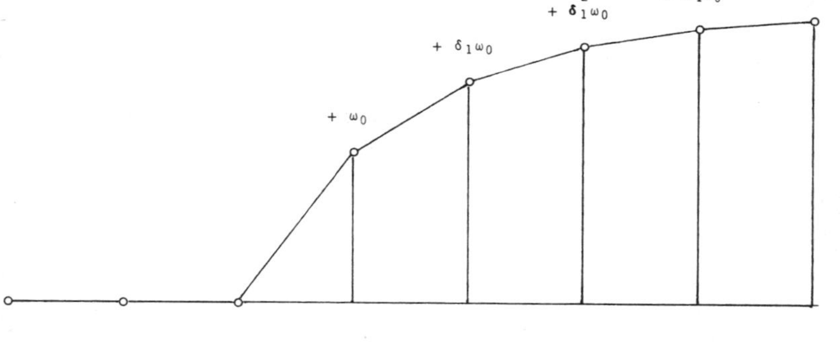

FIGURE 3.2(a) Pattern of Impact Expected of a First-Order Transfer Function

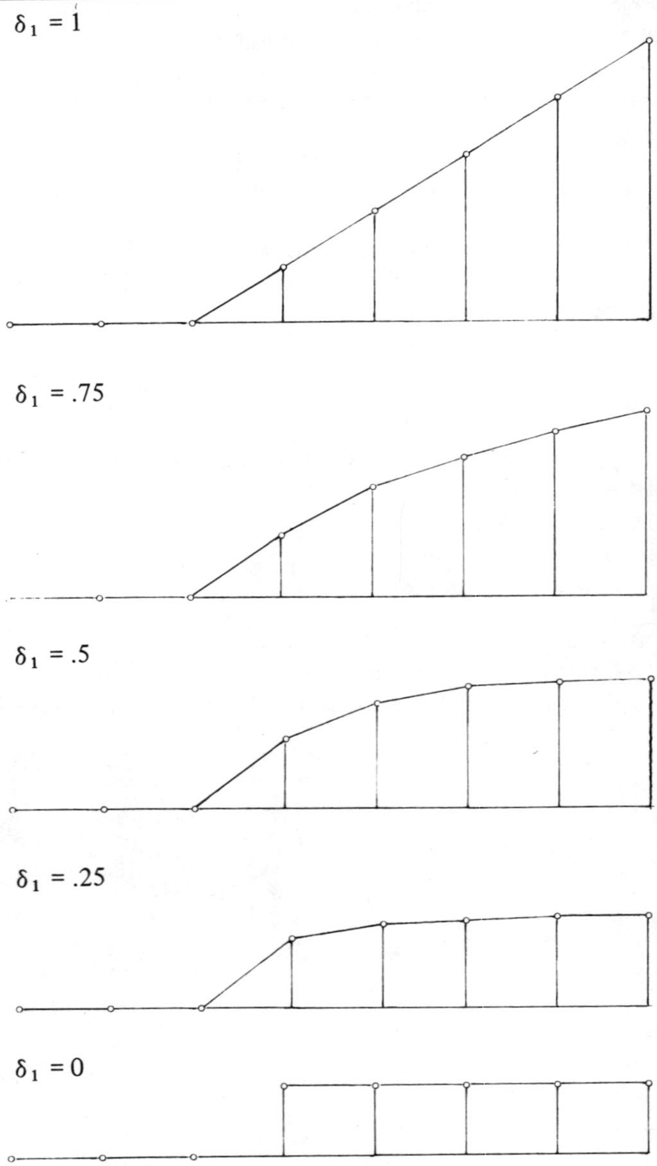

FIGURE 3.2(b) Pattern of Impact Expected for Several Values of δ_1

asymptotic impact is realized quickly. When δ_1 is larger, however, near unity for example, the asymptotic impact is realized slowly. The parameter δ_1 is thus interpreted as a *rate* parameter.

It is of some interest to note the behavior of this first-order transfer function at two values of δ_1. First, when $\delta_1 = 1$, the level of the Y_t^* series changes by the quantity ω_0 in each postintervention moment. This is so because when $\delta_1 = 1$, the first-order transfer function becomes:

$$Y_t^* = \frac{\omega_0}{1 - B} I_t.$$

Prior to the event, when $I_t = 0$, Y_t^* is an ARIMA (0,0,0) process, but when $I_t = 1$, Y_t^* becomes an ARIMA(0,1,0) process. The interpretation here is that, prior to intervention, the series is trendless, whereas postintervention, the series follows a trend with the parameter ω_0 interpreted as the slope.

Intuitively, the case where $\delta_1 = 1$ may seem useful. When examined more closely, however, the substantive implications of this model would seem to limit its utility. In its simplest form, the model describes a fixed-level (or stationary) process which, at the moment of intervention, begins to grow at a constant rate. Such a radical change (from a state of equilibrium to a state of growth) would rarely be observed in the social sciences in our opinion; and if observed, it is unlikely that this change would be associated with a manipulable social intervention.

Nevertheless, it is possible to observe an impact of this sort *when the postintervention time series segment is too short to encompass the equilibrium state of the process*. For example, the intervention may be such that the postintervention process reaches its equilibrium level slowly (the value of δ_1 may be quite large, that is, near unity). If the postintervention time series is too short, however, the postintervention change in level may have the appearance of a change in trend and the analyst may mistakenly conclude that the postintervention process is nonstationary and trending.

The only real solution to this dilemma is to wait for more postintervention data to become available. As these data become available, the analyst will be better able to decide whether the value of δ_1 is unity (and thus, that the postintervention process is trending) or slightly less than unity (in which case the postintervention process is *not* trending). Lacking these data, the analyst must depend upon informed substantive knowledge of the social process under analysis. If a change in slope seems to be a substantively reasonable impact, the parameter ω_0 is interpreted as the postintervention slope.

A second case of interest occurs when $\delta_1 = 0$. As shown in Figure 3.2(b), asymptotic impact is realized instantaneously in this case. This is so because when $\delta_1 = 0$, the first-order transfer function reduces to

$$Y_t^* = \frac{\omega_0}{1 - (0)B} = \omega_0 I_t,$$

the zero-order transfer function which we developed in the preceding section. We will make use of this relationship at a later point.

3.2.1 Chlorpromazine Impacts on Perceptual Speed

We are now concerned with the problem of selecting an appropriate intervention component. What are the consequences of using a zero-order transfer function to measure an impact that is not abrupt? Figure 3.2.1(a) shows a time series of 120 daily "perceptual speed" scores for a single schizophrenic patient as reported by Holtzman (1963; see also, Glass et al., 1975). On the 61st day, the patient was placed on a chlorpromazine regimen. Chlorpromazine is a radical tranquilizer, so one might expect a drop in perceptual speed for this patient coincident with the regimen. But is the impact abrupt or gradual? Figure 3.2.1(a), in our opinion, shows a gradual impact with perceptual speed dropping for several days before a new level is realized.

Using only the first 60 observations, we have estimated the ACF and PACF shown in Figure 3.2.1(b). These statistics would seem to indicate an ARIMA (1,0,1) model for the noise component. The basis of this identification is decay in both the ACF and the PACF. As noted in Section 2.9, ARIMA(p,0,q) processes are rarely encountered in social science time series. The ACF and PACF shown in Figure 3.2.1(b) nevertheless support identification of a mixed process.

Although the visual evidence supports a first-order transfer function for the intervention component, we will use a zero-order transfer function to demonstrate our point. The impact assessment model is:

$$Y_t = \omega_0 I_{61} + \frac{1 - \Theta_1 B}{1 - \phi_1 B} a_t.$$

Parameter estimates for this model are:

$$\begin{aligned} \hat{\phi}_1 &= .96 \text{ with t statistic} = 32.43 \\ \hat{\Theta}_1 &= .75 \text{ with t statistic} = 9.75 \\ \hat{\omega}_0 &= -27.09 \text{ with t statistic} = -4.40. \end{aligned}$$

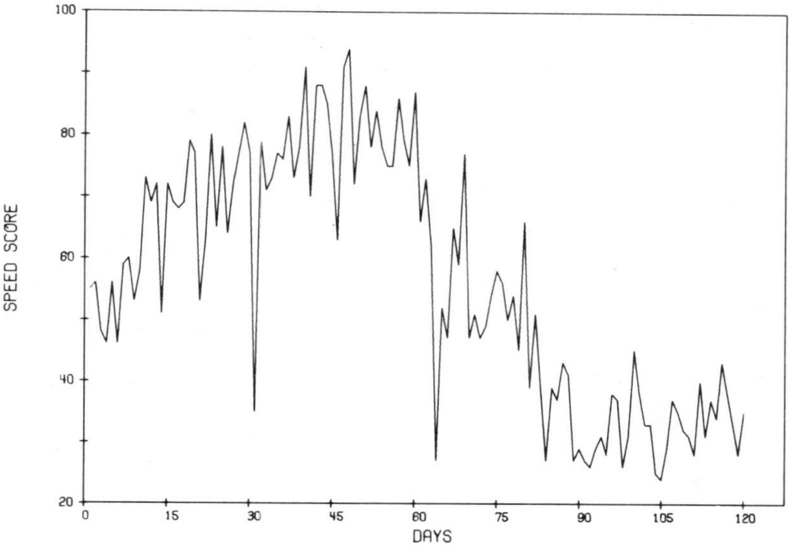

FIGURE 3.2.1(a) Daily Perceptual Speed Scores for a Schizophrenic Patient

While all three parameter estimates are statistically significant, the estimate of ϕ_1 is dangerously close to the bounds of stationarity. This indicates that an ARIMA(1,0,1) model is inappropriate. The 95% confidence interval around this estimate includes the value of $\hat{\phi}_1 = 1$, but, more important, because the estimates of ϕ_1 and Θ_1 are both large and positive, a problem with parameter redundancy is indicated.

In fact, the villain here is an outlier. Returning to Figure 3.2.1(a), we note that the 31st observation of the series is an order of magnitude smaller than neighboring observations. This outlier explains the aberrant ACF and PACF shown in Figure 3.2.1(b). The outlier has exaggerated the variance estimate used in these statistics and thus has biased the estimates of the ACF and PACF downward.

On the basis of the initial estimation, the noise component can be respecified as ARIMA(0,1,1). This leads to the impact assessment model

$$Y_t = \omega_0 I_{61} + \frac{\Theta_0 + (1 - \Theta_1 B)}{1 - B} a_t.$$

```
SERIES.. SPEED     (NOBS=  60)  PERCEPTUAL SPEED - PRE INTERVENTION
NO. OF VALID OBSERVATIONS   =    60.

AUTOCORRELATIONS OF LAGS 1 - 25.
Q( 25,   60) =  84.676       SIG =      .000

  LAG    CORR   SE -1  -.8  -.6  -.4  -.2    0   .2   .4   .6   .8    +
                    +----+----+----+----+----+----+----+----+----+----
    1    .427 .129                     (     IXXXXX)XXXXX
    2    .386 .151                   (       IXXXXXXXX)XX
    3    .471 .167                   (       IXXXXXXXX)XXXX
    4    .404 .187                  (        IXXXXXXXXX)X
    5    .428 .201                 (         IXXXXXXXXXX)X
    6    .239 .216                (          IXXXXXX    )
    7    .347 .220                (          IXXXXXXXXX )
    8    .267 .229                (          IXXXXXXX   )
    9    .166 .234               (           IXXXX       )
   10    .324 .236               (           IXXXXXXXX   )
   11    .109 .244               (           IXXX        )
   12    .155 .244               (           IXXXX       )
   13    .150 .246               (           IXXXX       )
   14    .090 .248               (           IXX         )
   15    .180 .248               (           IXXXXX      )
   16   -.000 .250              (            I            )
   17    .128 .250              (            IXXX         )
   18    .090 .251              (            IXX          )
   19    .006 .252              (            I            )
   20    .039 .252              (            IX           )
   21    .008 .252              (            I            )
   22   -.018 .252              (            I            )
   23   -.016 .252              (            I            )
   24    .011 .252              (            I            )
   25    .065 .252              (            IXX          )
                              -2SE                      +2SE
```

```
PARTIAL AUTOCORRELATIONS OF LAGS 1 - 25.

  LAG    CORR   SE -1  -.8  -.6  -.4  -.2    0   .2   .4   .6   .8    +
                    +----+----+----+----+----+----+----+----+----+----
    1    .427 .129                     (     IXXXXX)XXXXX
    2    .249 .129                     (     IXXXXX*
    3    .314 .129                     (     IXXXXX)XX
    4    .142 .129                     (     IXXXX )
    5    .175 .129                     (     IXXXX )
    6   -.159 .129                     ( XXXXI     )
    7    .111 .129                     (     IXXX  )
    8   -.080 .129                     (   XXI     )
    9   -.074 .129                     (   XXI     )
   10    .166 .129                     (     IXXXX )
   11   -.158 .129                     ( XXXXI     )
   12    .012 .129                     (     I     )
   13   -.008 .129                     (     I     )
   14   -.027 .129                     (    XI     )
   15    .086 .129                     (     IXX   )
   16   -.089 .129                     (   XXI     )
   17    .068 .129                     (     IXX   )
   18   -.004 .129                     (     I     )
   19   -.032 .129                     (    XI     )
   20   -.100 .129                     (  XXXI     )
   21    .074 .129                     (     IXX   )
   22   -.157 .129                     ( XXXXI     )
   23    .069 .129                     (     IXX   )
   24    .073 .129                     (     IXX   )
   25    .053 .129                     (     IX    )
                                     -2SE         +2SE
```

FIGURE 3.2.1(b) ACF and PACF for the Raw Preintervention Series

Parameter estimates for this model are:

$$\hat{\Theta}_0 = .01 \text{ with t statistic} = .06$$
$$\hat{\Theta}_1 = .77 \text{ with t statistic} = 12.93$$
$$\hat{\omega}_0 = -22.13 \text{ with t statistic} = -\ 3.37.$$

As the estimate of Θ_0 is not statistically significant, it must be dropped from the model. All other parameters are statistically significant and otherwise acceptable. Diagnostic checks of the model residuals indicate that they are not different than white noise, so this model is accepted.

The results of this impact assessment lead to the conclusion that the chlorpromazine regimen affected a drop of over 22 units in perceptual speed. This result is consistent at least with what is known about the physiological effects of chlorpromazine. However, the model implies an *abrupt* or instantaneous drop to the new level and this may be inconsistent with what is known about treatments in general.

Using the same ARIMA(0,1,1) model for a noise component, we now specify a first-order transfer function for the intervention component. This leads to the impact assessment model

$$Y_t = \frac{\omega_0}{1 - \delta_1 B} I_{61} + \frac{1 - \Theta_1 B}{1 - B} a_t.$$

Parameter estimates for this model are:

$$\hat{\Theta}_1 = .77 \text{ with t statistic} = 13.01$$
$$\hat{\delta}_1 = .53 \text{ with t statistic} = 2.37$$
$$\hat{\omega}_0 = -13.39 \text{ with t statistic} = -\ 2.23.$$

All parameter estimates are statistically significant and otherwise acceptable. Diagnostic checks of the model residuals indicate that they are not different than white noise, so this model is accepted.

As a first step in interpreting the results of this impact assessment analysis, the asymptotic impact is estimated as

$$\text{asymptotic change} = \frac{\hat{\omega}_0}{1 - \hat{\delta}_1} = \frac{-13.39}{1 - .53}$$
$$= -28.49 \text{ units}.$$

The model implies an impact amounting to a drop of over 28 perceptual speed score units. This impact is realized gradually, however. On the sixth day of the regimen, for example, the change from preintervention is:

$$\begin{aligned} \text{change} &= \hat{\omega}_0(1 + \hat{\delta}_1 + \hat{\delta}_1^2 + \hat{\delta}_1^3 + \hat{\delta}_1^4 + \hat{\delta}_1^5) \\ &= -13.39(1 + .53 + .281 + .149 + .079 + .042) \\ &= -27.86 \text{ units,} \end{aligned}$$

which is 98% of the asymptotic change. On the seventh and successive days, this patient's perceptual speed continues to drop but by negligible amounts.

Now having conducted two impact assessment analyses of the same time series data, we must note that the analyses lead to slightly different conclusions. To be sure, both models imply a drop in perceptual speed as a result of the chlorpromazine regimen. Assuming an *abrupt* response to treatment, however, the estimated reduction is 22.13 units while assuming a *gradual* response to treatment, the estimated reduction is 28.49 units. This is a substantial difference which must be reconciled. One of these estimated impacts must be judged more correct than the other.

If there were no substantive issues involved, the analyst could decide between these two estimates by statistical criteria alone. But whereas purely statistical criteria (such as the RMS statistics) can be used to compare two ARIMA noise models, these same criteria are less important in comparing two impact assessment models. If a model makes the best substantive sense, the analyst may judge it the "best" model regardless of its relative statistical properties. In this case, the first-order transfer function model is "better" than the zero-order transfer function model in both the statistical sense (its RMS statistic is the lower of the two; it fits the time series better) *and* the substantive sense.

At a later point, we will develop a strategy for comparing various low-order transfer function components by statistical criteria. For the time being, however, the analyst should understand that an estimate of impact will vary in quality according to how well the model represents the substantive process.

3.2.2 Sutter County Workforce: Temporary Impacts

A useful model of impact can be generated by applying the first-order transfer function to a *differenced* step function. For the step function I_t,

$$\ldots 0,\quad 0,\quad 0,\quad 0,\quad 1,\quad 1,\quad 1,\quad 1, \ldots$$

differencing results in

$$\begin{array}{lllllll} \ldots (0-0), & (0-0), & (0-0), & (1-0), & (1-1), & (1-1), \ldots \\ \ldots \quad 0, & 0, & 0, & 1, & 0, & 0, \ldots \end{array}$$

a pulse function, $(1 - B)I_t$, defined such that

$$(1 - B)I_t = 0 \text{ prior to the event}$$
$$= 1 \text{ at the onset of the event}$$
$$= 0 \text{ thereafter.}$$

Applying the first-order transfer function to $(1 - B)I_t$, the impact assessment model is:

$$Y_t = \frac{\omega_0}{1 - \delta_1 B}(1 - B)I_t + N_t$$

or

$$Y_t^* = \frac{\omega_0}{1 - \delta_1 B}(1 - B)I_t$$

$$(1 - \delta_1 B)Y_t^* = \omega_0(1 - B)I_t$$

$$Y_t^* = \delta_1 Y_{t-1}^* + \omega_0(1 - B)I_t.$$

This formulation may now be used to examine the behavior of the first-order transfer function applied to the differenced step function. Prior to the intervention, the step function and the Y_t^* series are both zero. If Y_i^* is the last preintervention observation, then $I_i = 0$ and $I_{i+1} = 1$. Hence,

$$(1 - B)I_{i+1} = I_{i+1} - I_i = 1 - 0 = 1$$

and the value of Y_{i+1}^* is:

$$Y_{i+1}^* = \delta_1 Y_i^* + \omega_0(1 - B)I_{i+1}$$
$$= \delta_1(0) + \omega_0(1) = \omega_0.$$

In the next postintervention moment, the differenced step function is equal to zero:

$$(1 - B)I_{i+2} = I_{i+2} - I_{i+1} = 1 - 1 = 0$$

and the value of Y_{i+2}^* is:

$$Y_{i+2}^* = \delta_1 Y_{i+1}^* + \omega_0(1 - B)I_{i+2}$$
$$= \delta_1(\omega_0) + \omega_0(0) = \delta_1\omega_0.$$

And in the next postintervention moment, the differenced step function is zero again and

$$Y_{i+3}^* = \delta_1 Y_{i+2}^* + \omega_0(1 - B)I_{i+3}$$
$$= \delta_1(\delta_1\omega_0) + \omega_0(0) = \delta_1^2\omega_0.$$

A progression begins to emerge. Continuing this procedure, it can be shown that the nth postintervention observation of the series, Y^*_{i+n}, is:

$$\begin{aligned} Y^*_{i+n} &= \delta_1 Y^*_{i+n-1} + \omega_0(1 - B)I_{i+n} \\ &= \delta_1(\delta_1^{n-2}\omega_0) + \omega_0(0) \\ &= \delta_1^{n-1}\omega_0. \end{aligned}$$

And as the value of δ_1 is constrained to the bounds of system stability, this term will be very small, nearly zero.

Figure 3.2.2 shows the expected impacts for various values of δ_1. The pulse function is distributed across the postintervention time series as a decaying spike. The value of the parameter δ_1 determines the rate at which the process returns to its preintervention equilibrium level. When δ_1 is large, near unity, return to the preintervention equilibrium level is slow. When δ_1 is small, return is rapid.

Friesema et al. (1979) used this abrupt, temporary impact model to assess the economic recovery of small communities from natural disasters. Like the pulse function, disasters are abrupt in onset and short in duration. Even though a natural disaster is short-lived, however, its impact remains for some time afterward. Using a temporary impact model, the parameter δ_1 can be interpreted as the *rate* of recovery during the disaster aftermath.

In Section 2.12.1, we built an ARIMA(0,1,1) $(0,1,1)_{12}$ model for the Sutter County Workforce time series. In December 1955 the 120th month of this series, a flood forced the evacuation of Sutter County. To assess the impact of the flood on the Workforce time series, we can use the ARIMA(0,1,1) $(0,1,1)_{12}$ model as the noise component. The impact assessment model is thus

$$Y_t = \frac{\omega_0}{1 - \delta_1 B}(1 - B)I_{121} + \frac{\Theta_0 + (1 - \Theta_1 B)(1 - \Theta_{12}B^{12})}{(1 - B)(1 - B^{12})} a_t.$$

Parameter estimates for this model are:

$$\begin{aligned} \hat{\Theta}_0 &= -\ \ .52 \text{ with t statistic} = -\ \ .22 \\ \hat{\Theta}_1 &= .60 \text{ with t statistic} = 11.38 \\ \hat{\Theta}_{12} &= .68 \text{ with t statistic} = 13.33 \\ \hat{\delta}_1 &= .84 \text{ with t statistic} = 2.64 \\ \hat{\omega}_0 &= -276.44 \text{ with t statistic} = -\ 1.36. \end{aligned}$$

The estimates of Θ_0 and ω_0 are not statistically significant. All parameter estimates are otherwise acceptable and a diagnostic check of the residuals indicates that they are not different than white noise.

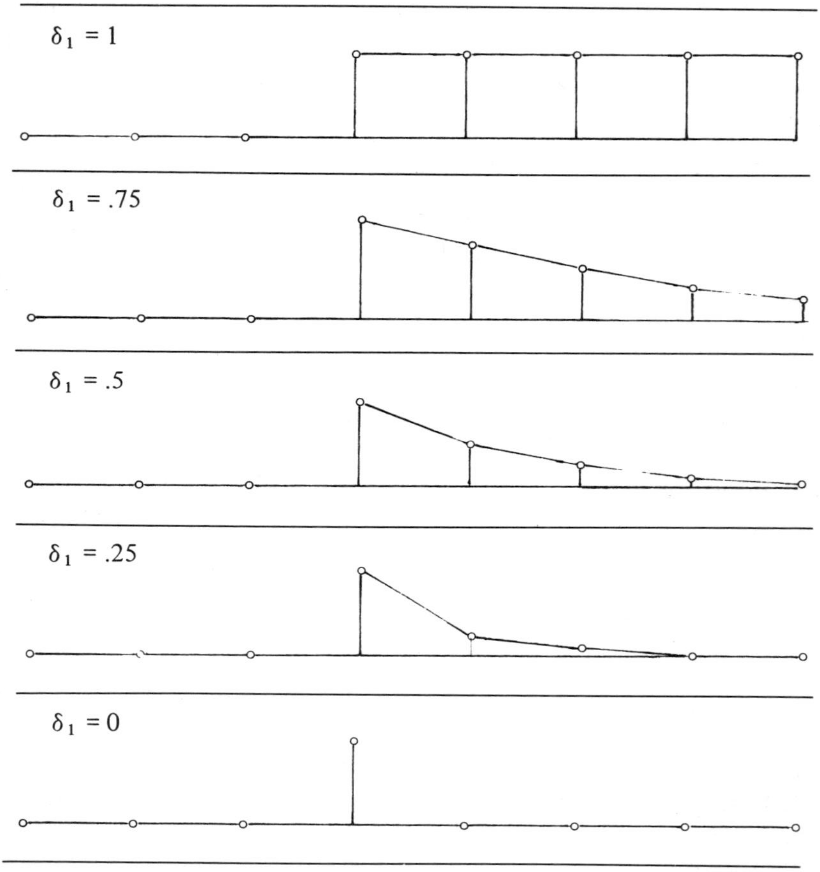

FIGURE 3.2.2 Pattern of Impact Expected for Several Values of δ_1

The parameter Θ_0 must be dropped from the model. A decision on the parameter ω_0 is not so easily made. Because the estimate of this parameter is not statistically different than zero, one might conclude that the flood had no effect whatsoever on the economy of Sutter County. This conclusion would be unsatisfactory, however. It is known in this case that there was an interruption in the local economy and, given this, the analysis must come up with a "best estimate" of the interruption.

According to Friesema et al., the economy of Sutter County is largely agricultural, and as the flood struck in December, after the normal growing season, there was little disruption. By the time of the next growing season, there was little unrepaired damage to local farmlands. Using the estimated values of δ_1 and ω_0 from this analysis, a "best estimate" of the flood's impact is:

$$\begin{array}{lll} \text{January 1956:} & \text{displacement} = \hat{\omega}_0 & = -276.44 \\ \text{February 1956:} & \text{displacement} = \hat{\delta}_1\hat{\omega}_0 & = -232.21 \\ \text{March 1956:} & \text{displacement} = \hat{\delta}_1^2\hat{\omega}_0 & = -195.06 \\ \text{April 1956:} & \text{displacement} = \hat{\delta}_1^3\hat{\omega}_0 & = -163.85 \\ \cdot & & \\ \cdot & & \\ \cdot & & \\ \text{December 1956:} & \text{displacement} = \hat{\delta}_1^{11}\hat{\omega}_0 & = -\ 40.61 \end{array}$$

and so forth. A year after the flood, the economy of Sutter County had returned to its normal condition. The displacement figures are given in worker-months. To estimate the total number of worker-months lost due to the flood, the infinite series

$$\sum_{k=0}^{\infty} \delta_1^k \omega_0$$

can be evaluated with the formula

$$\text{total displacement} = \frac{\hat{\omega}_0}{1 - \hat{\delta}_1} = -1727.75 \text{ worker-months.}$$

This total is interpreted geometrically as the area under the decaying spike. As there are approximately 36,000 worker-months in an average year, the impact of the flood on the Sutter County Workforce time series is substantively trivial.

Note finally that, because the interruption in this time series was relatively small, the noise component was identified with an ACF estimated from the entire time series.

3.2.3 Testing Rival Impact Hypotheses

In preceding sections, we developed three intervention components, each associated with a distinct pattern of impact. These include (1) an *abrupt, constant* pattern of impact determined by the zero-order transfer function

$$f(I_t) = \omega_0 I_t;$$

(2) a *gradual, constant* pattern of impact determined by the first-order transfer function

$$f(I_t) = \frac{\omega_0}{1 - \delta_1 B} I_t;$$

and (3) an *abrupt, temporary* pattern of impact determined by applying the first-order transfer function to a differenced I_t

$$f(I_t) = \frac{\omega_0}{1 - \delta_1 B}(1 - B)I_t.$$

In an ideal situation, the analyst works from a body of theory which points to one of these three patterns of impact, and hence, to one of these three intervention components. In many cases, for example, theory will define the impact as abrupt *or* gradual, permanent *or* temporary, and so forth. When theory is lacking, however, logical relationships between these three intervention component models (and between the three patterns of impact) will permit a test of rival hypotheses. This is a crucial aspect of impact modeling because, as demonstrated in the Perceptual Speed example of Section 3.2.1, two different intervention component models may often lead to two substantially different estimates of impact.

To illustrate the logical relationships between the three patterns of impact, consider the behavior of the abrupt, temporary impact pattern at the bounds of system stability. Referring to Figure 3.2.2, recovery is instantaneous when $\delta_1 = 0$. When $\delta_1 = 1$, however, there is no recovery at all. Writing out the first-order transfer function associated with this pattern of impact,

$$f(I_t) = \frac{\omega_0}{1 - \delta_1 B}(1 - B)I_t = \frac{1 - B}{1 - \delta_1 B}\omega_0 I_t.$$

Whenever $\delta_1 = 1$, the operator terms cancel out and the first-order transfer function reduces to

$$f(I_t) = \omega_0 I_t,$$

the zero-order transfer function.

This relationship suggests a rather simple method for checking the appropriateness of an intervention component. First, if the analyst has no a priori notions about the expected impact, an abrupt, temporary pattern of impact is hypothesized. If the estimated value of δ_1 is too large, near unity, a temporary impact is ruled out.

Next, the analyst hypothesizes a *permanent* but *gradual* pattern of impact based on the first-order transfer function

$$f(I_t) = \frac{\omega_0}{1 - \delta_1 B} I_t.$$

If the estimated value of δ_1 in this model is too small, near zero, a *gradual* pattern of impact is ruled out. When $\delta_1 = 0$, in fact, the first-order transfer function reduces to the zero-order transfer function associated with an *abrupt, constant* impact pattern.

To illustrate this procedure, we return to the Directory Assistance time series analyzed in Section 3.1. Eyeballing the plotted time series, Figure 3.1(a), there is no question but that an *abrupt, constant* pattern of impact is appropriate. But suppose now that the pattern of impact is unknown. The "blind" analysis begins with the model

$$Y_t = \frac{\omega_0}{1 - \delta_1 B}(1 - B)I_t + \frac{1 - \Theta_{12}B^{12}}{(1 - B)(1 - B^{12})} a_t.$$

The noise component is the one identified in Section 3.1. The intervention component hypothesizes an *abrupt, temporary* pattern of impact. Estimates for the transfer function parameters are:

$$\hat{\delta}_1 = .99295 \text{ with t statistic } = 70.64$$
$$\hat{\omega}_0 = -38{,}034 \text{ with t statistic } = -13.47.$$

The 95% confidence intervals about this estimate of δ_1 lie well outside the bounds of system stability. The parameter estimate is clearly "too large" to support the temporary effect hypothesis.

As a second step in the "blind" analysis, a permanent but gradual pattern of impact is hypothesized based on the model

$$Y_t = \frac{\omega_0}{1 - \delta_1 B} I_t + \frac{1 - \Theta_{12}B^{12}}{(1 - B)(1 - B^{12})} a_t.$$

Estimates for the transfer function parameters are:

$$\hat{\delta}_1 = -.0396 \text{ with t statistic } = -.56$$
$$\hat{\omega}_0 = -37{,}900 \text{ with t statistic } = -13.38.$$

The estimated value of δ_1 is clearly "too small" to support the hypothesis of a gradual effect. The only alternative remaining is the abrupt, constant pattern of impact associated with the zero-order transfer function.

The procedure by which competing patterns of impact are ruled out requires a theory in which only a limited number of distinct impact patterns are plausible. In almost all situations, the analyst can invoke the rule of parsimony, "Occam's razor," to make such a theory plausible. In the general case, there are infinitely many possible patterns of impact. There is no logical reason, for example, why second-, third-, and n^{th}-order transfer functions (and their associated patterns of impact) should not be considered by the analyst. The differences among these many patterns of impact are small, however, so for the purposes of ruling out alternative hypotheses, it will be helpful to collapse the infinitely many *possible* effects into two or three distinct classes of impact.

Figure 3.2.3 shows a scheme which we have found useful in practice. An impact is assumed to be *either* permanent *or* temporary, *either* gradual *or* abrupt. With this simple theory of impact, four distinct effects are possible. Three of these four are associated with zero- and first-order transfer functions. If our experiences are typical, almost all social interventions will have impacts reasonably well represented as one of these three models. Because the zero- and first-order transfer functions are related in the extreme, two of the three models can almost always be ruled out through analysis.

The fourth pattern of impact in Figure 3.2.3 is a *gradual, temporary* effect. The impact pattern cannot be easily modeled with a low-order transfer function applied to a step or pulse.[1] This model would seem to be the least useful of the four, so we will not develop it here.

When theory demands it, of course, the impact analysis should not be restricted to the lower order transfer functions. At a later point, we will develop compound lower order and higher order transfer functions which enable the analyst to model virtually any pattern of impact. We will also demonstrate the techniques of impact *fitting* (rather than modeling) which lead to uninterpretable impact assessments. In almost all situations, however, it will be possible to restrict the impact assessment analysis to the patterns of impact shown in Figure 3.2.3 and this restriction will generally pay off in terms of interpretability.

3.3 Interpreting Impact Parameters in the Natural Log Metric

In Chapter 2, we noted that a log transformation of a time series was appropriate when the series variance was proportional to change in the

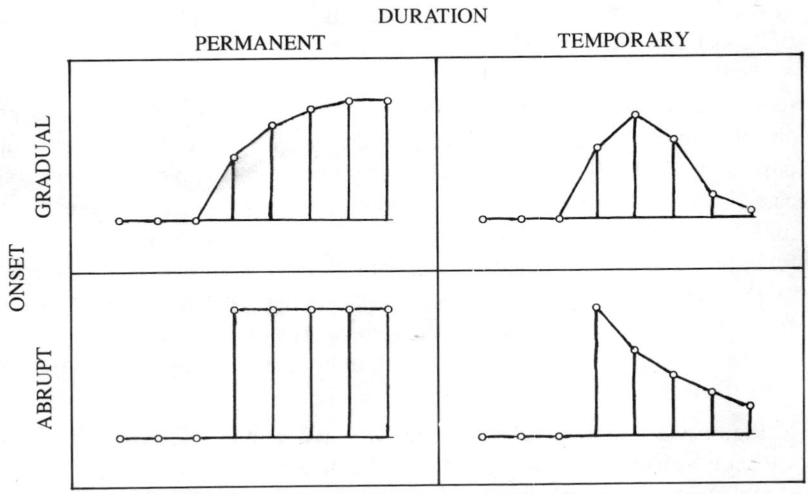

FIGURE 3.2.3 A Simple Theory of Impact

series level. The log-transformed series was then stationary in the larger sense and an appropriate ARIMA model could be fit. In estimating an impact model of such a transformed series, the ω_0 coefficient is interpreted as the pre- to postintervention change in the natural logarithm of the time series.

However precise and correct this interpretation may be, it lacks the easy interpretability of parameters estimated in the raw metric. Any social scientist who has tried to explain transformations and transformed effects to an audience of policy makers will immediately see the practical issue raised here. On one hand, in order to make a proper and correct assessment of impact, we must work in the natural log metric. Yet, on the other hand, by working in the log metric, we lose the easy interpretability of the model parameters.

Fortunately, a simple convention allows us to perform the analysis in the log metric but state our findings in terms of the raw metric. To demonstrate this convention, we must first develop the relationship of impact and ARIMA components in the log metric. Using the simplest intervention component in the log metric, we have

$$\mathrm{Ln}(Y_t) = \omega_0 I_t + \mathrm{ARIMA}.$$

The inverse procedure of the natural log operator is exponentiation, that is,

$$\mathrm{Ln}(x) = k$$
$$e^{\mathrm{Ln}(x)} = e^{k} = x.$$

To transform the model back into the raw metric then, we exponentiate it:

$$e^{\mathrm{Ln}(Y_t)} = e^{(\omega_0 I_t + \mathrm{ARIMA})}$$
$$Y_t = e^{(\omega_0 I_t)}\, e^{(\mathrm{ARIMA})}.$$

The term $e^{(\mathrm{ARIMA})}$ merely denotes a multiplicative shock form of the ARIMA model. For example, an ARIMA(0,1,0) process in the log metric is:

$$\mathrm{Ln}(Y_t) = Y_0 + a_1 + a_2 + \ldots + a_{t-1} + a_t$$

and exponentiating this,

$$Y_t = e^{(Y_0 + a_1 + a_2 + \ldots + a_{t-1} + a_t)}$$

but let $a_t^* = e^{(a_t)}$ and then

$$Y_t = Y_0^* (a_1^*)(a_2^*) \ldots (a_{t-1}^*)(a_t^*),$$

which is a multiplicative shock model. Similarly, an ARIMA(1,0,0) process in the natural log metric is:

$$(1 - \phi_1 B)\, \mathrm{Ln}(Y_t) = a_t.$$

As we demonstrated in the previous chapter, the ARIMA(1,0,0) process can be written as an infinite series of past shocks,

$$\mathrm{Ln}(Y_t) = a_t + \phi_1 a_{t-1} + \phi_1^2 a_{t-2} + \ldots + \phi_1^{n-1} a_{t-n-1} + \phi_1^n a_{t-n} + \ldots$$

and exponentiating this process, we have

$$Y_t = a_t^* (e^{\phi_1 a_{t-1}}) \ldots (e^{\phi_1^n a_{t-n}}),$$

which is a multiplicative shock model. Finally, an ARIMA(0,0,1) process in the log metric is:

$$\mathrm{Ln}(Y_t) = a_t - \Theta_1 a_{t-1}$$

$$\text{so} \qquad Y_t = \frac{a_t^*}{e^{(\Theta_1 a_{t-1})}}$$

which is a multiplicative shock model. By exponentiating the ARIMA model, then, we merely change from additive to multiplicative shocks. The exponentiated model still describes the preintervention equilibrium state of the process.

The impact component of the model multiplies the equilibrium state. But prior to the intervention, when $I_t = 0$, the model is:

$$Y_t = e^{(\omega_0 I_t)} e^{(\text{ARIMA})} = e^{(0)} e^{(\text{ARIMA})} = e^{(\text{ARIMA})}.$$

After the intervention, when $I_t = 1$, the model is:

$$Y_t = e^{(\omega_0 I_t)} e^{(\text{ARIMA})} = e^{(\omega_0)} e^{(\text{ARIMA})}.$$

It is convenient to think of pre- and postintervention equilibrium levels of the time series process. The ratio of post- to preintervention equilibrium is:

$$\frac{\text{postintervention equilibrium}}{\text{preintervention equilibrium}} = \frac{e^{(\omega_0)} e^{(\text{ARIMA})}}{e^{(\text{ARIMA})}} = e^{(\omega_0)}.$$

In fact, while the parameter ω_0 is not easily interpreted in the log metric, the term $e^{(\omega_0)}$ can be interpreted as *the ratio of the postintervention series level to the preintervention series level*. This ratio can be expressed as the percent change in the expected value of the process associated with the intervention:

$$\text{percent change} = (e^{(\omega_0)} - 1)\,100.$$

For example, in Section 2.12.2 we identified a noise model for the log-transformed Boston Armed Robbery series. Using the abrupt, permanent pattern of impact proposed by Deutsch and Alt (1977), we specify the tentative impact model

$$\text{Ln}(Y_t) = \omega_0 I_t + \frac{(1-.43B)(1+.19B^{12})}{1-B}\, a_t.$$

The estimated impact parameter is:

$$\hat{\omega}_0 = -.2070 \text{ with t statistic} = -1.33,$$

so $e^{-.2070} = .8130$

and percent change $= (.8130 - 1)\,100 = -18.7\%$.

Thus, we find that introduction of a gun control law in the 112th month is associated with an 18.7% reduction in armed robberies; however, this reduction is not significantly different from zero at the .05 level. This result contrasts sharply with that of Deutsch and Alt who, using an inappropriate noise model as previously discussed, found a statistically significant decline. The reader is referred to Hay and McCleary (1979) for a detailed discussion of the effect of inappropriate noise models on impact parameter estimates in this and other related series.

Interpreting log impact as percent change can be easily applied to dynamic models of impact. To illustrate this, consider the time series plotted in Figure 3.3(a). These are monthly public drunkenness arrests for Minneapolis. In June 1971, the 66th month of this series, public drunkenness was decriminalized in Minnesota. Aaronson et al. (1978; McCleary and Musheno, 1980) claim that decriminalization affected an abrupt and profound drop in the level of arrests and, given the visual appearance of the data, their claim seems reasonable. Starting in the 67th month, the level of this series appears to drop substantially.

A less notable impact (but an impact which is nevertheless noticeable) concerns the series variance. Prior to the intervention, month-to-month fluctuations are relatively large, while postintervention, month-to-month fluctuations are relatively small. Of course, this postintervention change in variance is not a unique impact of decriminalization, but rather is due to a "floor" effect of the sort we alluded to in Section 2.4. As a result of decriminalization, the process drops to a new equilibrium level near the "floor." At this new equilibrium level, the series variance is constrained (see McCleary and Musheno, 1980). The log-transformed series is shown in Figure 3.3(b). In the natural logarithm metric, variance is more nearly constant throughout the length of the series.

As a preliminary step to impact analysis, we will build an ARIMA model for this time series. Due to the magnitude of the intervention, only the 66 preintervention observations will be used for identification.

Identification

An ACF and PACF estimated from the first 66 observations of the log-transformed series, Figure 3.3(c), indicate that the series is seasonally nonstationary. The ACF and PACF for the seasonally differenced log series, Figure 3.3(d), suggest an ARIMA $(0,0,0)\ (0,1,1)_{12}$ model for the noise component.

Estimation

To this noise component, we add an impact component to reflect a gradual, permanent effect pattern. The full model is thus

$$\text{Ln}\,(Y_t) = \frac{\omega_0}{1-\delta_1 B} I_t + \frac{\Theta_0 + (1-\Theta_{12}B^{12})}{1-B^{12}} a_t \,.$$

(text continued on p. 181)

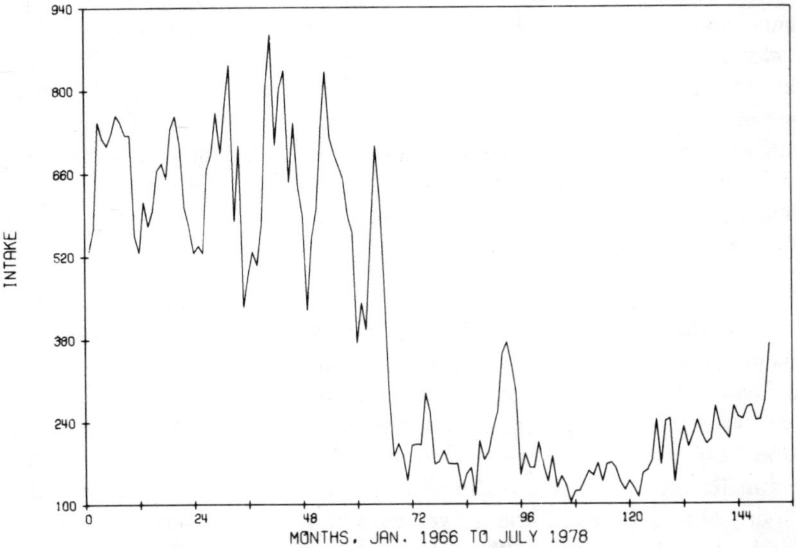

FIGURE 3.3(a) Minneapolis Public Drunkenness Intakes

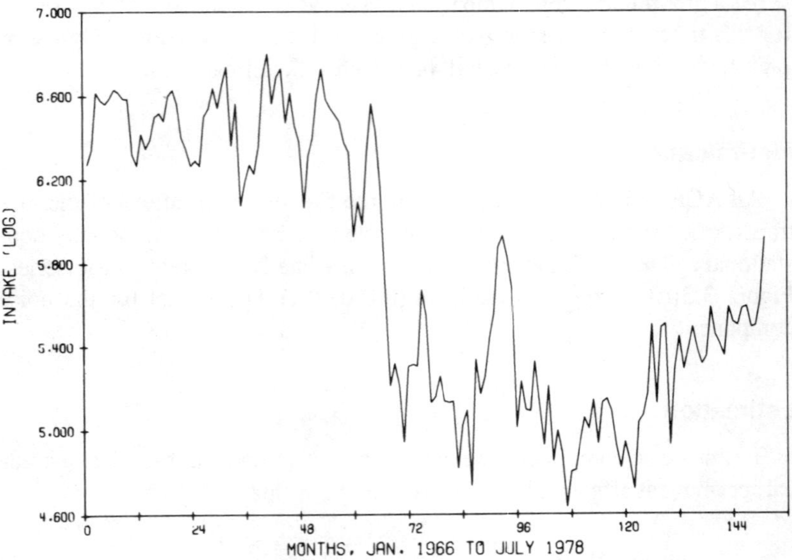

FIGURE 3.3(b) Minneapolis Public Drunkenness Intakes (Logged)

```
SERIES.. DRUNK     (NOBS=  66)  PUBLIC DRUNKENESS - PRE INTERVENTION
TRANSFORMED BY LOG.
NO. OF VALID OBSERVATIONS  =    66.

AUTOCORRELATIONS OF LAGS 1 - 25.
Q( 25,   66) =  167.30      SIG =      .000

  LAG    CORR     SE  -1    -.8   -.6   -.4   -.2    0    .2    .4    .6    .8   +1
                       +----+----+----+----+----+----+----+----+----+----+
   1    .585  .123                       (     IXXXXX)XXXXXXXXX
   2    .327  .160                     (       IXXXXXXX*
   3    .061  .170                     (       IXX      )
   4   -.176  .170                     (   XXXXI        )
   5   -.346  .173                   *XXXXXXXXI          )
   6   -.371  .183                   *XXXXXXXXI          )
   7   -.364  .194                  (XXXXXXXXXI           )
   8   -.335  .204                  ( XXXXXXXXI           )
   9   -.068  .212                 (         XXI            )
  10    .195  .213                 (           IXXXXX       )
  11    .407  .215                 (           IXXXXXXXXXX)
  12    .486  .227                 (           IXXXXXXXXXX)X
  13    .371  .242                (            IXXXXXXXXX  )
  14    .155  .250              (              IXXXX          )
  15   -.083  .252              (            XXI            )
  16   -.240  .252              (        XXXXXXI            )
  17   -.350  .256              (    XXXXXXXXXI             )
  18   -.419  .263              (  XXXXXXXXXXXI             )
  19   -.373  .273             (     XXXXXXXXXI              )
  20   -.263  .260             (       XXXXXXXI              )
  21   -.027  .284             (            XI               )
  22    .097  .284             (             IXX             )
  23    .301  .285             (             IXXXXXXXXX      )
  24    .415  .289             (             IXXXXXXXXXXX    )
  25    .314  .298            (              IXXXXXXXXX        )
                            -2SE                               +2SE

PARTIAL AUTOCORRELATIONS OF LAGS 1 - 25.

  LAG    CORR     SE  -1    -.8   -.6   -.4   -.2    0    .2    .4    .6    .8   +1
                       +----+----+----+----+----+----+----+----+----+----+
   1    .585  .123                       (     IXXXXXX)XXXXXXXXX
   2   -.024  .123                       (     XI      )
   3   -.184  .123                       (XXXXXI       )
   4   -.209  .123                       (XXXXXI       )
   5   -.186  .123                       (XXXXXI       )
   6   -.051  .123                       (     XI      )
   7   -.110  .123                       (  XXXI       )
   8   -.158  .123                       ( XXXXI       )
   9    .219  .123                       (      IXXXXX)
  10    .216  .123                       (      IXXXXX)
  11    .188  .123                       (      IXXXXX)
  12    .071  .123                       (      IXX    )
  13   -.136  .123                       (  XXXI       )
  14   -.139  .123                       (  XXXI       )
  15   -.154  .123                       ( XXXXI       )
  16   -.065  .123                       (   XXI       )
  17    .013  .123                       (      I      )
  18   -.046  .123                       (     XI      )
  19    .037  .123                       (      IX     )
  20   -.010  .123                       (      I      )
  21    .054  .123                       (      IX     )
  22   -.213  .123                       (XXXXXI       )
  23    .008  .123                       (      I      )
  24    .166  .123                       (      IXXXXX )
  25   -.023  .123                       (     XI      )
                                       -2SE         +2SE
```

FIGURE 3.3(c) ACF and PACF for the Raw (Logged) Preintervention Series

```
SERIES.. DRUNK     (NOBS=  66)  PUBLIC DRUNKENESS - PRE INTERVENTION
TRANSFORMED BY LOG.
DIFFERENCED  1 TIME(S) OF ORDER 12.
NO. OF VALID OBSERVATIONS  =    54.

AUTOCORRELATIONS OF LAGS 1 - 25.
Q( 25,  54) =  31.360      SIG =      .177

LAG    CORR    SE  -1  -.8  -.6  -.4  -.2   0   .2   .4   .6   .8   +1
                   +----+----+----+----+----+----+----+----+----+----+
  1    .154  .136                    (      IXXXX  )
  2    .218  .139                    (      IXXXXX )
  3    .252  .145                    (      IXXXXXX)
  4    .259  .153                   (       IXXXXXX )
  5    .029  .161                   (       IX      )
  6    .200  .161                   (       IXXXXX  )
  7    .226  .166                   (       IXXXXXX )
  8   -.101  .171                  (     XXXI        )
  9   -.067  .173                  (      XXI        )
 10    .084  .173                  (        IXX      )
 11    .028  .174                  (        IX       )
 12   -.307  .174                  (XXXXXXXXI        )
 13   -.008  .184                  (        I        )
 14   -.005  .184                  (        I        )
 15   -.270  .184                  ( XXXXXXXI        )
 16    .040  .191                 (         IX        )
 17    .031  .191                 (         IX        )
 18   -.118  .191                 (      XXXI         )
 19   -.188  .192                 (    XXXXXI         )
 20   -.011  .196                 (         I         )
 21    .002  .196                 (         I         )
 22   -.168  .196                 (     XXXXI         )
 23    .031  .198                 (         IX        )
 24   -.047  .199                 (        XI         )
 25   -.121  .199                 (      XXXI         )
                                -2SE                 +2SE


PARTIAL AUTOCORRELATIONS OF LAGS 1 - 25.

LAG    CORR    SE  -1  -.8  -.6  -.4  -.2   0   .2   .4   .6   .8   +1
                   +----+----+----+----+----+----+----+----+----+----+
  1    .154  .136                    (      IXXXX  )
  2    .199  .136                    (      IXXXXX )
  3    .208  .136                    (      IXXXXX )
  4    .188  .136                    (      IXXXXX )
  5   -.102  .136                    (   XXXI      )
  6    .092  .136                    (      IXX    )
  7    .145  .136                    (      IXXXX  )
  8   -.238  .136                    (XXXXXXI      )
  9   -.171  .136                    (  XXXXI      )
 10    .041  .136                    (      IX     )
 11    .084  .136                    (      IXX    )
 12   -.303  .136                   X(XXXXXXI      )
 13   -.037  .136                    (     XI      )
 14    .149  .136                    (      IXXXX  )
 15   -.133  .136                    (   XXXI      )
 16    .198  .136                    (      IXXXXX )
 17    .016  .136                    (      I      )
 18   -.067  .136                    (    XXI      )
 19   -.017  .136                    (      I      )
 20   -.155  .136                    (  XXXXI      )
 21    .071  .136                    (      IXX    )
 22   -.059  .136                    (     XI      )
 23    .022  .136                    (      IX     )
 24   -.133  .136                    (   XXXI      )
 25   -.012  .136                    (      I      )
                                   -2SE           +2SE
```

FIGURE 3.3(d) ACF and PACF for the Seasonally Differenced (Logged) Preintervention Series

```
SERIES.. RESIDUAL (NOBS= 138)  ESTIMATED MODEL RESIDUALS
NO. OF VALID OBSERVATIONS   =  138.

AUTOCORRELATIONS OF LAGS 1 - 25.
Q( 24, 138) =  131.31      SIG =      .000

 LAG    CORR     SE  -1    -.8  -.6  -.4  -.2   0    .2   .4   .6   .8  +1
                       +----+----+----+----+----+----+----+----+----+----+
   1   .490  .090                           (     IXXXX)XXXXXXX
   2   .420  .100                           (     IXXXX)XXXXXX
   3   .440  .120                          (      IXXXXX)XXXXX
   4   .290  .130                         (       IXXXXXX*
   5   .230  .130                         (       IXXXXXX)
   6   .240  .130                         (       IXXXXXX)
   7   .120  .140                         (       IXXX   )
   8   .070  .140                         (       IXX    )
   9   .050  .140                         (       IX     )
  10   .010  .140                         (       I      )
  11  -.020  .140                         (      XI      )
  12   .010  .140                         (       I      )
  13  -.140  .140                         (  XXXXI       )
  14  -.130  .140                         (   XXXI       )
  15  -.060  .140                         (     XXI      )
  16  -.070  .140                         (     XXI      )
  17  -.020  .140                         (      XI      )
  18  -.030  .140                         (      XI      )
  19   .040  .140                         (       IX     )
  20  -.020  .140                         (      XI      )
  21  -.080  .140                         (     XXI      )
  22   .020  .140                         (       IX     )
  23  -.120  .140                         (   XXXI       )
  24  -.180  .140                         ( XXXXXI       )
  25  -.150  .140                         (  XXXXI       )
                                        -2SE          +2SE

PARTIAL AUTOCORRELATIONS OF LAGS 1 - 25.

 LAG    CORR     SE  -1    -.8  -.6  -.4  -.2   0    .2   .4   .6   .8  +1
                       +----+----+----+----+----+----+----+----+----+----+
   1   .490  .090                           (     IXXXX)XXXXXXX
   2   .240  .090                           (     IXXXX)X
   3   .230  .090                           (     IXXXX)X
   4  -.040  .090                           (    XI    )
   5  -.030  .090                           (    XI    )
   6   .050  .090                           (     IX   )
   7  -.080  .090                           (   XXI    )
   8  -.060  .090                           (   XXI    )
   9  -.030  .090                           (    XI    )
  10  -.010  .090                           (     I    )
  11  -.020  .090                           (    XI    )
  12   .050  .090                           (     IX   )
  13  -.180  .090                           *XXXXI     )
  14  -.040  .090                           (    XI    )
  15   .060  .090                           (     IXX  )
  16   .070  .090                           (     IXX  )
  17   .090  .090                           (     IXX  )
  18  -.040  .090                           (    XI    )
  19   .120  .090                           (     IXXX )
  20  -.100  .090                           ( XXXI     )
  21  -.140  .090                           (XXXXI     )
  22   .050  .090                           (     IX   )
  23  -.170  .090                           (XXXXI     )
  24  -.130  .090                           ( XXXI     )
  25  -.030  .090                           (    XI    )
                                          -2SE      +2SE
```

FIGURE 3.3(e) *Diagnosis:* ACF and PACF for the Residuals of an ARIMA (0,0,0) $(0,1,1)_{12}$ Model and Intervention Component

```
SERIES.. RESIDUAL (NOBS= 149)  ESTIMATED MODEL RESIDUALS
NO. OF VALID OBSERVATIONS   =  149.

AUTOCORRELATIONS OF LAGS 1 - 25.
Q( 24, 149) =  46.086      SIG =      .004

  LAG   CORR    SE      -1  -.8  -.6  -.4  -.2   0   .2   .4   .6   .8   +1
                        +----+----+----+----+----+----+----+----+----+----+
   1   -.340  .080                      XXXXX(XXXI   )
   2   -.050  .090                          (   XI    )
   3    .060  .090                          (    IXX  )
   4   -.040  .090                          (   XI    )
   5   -.140  .090                          (XXXXI    )
   6    .140  .090                          (    IXXXX)
   7   -.100  .090                          ( XXXI    )
   8   -.060  .100                          (  XXI    )
   9   -.010  .100                          (    I    )
  10    .010  .100                          (    I    )
  11    .100  .100                          (    IXXX )
  12   -.020  .100                          (   XI    )
  13    .000  .100                          (    I    )
  14   -.020  .100                          (   XI    )
  15   -.050  .100                          (   XI    )
  16   -.040  .100                          (   XI    )
  17    .040  .100                          (    IX   )
  18   -.070  .100                          (  XXI    )
  19    .100  .100                          (    IXXX )
  20    .010  .100                          (    I    )
  21   -.190  .100                          *XXXXI    )
  22    .220  .100                          (    IXXXX)X
  23    .000  .100                          (    I    )
  24   -.070  .100                          (  XXI    )
  25    .110  .100                          (    IXXX )
                                          -2SE      +2SE

PARTIAL AUTOCORRELATIONS OF LAGS 1 - 25.

  LAG   CORR    SE      -1  -.8  -.6  -.4  -.2   0   .2   .4   .6   .8   +1
                        +----+----+----+----+----+----+----+----+----+----+
   1   -.340  .080                      XXXXX(XXXI   )
   2   -.190  .080                          X(XXXI   )
   3   -.030  .080                           (  XI   )
   4   -.030  .080                           (  XI   )
   5   -.190  .080                          X(XXXI   )
   6    .020  .080                           (   IX  )
   7   -.090  .080                           ( XXI   )
   8   -.120  .080                           (XXXI   )
   9   -.140  .080                           *XXXI   )
  10   -.100  .080                           (XXXI   )
  11    .080  .080                           (   IXX )
  12    .010  .080                           (   I   )
  13    .000  .080                           (   I   )
  14   -.040  .080                           (  XI   )
  15   -.090  .080                           ( XXI   )
  16   -.100  .080                           (XXXI   )
  17   -.070  .080                           ( XXI   )
  18   -.100  .080                           (XXXI   )
  19    .050  .080                           (   IX  )
  20    .060  .080                           (   IXX )
  21   -.210  .080                          X(XXXI   )
  22    .040  .080                           (   IX  )
  23    .040  .080                           (   IX  )
  24   -.010  .080                           (   I   )
  25    .070  .080                           (   IXX )
                                           -2SE    +2SE
```

FIGURE 3.3(f) *Diagnosis:* ACF and PACF for the Residuals of an ARIMA (0,1,0) $(0,0,1)_{12}$ Model and Intervention Component

Parameter estimates for this tentative model are:

$$\hat{\omega}_0 = -\ .992 \text{ with t statistic} = -\ 4.17$$
$$\hat{\delta}_1 = .190 \text{ with t statistic} = .99$$
$$\hat{\Theta}_0 = -\ .007 \text{ with t statistic} = -\ .77$$
$$\hat{\Theta}_{12} = 1.039 \text{ with t statistic} = 24.73.$$

This tentative model is clearly unacceptable. The estimate of Θ_{12} lies outside the bounds of invertibility.

Diagnosis

There is little to be gained returning to the ACFs and PACFs used to identify the ARIMA (0,0,0) $(0,1,1,)_{12}$ noise component. The problem here is that a model identified from the preintervention series only will not adequately reflect the stochastic behavior of the entire series. The noise component in this example appears to be too complicated to be identified with such a short series. We can nevertheless make some educated guess as to the appropriate noise component on the basis of the information available at this point.

First, the unacceptable estimate of Θ_{12} suggests that seasonal differencing is *not* required. Second, the ACF and PACF of the model residuals, Figure 3.3(e), indicate a nonstationary process. We thus tentatively specify an ARIMA (0,1,0) $(0,0,1)_{12}$ model for the noise component.

Estimation

For the tentative model

$$\text{Ln}\,(Y_t) = \frac{\omega_0}{1 - \delta_1 B} I_t + \frac{\Theta_0 + (1 - \Theta_{12}B^{12})}{1 - B} a_t,$$

parameter estimates are:

$$\hat{\omega}_0 = -.543 \text{ with t statistic} = -2.76$$
$$\hat{\delta}_1 = .451 \text{ with t statistic} = 1.71$$
$$\hat{\Theta}_0 = .004 \text{ with t statistic} = .21$$
$$\hat{\Theta}_{12} = -.153 \text{ with t statistic} = -1.80.$$

Two parameter estimates are only marginally significant while the estimate of Θ_0 is clearly insignificant. The estimate of Θ_{12} now lies within the bounds of invertibility.

Diagnosis

The residual ACF and PACF, Figure 3.3(f) now indicate stationarity. Spikes at the first and fourth lags, however, indicate unmodeled moving average terms of those orders. A $\Theta_1 B$ and a $\Theta_4 B^4$ term must be incorporated into the noise component but there is no indication in the ACF or PACF as to how this should be done. There are three possibilities including the two-factor models

$$\text{Ln}\,(Y_t) = \frac{\omega_0}{1 - \delta_1 B} I_t + \frac{\Theta_0 + (1 - \Theta_1 B - \Theta_4 B^4)(1 - \Theta_{12} B^{12})}{1 - B} a_t$$

and

$$\text{Ln}\,(Y_t) = \frac{\omega_0}{1 - \delta_1 B} I_t + \frac{\Theta_0 + (1 - \Theta_1 B)(1 - \Theta_4 B^4 - \Theta_{12} B^{12})}{1 - B} a_t$$

and the three factor model

$$\text{Ln}\,(Y_t) = \frac{\omega_0}{1 - \delta_1 B} I_t + \frac{\Theta_0 + (1 - \Theta_1 B)(1 - \Theta_4 B^4)(1 - \Theta_{12} B^{12})}{1 - B} a_t.$$

To be sure, these three models are nearly identical whenever Θ_4 is small; when Θ_4 is zero, they are identical. Judging from the size of the lag-4 ACF in Figure 3.3(f), however, the estimated value of Θ_4 will not be small. The differences in these three models are in their cross-product terms. Expanding the first two-factor noise component,

$$(1 - \Theta_1 B - \Theta_4 B^4)(1 - \Theta_{12} B^{12}) =$$
$$(1 - \Theta_1 B - \Theta_4 B^4 - \Theta_{12} B^{12} + \Theta_1 \Theta_{12} B^{13} + \Theta_4 \Theta_{12} B^{16}).$$

Expanding the second two-factor noise component

$$(1 - \Theta_1 B)(1 - \Theta_4 B^4 - \Theta_{12} B^{12}) =$$
$$(1 - \Theta_1 B - \Theta_4 B^4 - \Theta_{12} B^{12} + \Theta_1 \Theta_4 B^5 + \Theta_1 \Theta_{12} B^{13}).$$

And expanding the three-factor noise component

$$(1 - \Theta_1 B)(1 - \Theta_4 B^4)(1 - \Theta_{12} B^{12}) =$$
$$(1 - \Theta_1 B - \Theta_4 B^4 - \Theta_{12} B^{12} + \Theta_1 \Theta_4 B^5 + \Theta_1 \Theta_{12} B^{13}$$
$$+ \Theta_4 \Theta_{12} B^{16} - \Theta_1 \Theta_4 \Theta_{12} B^{17}).$$

The analyst might ordinarily look to the residual ACF in Figure 3.3(f) for evidence favoring one of these three components. A spike at ACF(17), for example, would argue for the three-factor model. We see nothing in the ACF to inform this decision, however. Lacking information from this source, the analyst is advised to estimate all three models, deciding the competition with a comparison of residual statistics. Following this advice, the "best" model of the three is:

$$\text{Ln}(Y_t) = \frac{\omega_0}{1 - \delta_1 B} I_t + \frac{\Theta_0 + (1 - \Theta_1 B - \Theta_4 B^4)(1 - \Theta_{12} B^{12})}{1 - B} a_t.$$

This model is the "best" because, of the three competing models, it has the lowest residual mean square (RMS) statistic. To learn this fact, of course, the analyst must estimate all three models.

Estimation

Parameter estimates for this tentative model are:

$$\hat{\omega}_0 = -.6116 \text{ with t statistic} = -4.16$$
$$\hat{\delta}_1 = .5186 \text{ with t statistic} = 4.30$$
$$\hat{\Theta}_0 = .0040 \text{ with t statistic} = 1.09$$
$$\hat{\Theta}_1 = .5052 \text{ with t statistic} = 8.14$$
$$\hat{\Theta}_4 = .5727 \text{ with t statistic} = 8.94$$
$$\hat{\Theta}_{12} = -.2384 \text{ with t statistic} = -2.74.$$

With the exception of the Θ_0 estimate, all parameter estimates are statistically significant and otherwise acceptable. A diagnostic check of the model residuals indicates that they are not different than white noise.

Whenever a statistically insignificant parameter is dropped from a tentative model, the remaining parameters must be reestimated. The estimate of Θ_0 is not statistically significant and must be dropped from the model for this series. The new estimates of the impact component parameters are:

$$\hat{\omega}_0 = -.6070 \quad \text{with t statistic} = -4.05$$
$$\hat{\delta}_1 = \ .41287 \ \text{with t statistic} = \ 2.83,$$

which is a substantial change. This should warn the reader to always estimate all parameters of a model simultaneously. In all cases, when a parameter is dropped from a model, estimates of the remaining parameters will change if only slightly.

We can now interpret this result. In the first month following decriminalization, the level of the log-transformed time series process changed by the quantity $\hat{\omega}_0$. This amounted to a decrease and, in successive months, the process level continued to drop. Log levels for the first six months following decriminalization are expected to be:

$$\begin{array}{lll}
\text{first month:} & \hat{\omega}_0 & = -\ .607 \\
\text{second month:} & \hat{\omega}_0\,(1 + \hat{\delta}_1) & = -\ .858 \\
\text{third month:} & \hat{\omega}_0\,(1 + \hat{\delta}_1 + \hat{\delta}_1^2) & = -\ .961 \\
\text{fourth month:} & \hat{\omega}_0\,(1 + \hat{\delta}_1 + \hat{\delta}_1^2 + \hat{\delta}_1^3) & = -1.004 \\
\text{fifth month:} & \hat{\omega}_0\,(1 + \hat{\delta}_1 + \hat{\delta}_1^2 + \hat{\delta}_1^3 + \hat{\delta}_1^4) & = -1.021 \\
\text{sixth month:} & \hat{\omega}_0\,(1 + \hat{\delta}_1 + \hat{\delta}_1^2 + \hat{\delta}_1^3 + \hat{\delta}_1^4 + \hat{\delta}_1^5) & = -1.029
\end{array}$$

and so forth. The postintervention level of the log process continues to drop but by smaller and smaller increments. The asymptotic change in log level is:

$$\text{asymptotic change} = \frac{\hat{\omega}_0}{1 - \hat{\delta}_1} = -1.034.$$

And thus, by the end of the sixth postintervention month, the log process has achieved 99.52% of its asymptotic change in level. While the process level continues to drop after the sixth month, this change is negligible.

To translate this finding into the raw metric, the analyst needs only to exponentiate the asymptotic change in level:

$$\frac{\text{postintervention equilibrium}}{\text{preintervention equilibrium}} = e^{-1.034} = .35558.$$

This result in turn is translated into a percent change estimate of impact:

$$\text{percent change} = .35558 - 1.0 = -64.44\%.$$

As the preintervention mean for this series is approximately 651 arrests per month, the percent change represents a reduction of approximately 420 arrests per month. This interpretation is consistent with the visual evidence in the plotted time series, Figure 3.3(a).

We urge the reader to replicate this analysis. The manner in which a full model was built is typical of the iterative procedure except, of course, that the preintervention series was not long enough to permit a confident identification of the noise component. Through a conservative series of model-building steps, we were nevertheless able to arrive at an adequate but parsimonious representation of the process underlying this time series.

3.4 Higher Order and Compound Intervention Components

In Section 3.2.3, we developed a modeling strategy based on two characteristics of an impact: onset and duration. An impact can be *either* abrupt *or* gradual in onset and *either* permanent *or* temporary in duration. Using the zero- and first-order transfer functions associated with three distinct patterns of impact, almost any problem in any substantive area can be analyzed. There are two advantages to working from this perspective. First, the results of the analysis are easily interpreted and, second, because the transfer functions are related at the bounds of system stability, alternative impact hypotheses can ordinarily be ruled out through analysis. But should an impact assessment require a more complicated effect model, higher order and compound intervention components can be used. Unfortunately, interpretability of the impact assessment may suffer in many cases.

A higher order intervention component is defined as one with powers of B higher than zero or one. For example,

$$f(I_t) = \frac{\omega_0}{1 - \delta_1 B - \delta_2 B^2} I_t$$

is a second-order transfer function. The bounds of system stability for higher order transfer functions are identical with the bounds of invertibility for higher order autoregressive operators (see Section 2.5). For this second-order transfer function then, the bounds are:

$$\begin{aligned} -1 &< \delta_2 < +1 \\ \delta_1 + \delta_2 &< +1 \\ \delta_2 - \delta_1 &< +1. \end{aligned}$$

The behavior of this or any higher order transfer function can be determined by expanding the inverse operator as an infinite series. A simpler recursive method is more practical, however. Working with the Y_t^* time series in this case

$$Y_t^* = \frac{\omega_0}{1 - \delta_1 B - \delta_2 B^2} I_t$$

$$(1 - \delta_1 B - \delta_2 B^2)\, Y_t^* = \omega_0 I_t$$

$$Y_t^* = \delta_1 Y_{t-1}^* + \delta_2 Y_{t-2}^* + \omega_0 I_t.$$

Prior to onset of the event, when $I_t = 0$, $Y_t^* = 0$. At $i + 1$, however, the step function changes from zero to one and the value of Y_{i+1}^* is expected to be:

$$\begin{aligned} Y_{i+1} &= \delta_1 Y_i^* + \delta_2 Y_{i-1}^* + \omega_0 I_{i+1} \\ &= \delta_1 (0) + \delta_2 (0) \quad + \omega_0 (1) = \omega_0. \end{aligned}$$

In the next moment,

$$\begin{aligned} Y_{i+2}^* &= \delta_1 Y_i^* + \delta_2 Y_i^* + \omega_0 I_{i+2} \\ &= \delta_1 (\omega_0) \quad + \delta_2 (0) \quad + \omega_0 (1) \quad = \delta_1 \omega_0 + \omega_0. \end{aligned}$$

And in the next moment,

$$\begin{aligned} Y_{i+3}^* &= \delta_1 Y_{i+2}^* + \delta_2 Y_{i+1}^* + \omega_0 I_{i+3} \\ &= \delta_1 (\delta_1 \omega_0 + \omega_0) + \delta_2 (\omega_0) + \omega_0 (1), \\ &= \delta_1^2 \omega_0 + \delta_1 \omega_0 + \delta_2 \omega_0 + \omega_0, \end{aligned}$$

and so forth. The reader may use this recursive method to determine the behavior of any higher order transfer function. In this case, *two* rate parameters, δ_1 and δ_2, determine the rate at which the process achieves a new equilibrium level. The interpretability of this model is somewhat lessened by including an extra rate parameter, however.

As a general strategy, the analyst can *fit* a higher order transfer function to the time series: δ- and ω-parameters are added to the model until a desired degree of fit is achieved. In model *fitting,* however, the analyst must be conscious of what has been lost. First, the impact assessment is no longer a *confirmatory* analysis based on a null hypothesis. Second, the estimate of effect is likely to be biased. Model fitting takes advantage of properties of the specific realization and it may be incorrect to attribute those properties to the generating process itself. We will demonstrate a model-fitting procedure in the next section.

While the use of higher order transfer functions often reduces model interpretability, the use of *compound* intervention components may increase model interpretability. A *compound* intervention component is defined for our purposes as a sum of two low-order transfer functions. Suppose, for example, that a time series has been impacted by two distinct interventions at two different times. A model reflecting this compound effect would be:

$$Y_t = f(I_t) + f(I_{t+n}) + N_t.$$

The sense of this compound intervention component is that the generating process is impacted once and then is impacted again n observations later.

The distinct impacts of a compound intervention component need not occur at different times, of course. There are many situations in which, on the basis of theory, a single intervention may have two distinct impacts, and in these situations, a compound intervention component can be used as a model of the effect. One of the most useful compound components in our experience is the one composed of the zero-order transfer function

$$f(I_t) = \omega_0 I_t$$

and the first-order transfer function applied to a pulse

$$f(I_t) = \frac{\omega_0}{1 - \delta_1 B}(1 - B)\, I_t.$$

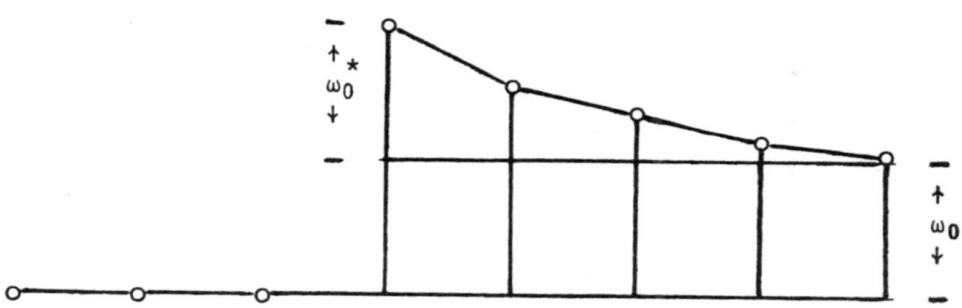

FIGURE 3.4 Pattern of Impact Expected of a Compound Intervention Component

Combining these two components, a compound component is formed as

$$f(I_t) = \omega_0 I_t + \frac{\overset{*}{\omega_0}}{1 - \delta_1 B}.$$

An asterisk (*) is used only to denote the fact that these two zero-order ω-parameters are not the same. Figure 3.4 shows the pattern of impact determined by this compound model. In a literal sense, this impact is the sum of the impacts expected of the two component elements.

We have found this compound intervention model useful in many situations in which an intervention may produce an abrupt but short-lived demonstration effect as well as a larger permanent effect. By accounting for the temporary demonstration effect with a decaying spike, this model permits a more precise estimate of the permanent effect.

To illustrate the use of this compound model, we return to the Perceptual Speed time series. In Section 3.2.1, we used a first-order transfer function to estimate a *gradual, permanent* reduction in perceptual speed scores due to the onset of a chlorpromazine regimen. In many psychological phenomena such as learning, habit formation, and so forth, gradual impacts are expected theoretically as a result of a simple growth process. A "learning curve," for example, may result from a gradual accumulation of correct responses or, alternatively, from a gradual decay of incorrect responses.

There is another mechanism which can result in a gradual, permanent impact, however. First, a treatment itself can result in an abrupt, constant impact of the sort

Before o o o o o o

o o o o o o After.

Associated with the treatment, however, might be a *placebo* (or *trauma; novelty; reactivity;* and so forth) effect of the sort

o

o

o

Before o o o o o o o o o After.

This temporary effect is due only to the novelty (from the perspective of the single schizophrenic patient in this case) of the change in state from "no treatment" to "treatment." After a few days, the effect wears off, leaving only the true physiological effect. When these two distinct impacts are added together, a *gradual, permanent* impact is realized. For all practical purposes, of course, the analyst cannot tell whether the observed impact is due to a single process (attenuation of a natural resistance to chlorpromazine in this case) or due to two distinct processes (physiological impact plus novelty).

Apologizing for this simplification of theory, we can reanalyze the Perceptual Speed time series with an assumption that there are two distinct impacts of the chlorpromazine regimen. Using the same noise component selected in Section 3.2.1, the impact assessment model corresponding to a double impact theory is:

$$Y_t = [\omega_0 + \frac{1 - B}{1 - \delta_1 B} \omega_0^*] I_{61} + \frac{1 - \Theta_1 B}{1 - B} a_t.$$

Parameter estimates for this model are:

$$\begin{aligned}
\hat{\Theta}_1 &= .77 \text{ with t statistic} = 12.97 \\
\hat{\omega}_0 &= -28.8 \text{ with t statistic} = -3.50 \\
\hat{\delta}_1 &= .51 \text{ with t statistic} = .56 \\
\hat{\omega}_0^* &= 18.05 \text{ with t statistic} = 1.72.
\end{aligned}$$

The estimates of δ_1 and ω_0^* are not statistically significant so, strictly speaking, these parameters should be dropped from the model. There are a number of reasons which have nothing to do with the null hypothesis which might explain the lack of statistical significance, however. We will address these reasons shortly.

Assuming that statistical significance is not an issue here, and accepting the estimated parameters of this model at face value, the impact on the nth day of the chlorpromazine regimen is given by

$$\begin{aligned}
\text{impact on day n} &= \hat{\omega}_0 + \hat{\delta}_1^{n-1} \hat{\omega}_0^* \\
&= -28.8 + (.51)^{n-1} (18.05),
\end{aligned}$$

which is merely the sum of the two simple impacts. In Section 3.2.1, on the other hand, impact on the nth day of the chlorpromazine regimen was given by

$$\text{impact on day n} = \sum_{k=0}^{n} \hat{\delta}_1^k \hat{\omega}_0$$

$$= \sum_{k=0}^{n} (.53)^k (-28.49).$$

If the results of these two models are compared for the first seven days of the regimen,

	Compound Model	Gradual Model
First day	−10.75 units	−13.30 units
Second day	−19.64 units	−20.08 units
Third day	−24.15 units	−23.87 units
Fourth day	−26.44 units	−25.89 units
Fifth day	−27.60 units	−26.96 units
Sixth day	−28.19 units	−27.53 units
Seventh day	−28.49 units	−27.83 units,

it is clear that these two models give similar estimates of the impact. Estimates of asymptotic effect (−28.8 units for the compound model *versus* −28.49 units for the gradual model) are nearly identical. *There is nevertheless a substantive difference between the impact estimates of these two models.* Whereas the gradual model predicts the effect based on a single mechanism, the compound model predicts the effect based on two distinct mechanisms. Unique treatment and novelty effects are untangled and estimated.

Because the transfer function parameter estimates for the compound model were statistically insignificant, one might argue that the compound impact theory is unjustified. Multicollinearity is always a problem with compound models, however; we will discuss this problem at some length in Chapter 6. Another problem affecting this particular time series is the outlier in the 31st observation. We will discuss the problem of outliers generally in impact assessments at the end of this chapter but, in all cases, transfer function parameter estimates will be sensitive to outliers. Finally, depending upon theoretical perspective, statistical significance may be a minor concern. Given a theory where the compound impact model is "true," the transfer function parameter estimates will be interpreted without qualification. In the next section, we will demonstrate the procedures of model *fitting* (as opposed to model building). A comparison of these two sections will illuminate the role of theory in impact assessment.

3.5 U.S. Suicides

The examples preceding this have been straightforward impact assessment analyses, aimed at estimating the effects of discrete interventions (events) on time series. We will now address a related use of impact assessment models. Figure 3.5(a) shows a time series of annual U.S. suicide rates for the 1920–1969 period. The series begins at a level of approximately seventeen suicides (per 100,000 total population), jumps up to approximately 29, and then returns gradually to the starting level. A striking feature of this plot is the set of observations for the 1930s. These are outliers but, in this case, recording errors are not suspected as in the Hyde Park Purse

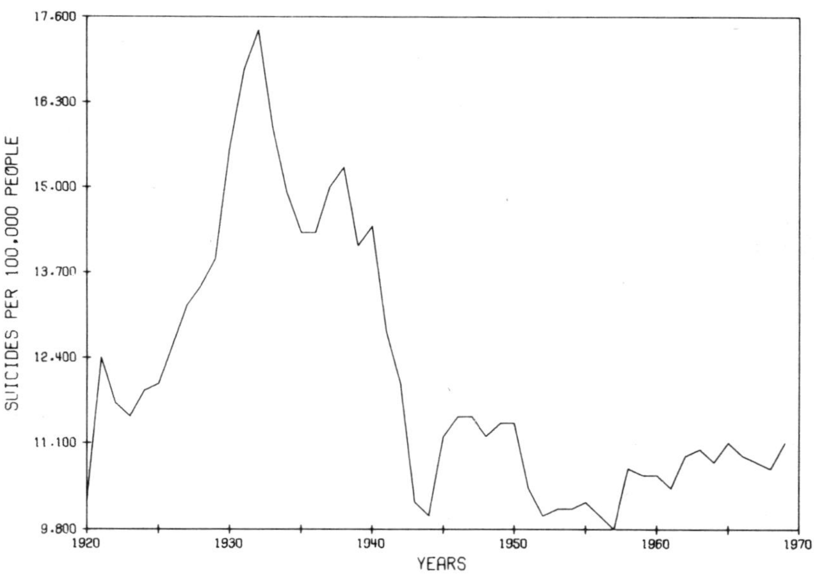

FIGURE 3.5(a) U.S. Suicide Rate, 1920–1969

```
SERIES.. SUICIDES (NOBS=  50)  U.S. SUCIDE RATE, 1920-1969
NO. OF VALID OBSERVATIONS  =    50.

AUTOCORRELATIONS OF LAGS 1 - 10.
Q( 10,   50) =  164.72       SIG =     0.000

  LAG    CORR     SE  -1    -.8   -.6   -.4   -.2   0    .2    .4    .6    .8   +1
                      +----+----+----+----+----+----+----+----+----+----+
   1    .917  .141                          (       IXXXXXX)XXXXXXXXXXXXXXXX
   2    .822  .232                     (            IXXXXXXXXXXX)XXXXXXXXX
   3    .715  .284                   (              IXXXXXXXXXXXXX)XXXX
   4    .625  .318                 (                IXXXXXXXXXXXXXXXX*
   5    .551  .342               (                  IXXXXXXXXXXXXXX   )
   6    .473  .359              (                   IXXXXXXXXXXXX       )
   7    .403  .371            (                     IXXXXXXXXXX          )
   8    .334  .380            (                     IXXXXXXXX            )
   9    .241  .386            (                     IXXXXXX              )
  10    .127  .389            (                     IXXX                 )
                            -2SE                                      +2SE


    PARTIAL AUTOCORRELATIONS OF LAGS 1 - 10.

    LAG    CORR     SE  -1    -.8   -.6   -.4   -.2   0    .2    .4    .6    .8   +1
                        +----+----+----+----+----+----+----+----+----+----+
     1    .917  .141                          (       IXXXXXX)XXXXXXXXXXXXXXXX
     2  -.115  .141                           (   XXXI        )
     3  -.126  .141                           (   XXXI        )
     4    .049  .141                          (       IX      )
     5    .050  .141                          (       IX      )
     6  -.111  .141                           (   XXXI        )
     7    .001  .141                          (       I       )
     8  -.027  .141                           (      XI       )
     9  -.222  .141                           (XXXXXXI        )
    10  -.198  .141                           ( XXXXXI        )
                                            -2SE          +2SE
```

FIGURE 3.5(b) ACF and PACF for the Raw Series

Snatchings time series. These extreme values are instead attributed to the obvious exogenous effect of the Great Depression.

Referring to these observations as "outliers" should not imply a substantive and/or theoretical argument about the sociological relevance of economic depression. Some would argue that the Depression was an unfortunate "accident" caused by the improbable intersection of several economic and political events. Others would argue that the Depression was merely an extreme instance of the periodic crises which characterize capitalist economic systems. Although time series analysis could certainly be used profitably to illuminate these questions, that is not our purpose here. We are concerned only with the statistical issues associated with these Depression outliers.

During the 1920–1969 period, the time series appears to be relatively flat, fluctuating about a mean of (approximately) 17 suicides. The exception

```
SERIES.. SUICIDES (NOBS=  50)  U.S. SUCIDE RATE, 1920-1969
DIFFERENCED  1 TIME(S) OF ORDER  1.
NO. OF VALID OBSERVATIONS  =     49.

AUTOCORRELATIONS OF LAGS 1 - 10.
Q( 10,  49) =  7.4952        SIG =      .678

 LAG   CORR    SE  -1    -.8   -.6   -.4   -.2   0     .2    .4    .6    .8   +1
                    +----+----+----+----+----+----+----+----+----+----+
  1    .216  .143                        (       IXXXXX )
  2    .043  .149                        (       IX      )
  3   -.121  .150                        (    XXXI       )
  4   -.068  .152                       (      XXI        )
  5    .032  .153                       (        IX       )
  6   -.012  .153                       (        I        )
  7    .035  .153                       (        IX       )
  8    .173  .153                       (        IXXXX    )
  9    .180  .157                       (        IXXXX    )
 10    .130  .161                       (        IXXX     )
                                      -2SE             +2SE
```

```
PARTIAL AUTOCORRELATIONS OF LAGS 1 - 10.

 LAG   CORR    SE  -1    -.8   -.6   -.4   -.2   0     .2    .4    .6    .8   +1
                    +----+----+----+----+----+----+----+----+----+----+
  1    .216  .143                        (       IXXXXX )
  2   -.004  .143                        (       I       )
  3   -.136  .143                        (    XXXI       )
  4   -.036  .143                        (      XI       )
  5    .072  .143                        (       IXX     )
  6   -.049  .143                        (      XI       )
  7    .028  .143                        (       IX      )
  8    .186  .143                        (       IXXXXX )
  9    .115  .143                        (       IXXX    )
 10    .058  .143                        (       IX      )
                                       -2SE           +2SE
```

FIGURE 3.5(c) ACF and PACF for the Regularly Differenced Series

to this, of course, is the period of time during and following the Depression onset. While the process as a whole may be stationary (or trendless at least), these observations change the appearance of the series. They are outliers and, in a statistical sense only, they violate the assumption of stationarity. The generating process itself may well be a stationary process but this realization of the process is not. From this series alone, the analyst could not infer a stationary property of the generating process.

The consequences of these outliers for identification and estimation are similar to those we discussed in Section 2.12.4 for the Hyde Park time series. Figure 3.5(b) shows the ACF and PACF estimated from the raw suicide time series. While these statistics suggest a nonstationary process, that appearance may be due only to the Depression outliers. Figure 3.5(c) shows the ACF and PACF estimated from the differenced suicide time

```
SERIES.. SUICIDES (NOBS=  50)  U.S. SUCIDE RATE, 1920-1969
SELECTED OBSERVATIONS    24 -    50.
NO. OF VALID OBSERVATIONS   =    27.

AUTOCORRELATIONS OF LAGS 1 - 10.
Q( 10,   27) =  31.691      SIG =      .000

  LAG    CORR     SE  -1    -.8  -.6  -.4  -.2   0    .2   .4   .6   .8  +1
                       +----+----+----+----+----+----+----+----+----+----+
   1     .653  .192               (          IXXXXXXXXXX)XXXXXX
   2     .319  .262            (             IXXXXXXXXX     )
   3     .165  .276          (               IXXXXX           )
   4     .051  .280          (               IX               )
   5    -.195  .280          (          XXXXXI                )
   6    -.437  .285          (  XXXXXXXXXXXXI                 )
   7    -.417  .309         (     XXXXXXXXXXI                  )
   8    -.305  .329       (          XXXXXXXXI                   )
   9    -.239  .339      (             XXXXXXI                    )
  10    -.251  .345      (             XXXXXXI                    )
                       -2SE                                   +2SE



PARTIAL AUTOCORRELATIONS OF LAGS 1 - 10.

  LAG    CORR     SE  -1    -.8  -.6  -.4  -.2   0    .2   .4   .6   .8  +1
                       +----+----+----+----+----+----+----+----+----+----+
   1     .653  .192               (          IXXXXXXXXXX)XXXXXX
   2    -.186  .192               (     XXXXXI          )
   3     .070  .192               (          IXX        )
   4    -.086  .192               (        XXI          )
   5    -.329  .192               ( XXXXXXXXXI          )
   6    -.259  .192               (   XXXXXXXI          )
   7     .076  .192               (          IXX        )
   8    -.020  .192               (         XI          )
   9    -.015  .192               (          I          )
  10    -.157  .192               (      XXXXI          )
                                -2SE                 +2SE
```

FIGURE 3.5(d) ACF and PACF for the Raw Series (1943–1969 Only)

series. These indicate white noise but this appearance too may be due only to the Depression outliers.

It is instructive to compare the effects of multiple outliers in this series with the effect of a single outlier in the Hyde Park series. Here the outliers give an overall trend to the series which is reflected in the ACF shown in Figure 3.5(b). Once this trendlike component is removed through differencing, however, the ACF, shown in Figure 3.5(c), indicates a white noise process. The differenced suicide series and the undifferenced Hyde Park series in fact have almost identical ACFs and this should not be surprising. *In both cases, outliers inflate the estimate of process variance and, as a result, understate the values of low-order ACFs.*

Since the Depression observations of this time series are not due to recording error (and hence cannot be "corrected" or adjusted), a noise component must be built around the outliers.

Identification

Figure 3.5(d) shows an ACF and PACF estimated from the 1943–1969 segment of the series. This series is too short to permit a confident identification. Although the ACF and PACF indicate an ARIMA (1,0,0) process, the large standard errors give us little confidence in estimates of the high-order ACF (k) and PACF (k). The reader who replicates our analysis will discover that deleting one or two observations from this segment changes the appearance of these statistics markedly. Acknowledging the weakness of this identification, we tentatively select an ARIMA (1,0,0) model for the noise component.

Estimation

Parameter estimates for the noise component model are:

$$\hat{\phi}_1 = .67 \text{ with t statistic} = 4.47.$$

Although this estimate is based only on the last 27 observations of the series, the estimate is statistically significant and otherwise acceptable. Diagnostic checks of the residuals indicate that the residuals of this model are not significantly different than white noise.

Accepting the ARIMA (1,0,0) model for a noise component, the exogenous shock of the Great Depression will be represented by an impact assessment model of the form

$$Y_t = f(I_t) + \frac{a_t}{1 - \phi_1 B}.$$

In this case, social theories suggest an abrupt, temporary pattern of impact associated with the transfer function

$$f(I_t) = \frac{\omega_0}{1 - \delta_1 B}(1 - B) I_t$$

$$\text{where} \quad I_t = 0 \text{ prior to } 1930$$
$$= 1 \text{ from } 1930 \text{ on.}$$

visual inspection of the plotted time series raises questions as to the adequacy of this model, however. *Note that the series does not reach its highest point in 1930, but continues to climb higher in 1931 and 1932 before starting to decay.* Also, decay from the 1932 zenith is interrupted by an increase in 1937. (By many accounts, 1937 was the worst year of the

Depression.) To incorporate these impacts into the model, we specify the compound model

$$f(I_t) = \frac{\omega_0 - \omega_1 B - \omega_2 B^2 - \omega_7 B^7}{1 - \delta_1 B}(1 - B) I_t.$$

As we have not encountered a higher order transfer function up to this point, this model requires some explanation.

The exogenous shock of the Great Depression is represented by a pulse function, $(1 - B)I_{1930}$, visited upon the economic-social system in 1930. The scalar weight, ω_0, translates this pulse into an increase in the suicide rate which begins immediately to decay. The *rate* of decay is determined by the parameter δ_1. The scalar weights ω_1, ω_2, and ω_7 translate the pulse into delayed spikes, one each occurring in 1931, 1932, and 1937. Each of these delayed spikes also begins immediately to decay at a rate determined by the parmeter δ_1. The seventh-order transfer function is thus identical to *four* decaying spikes. The transfer function in fact can be rewritten as

$$f(I_t) = \frac{\omega_0}{1 - \delta_1 B}(1 - B) I_t - \frac{\omega_1}{1 - \delta_1 B}(1 - B) I_{t-1}$$

$$- \frac{\omega_2}{1 - \delta_1 B}(1 - B) I_{t-2} - \frac{\omega_7}{1 - \delta_1 B}(1 - B) I_{t-7},$$

which makes this point clear. To draw the expected pattern of impact determined by the transfer function, the reader need only draw four decaying spikes (beginning in 1930, 1931, 1932, and 1937) one on top of the other.

Estimation

Parameter estimates for the full intervention model are:

$$\hat{\phi}_1 = .87 \text{ with t statistic } = 8.21$$
$$\hat{\delta}_1 = .80 \text{ with t statistic } = 8.00$$
$$\hat{\omega}_0 = 2.08 \text{ with t statistic } = 3.12$$
$$\hat{\omega}_1 = -1.94 \text{ with t statistic } = -3.21$$
$$\hat{\omega}_2 = -1.60 \text{ with t statistic } = -2.62$$
$$\hat{\omega}_7 = -1.26 \text{ with t statistic } = -2.19.$$

All coefficients are statistically significant and otherwise acceptable. Our estimate of ϕ_1 is dangerously close to the bounds of invertibility, however. When autoregressive parameters are so large, we prefer to respecify the noise component. The noise component is thus respecified as ARIMA (0,1,0) and parameters are reestimated as

$$\hat{\omega}_0 = 1.855 \text{ with t statistic} = 3.26$$

$$\hat{\omega}_1 = -1.851 \text{ with t statistic} = -3.29$$

$$\hat{\omega}_2 = -1.614 \text{ with t statistic} = -2.69$$

$$\hat{\omega}_7 = -1.117 \text{ with t statistic} = -2.04$$

$$\hat{\delta}_1 = .733 \text{ with t statistic} = 5.43.$$

All parameters are statistically significant at a .05 level and are otherwise acceptable.

To intrepret this finding, we begin with a zero level in the year preceding the Depression. Increases in successive years are expected to be:

$$1930\text{: } \hat{\omega}_0 = +1.855$$

$$1931\text{: } \hat{\delta}_1\hat{\omega}_0 - \hat{\omega}_1 = +3.210$$

$$1932\text{: } \hat{\delta}_1^2\hat{\omega}_0 - \hat{\delta}_1\hat{\omega}_1 - \hat{\omega}_2 = +3.967$$

$$1933\text{: } \hat{\delta}_1^3\hat{\omega}_0 - \hat{\delta}_1^2\hat{\omega}_1 - \hat{\delta}_1\hat{\omega}_2 = +2.908$$

$$1934\text{: } \hat{\delta}_1^4\hat{\omega}_0 - \hat{\delta}_1^3\hat{\omega}_1 - \hat{\delta}_1^2\hat{\omega}_2 = +2.131$$

$$1935\text{: } \hat{\delta}_1^5\hat{\omega}_0 - \hat{\delta}_1^4\hat{\omega}_1 - \hat{\delta}_1^3\hat{\omega}_2 = +1.562$$

$$1936\text{: } \hat{\delta}_1^6\hat{\omega}_0 - \hat{\delta}_1^5\hat{\omega}_1 - \hat{\delta}_1^4\hat{\omega}_2 = +1.145$$

$$1937\text{: } \hat{\delta}_1^7\hat{\omega}_0 - \hat{\delta}_1^6\hat{\omega}_1 - \hat{\delta}_1^5\hat{\omega}_2 - \hat{\omega}_7 = +1.956$$

$$1938\text{: } \hat{\delta}_1^8\hat{\omega}_0 - \hat{\delta}_1^7\hat{\omega}_1 - \hat{\delta}_1^6\hat{\omega}_2 - \hat{\delta}_1\hat{\omega}_7 = +1.434$$

and so forth. The pattern of impact from this transfer function is *isomorphic* to the effect seen in the plotted time series, Figure 3.5(a). Diagnostic checks of the model residuals indicate that they are not different than white noise.

This model must now be interpreted. The use of a seventh-order transfer function as a model of the Great Depression had only one justification: *it fit the data well.* There is no theoretical basis for assuming that depressions generally will have an impact of this sort on suicide rates. This use of a compound intervention component may be contrasted with the use of a compound intervention component in the analysis of the Perceptual Speed time series in Section 3.4. There a compound intervention component was justified by a theory which predicted that there would be *two* distinct impacts from a single intervention. The results of that analysis could be interpreted within the context of that theory. In this present example, however, no confident statement can be made about the impact of the Depression on suicides except this obvious one: There *appears* to be a substantial effect.

In a broader sense, this impact assessment model has no interpretation whatsoever. The model built here nevertheless may be seen as a "clean" picture of the suicide rate generating process and may be used for a number of purposes not related to impact assessment.

Vigderhous (1978) and Mark (1979), for example, have suggested causal models for this time series, using a time series of annual U.S. unemployment rates to predict suicide rates. A causal model requires a "clean" picture of both unemployment and suicide time series. In the simplest case, a causal model might be diagramed as

unemployment ———————————→ suicide.

Yet if the Depression outliers are not removed from both time series, a more complicated causal model,

unemployment ———————————→ suicide,

↑—————— depression ——————↑

might lead to spurious causal inferences. There is no doubt that both unemployment rates and suicide rates were at their highest levels during the Great Depression (and at their lowest levels during World War II). This does not mean that "unemployment causes suicides," however.

In Chapters 4 and 5, we develop the use of ARIMA models and methods for univariate forecasting and multivariate causal analysis. In both of these applications, the analyst must assume that the generating process of a time series in invariant; that the process will continue to generate realizations that are identical with one another within the limits of sampling variance. To be sure, wars and depressions are part of any social science generating process. When only one short realization of the process is available, however, and when that realization includes a war or depression, a number of practical problems arise.

Here it may be instructive to compare social and industrial processes. In the field of industrial control, generating processes are always invariant. Manufacturing processes are relatively constant over time. Error in the process is usually due only to slight variations in the quality of process inputs (electricity, water, oil, and so on), and as the quality of these inputs is tightly controlled, errors are slight. Input variance is so relatively small, in fact, that it is always well described as white noise.

In the social sciences, on the other hand, generating processes are subject to a wide range of exogenous (input) forces which are not tightly controlled. So long as no single exogenous force exerts a primary influence on the process, the analyst is justified in treating the sum of many exogenous forces as white noise. When a dormant force suddenly asserts itself, however, the effect of that variable cannot be treated ideally as "just another random shock," part of the white noise process. This is a *practical* dictum, of course, for in an infinitely long realization of the suicide rate process, the Great Depression would indeed be "just another random shock."

Wars and depressions commonly exert primary influences on most social indicators. Interaction among nations is ordinarily a stochastic process whose impact on social indicators is adequately described as white noise; except, as a practical matter, in the extreme case of a declared war. Economic fluctuations are similarly stochastic, exerting no strong deterministic influence on social indicators except, as a practical matter, in the extreme case. When only a finite realization of a social process is available, wars and depressions are best thought of as cataclysmic *events* rather than as points on a continuum.

This point is brought home by the analysis of the U.S. Suicide rate time series. While there is no problem *fitting* an intervention component to the series, it would be incorrect to say that this component is a *model* of the Great Depression impact. It is not. Nevertheless, the results of this analysis give a "clean" picture of the suicide rate generating process which may then be used in forecasting or causal modeling.

3.6 A Final Note on the Outlier Problem

In Section 2.12.4, we discussed the problem of outliers in identification. As illustrated by the Hyde Park Purse Snatchings time series, outliers inflate the estimate of process variance and thus understate low-order autocorrelation. This problem is analogous to the problem of estimating a noise component from an impacted series. If the impact is large, as in the Directory Assistance time series, for example, the estimated ACF is practically meaningless. A noise component must be estimated from the preintervention time series only. Identification notwithstanding, outliers cause another specific problem for impact assessment analyses: Transfer function parameter estimates may be unduly influenced by outliers.

The Hyde Park Purse Snatchings time series had an outlier (due to a recording error which was later corrected). In the 42nd observation of the series, a community whistle alert program was implemented (see Reed, 1978). We will estimate the impact of that program on purse snatchings both with and without the incorrect deviant observation.

First, with the outlier: In Section 2.12.4, we identified an ARIMA (0,0,0) model for this series; the raw series was not different than white noise. Reed proposed an abrupt, constant impact for the program, so the impact assessment model is set tentatively as

$$y_t = \omega_0 I_{42} + a_t,$$

$$\text{where } I_{42} = 0 \text{ for the first 41 observations}$$
$$= 1 \text{ for the next 30 observations.}$$

The parameter estimate for this model is:

$$\hat{\omega}_0 = -4.17 \text{ with t statistic} = -1.78.$$

The sign of the parameter estimate is negative, as one would expect if the program had any impact at all, and it is statistically significant at the .10 level. Diagnostic checks of the model residuals indicate that they are not different than white noise. However, given our discussion of ACF distortion associated with outliers, this is to be expected.

Second, with the corrected observation: Considering the effect of a large outlier on the estimation of impact parameters, one might suspect that, because the outlier is in the postintervention segment, it would *inflate* the estimate of postintervention level and thus *deflate* the estimate of impact. By this line of reasoning, one would expect an analysis of the corrected time series to show a larger drop in purse snatchings coincident with the intervention.

But in practice this line of reasoning is flawed. Outliers bias estimates of both the noise and impact component parameters, resulting in a *joint* effect. Replacing the incorrect deviant observation (66 purse snatchings) with its correct value (12) might nevertheless be expected to result in a larger estimate of impact. *Using the statistically inadequate ARIMA (0,0,0) noise component, this is indeed true*. For the model

$$y_t = \omega_0 I_{42} + a_t,$$

the parameter estimate is:

$$\hat{\omega}_0 = -5.97 \text{ with t statistic} = -3.50.$$

A diagnostic check of these model residuals would reveal that the model is not statistically adequate, however. Using the more appropriate ARIMA (2,0,0) for a noise component, parameter estimates are:

$$\hat{\phi}_1 = .28 \text{ with t statistic} = 2.37$$
$$\hat{\phi}_2 = .36 \text{ with t statistic} = 3.09$$
$$\hat{\omega}_0 = -3.07 \text{ with t statistic} = -.87.$$

Both autoregressive parameters are statistically significant and otherwise acceptable. The transfer function parameter is not statistically significant, however. Diagnostic checks of the model residuals indicate that they are not different than white noise, so the model is statistically adequate.

An interpretation of these results is that *there is no evidence to support the hypothesis that this program had an impact on purse snatchings.*

A more important result of this analysis concerns the effect of an outlier on the estimate of impact. Using the same statistically inadequate noise component, the estimate of effect changed by over 40% when the outlier was removed from the time series. The degree of distortion attributable to outliers will depend upon their size and number, the size of the effect, and upon the length of the time series. There will always be some distortion, however.

3.7 Conclusion

We have required a mental leap of the reader from Chapter 2 to Chapter 3. We developed an atheoretical, mechanical model-building strategy in Chapter 2 which leads to an adequate, parsimonious ARIMA model. In Chapter 3, however, we developed a subtler strategy which is not amenable to description as a rigid series of steps or as a flow chart, see Figure 2.11(a). Unlike univariate ARIMA modeling, the *use* of ARIMA models and methods in impact assessment requires a thoughtfully flexible strategy which may change from situation to situation.

The impact itself causes real practical problems for the identification of a noise component. In analyzing the Sutter County Workforce series, for example, the impact was so slight that noise component identification was a simple task. In analyzing the Directory Assistance time series, on the other hand, the impact was so profound that the noise model had to be identified from the 146 preintervention observations. The impact was equally profound in the analysis of the Minneapolis Public Drunkenness series, but in that analysis, only 66 preintervention observations were available. In effect, the noise component had to be identified from the residuals. The procedure in each case was determined by idiosyncracies of the time series under analysis. The decisions involved in a *general* modeling procedure are too many to be described in a simple flow chart.

The considerations involved in selecting an appropriate intervention component are even more complicated. The impact assessment model must be *interpretable,* so the "best" model is always the statistically adequate model whose substantive implications make the most sense. Whereas two competing ARIMA noise components can be compared absolutely by purely statistical criteria (their RMSs, for example), two impact assessment models cannot always be compared along purely statistical dimensions.

In cases in which the analyst can theoretically justify the simple *form* of the expected impact, an intervention component can be selected a priori. The results of the impact assessment analysis then constitute a testing of the theoretically generated null hypothesis. When theory does not point to a single expected impact, however, it is often possible to narrow the possibilities to a few alternatives. A logical system that we have found useful assumes that an impact will be *either* abrupt *or* gradual, *either* permanent *or* temporary. Given these possibilities, the analyst can select a zero- or first-order transfer function for the intervention component. Because these transfer functions are related at their extremes, alternative models can almost always be ruled out in the analysis.

When the expected pattern of impact cannot be limited to a few alternatives, the repertoire of intervention components can be expanded to include *compound* components. The compound intervention component is the sum of two low-order transfer functions and the expected impact of the compound component is simply the sum of the expected impacts of its element components. The compound intervention component is thus easily interpreted. At the extreme, the analyst may *fit* an intervention component to a time series by simply adding δ- and ω-parameters to the intervention component until the fit is complete. An impact assessment analysis based on a *fitted* model is uninterpretable, however. As demonstrated in our analysis of the U.S. Suicide Rate time series, fitted models may be useful but they are not generally interpretable.

These theoretical considerations make ARIMA impact assessment modeling a "confirmatory" analysis. In the analysis of the Sutter County Workforce time series, for example, only one pattern of impact was considered because, on theoretical grounds, only one pattern of impact was plausible. The findings of that analysis were interpretable only in the context of the theory. Similarly, the Perceptual Speed time series was analyzed from two different perspectives. Neither analysis was more or less correct than the other outside of the theoretical context. While the results of these analyses could be used to illuminate the theoretical context, the results will require a theoretical interpretation.

For Further Reading

The *design* of time series quasi-experiments is thoroughly discussed by Cook and Campbell (1979: Chapter 5) or Glass et al. (1975: Chapters 1–4). The Glass et al. work also has an extensive bibliography of published research. The *analysis* of time series quasi-experiments is developed by McCain and McCleary (1979), McDowall and McCleary (1980), Hibbs (1977), Glass et al. (1975: Chapters 5–7), or Box and Tiao (1975). Glass et al. do not develop seasonal ARIMA models or dynamic intervention components. This work is now outdated and is not generally recommended. The Box-Tiao article must be regarded as the source work for this field. Unfortunately, it may not be accessible to the mathematically unsophisticated reader. The Hibbs article has proved to be extremely influential and is highly recommended.

NOTE TO CHAPTER 3

1. A *gradual, temporary* pattern of impact can be determined by mapping any unimodal function (Normal, Poisson, and so forth) to a pulse. The function will ordinarily be determined by theory in a substantive area. As theory varies tremendously, we have not covered these methods. For an instructive example, however, the reader should see Izenman and Zabel (1980).

4 Univariate ARIMA Forecasts

In this chapter, the shortest one of the volume, we describe univariate ARIMA forecasting methods. While ARIMA methods give the "best" short-range forecasts for a wide variety of time series, there are other univariate forecasting methods which, for some data and in some situations, give "better" forecasts. Much of the forecaster's work involves preparing the "best" forecast in a particular situation. Preparing the forecast itself is not a difficult task and requires little experience. Recognizing the idiosyncracies of each situation, however, and accounting heuristically for these idiosyncracies in the forecast, requires some experience. The reader who is interested primarily in forecasting will not benefit greatly from our treatment of this area. We direct those readers to other sources, particularly to Makridakis and Wheelwright (1978) and to Pindyck and Rubinfeld (1976), where univariate forecasting methods are developed in a richer context.

Our decision to de-emphasize univariate ARIMA forecasting methods in this volume is based on two points. First, almost every book written on the topic of applied time series analysis is concerned exclusively with forecasting. We would have little original thought to add to this body of work. Second, univariate forecasts are usually reliable only in the short range (two or three periods into the future, that is), so univariate ARIMA forecasting is not itself likely to become a widely used method of social research. In the fields of business and management, short-range forecasts can be extremely useful. Managers use month-to-month forecasts to optimize control of inventories, to allocate and schedule salesmen, and so forth. Social scientists, of course, do not ordinarily have such well-defined problems. Unlike the firm, the economy or the social system is seldom representable by a few crucial indicators. More important, economies and social systems are so relatively cumbersome that it would be practically impossible for a policy maker to react to monthly changes in a social indicator.

In contrast to univariate ARIMA forecasts, *multivariate* ARIMA forecasts can be extremely useful in social research. A multivariate forecasting model will ideally account for the joint variation of several social indicators and, based on this structure, will give reliable *long*-range forecasts of a time series. We will develop multivariate ARIMA forecasting methods in the next chapter but that presentation will assume a knowledge of the univariate material developed in this chapter. Beyond this, by learning the algebra of univariate ARIMA forecasting models, the reader will gain a final crucial insight into the nature of the general ARIMA model. Finally, univariate ARIMA forecasts are useful as metadiagnostic tools in many situations. Other things equal, the relative utility and validity of two competing ARIMA models can be compared by contrasting the forecasting abilities of the two models. We will discuss this technique in the concluding section of this chapter.

4.1 Point and Interval Forecasts

All univariate forecasting methods (including ARIMA methods) are based on the same logic. First, the expected value of the time series process is calculated and, second, the expected value is extrapolated into the future. The underlying assumption of this logic is that the process is invariant and this may not always be a wise assumption. It is nevertheless an assumption which the forecaster must be willing to make.

If the current time series observation is Y_t, then we are interested in predicting the values of $Y_{t+1}, Y_{t+2}, \ldots, Y_{t+n}$. We will denote our ARIMA forecast of Y_{t+n} by $Y_t(n)$. We call $Y_t(n)$ *the origin-t forecast of Y with a lead time of n observations.*

As a first step in generating an estimate of $Y_t(n)$, we calculate the expected value of the Y_t process. Our calculations will be simplified considerably if we work in terms of the deviate process, y_t. Noting that the Y_t and y_t processes are related by

$$Y_t = y_t + \Theta_0,$$

we can translate our calculations back into the Y_t metric simply by adding a constant to our result.

Now there are actually *two* expected values of a time series process which can be used for univariate forecasts: the *un*conditional and the *conditional* process expectations. To illustrate the differences between these two expectations, consider the ARIMA(1,0,0) process

$$(1 - \phi_1 B)y_t = a_t.$$

As demonstrated in Chapter 2, this process can be expressed identically as an exponentially weighted sum of past shocks:

$$\begin{aligned} Y_t &= (1 - \phi_1 B)^{-1} a_t \\ &= (1 + \phi_1 B + \phi_1^2 B^2 + \ldots + \phi_1^n B^n + \ldots)\, a_t \\ &= a_t + \phi_1 a_{t-1} + \phi_1^2 a_{t-2} + \ldots + \phi_1^n a_{t-n} + \ldots. \end{aligned}$$

Taking the expected value of this expression,

$$Ey_t = Ea_t + \phi_1 Ea_{t-1} + \phi_1^2 Ea_{t-2} + \ldots + \phi_1^n Ea_{t-n} + \ldots.$$

Because the expected value of any random shock is zero, the expected value of the infinite series is zero and

$$\begin{aligned} Ey_t &= 0 + \phi_1 (0) + \phi_1^2 (0) + \ldots + \phi_1^n (0) + \ldots \\ &= 0 \end{aligned}$$

and thus

$$EY_t = Ey_t + \Theta_0 = \Theta_0.$$

This is the unconditional expectation of an ARIMA (1,0,0) process. Extrapolating this term into the future,

$$\begin{aligned} y_t(1) &= Ey_{t+1} = 0 \\ y_t(2) &= Ey_{t+2} = 0 \\ &\vdots \\ y_t(n) &= Ey_{t+n} = 0. \end{aligned}$$

When the unconditional expectation of the process is used as a univariate forecast, the process mean is the forecast regardless of lead time.

The problem with forecasts based on the unconditional expectation of a process is that much valuable information is ignored. While each past random shock of the process was *expected* to be zero, for example, almost all of these shocks will not be exactly zero. (A time series process in which each shock is zero will have zero variance.) In fact, we know the precise values of some of these shocks (indirectly) and this information could be useful if it were incorporated into the forecast model.

The *conditional* process expectation uses this information. The conditional expectation of y_{t+1} is:

$$E(y_{t+1} \mid y_t, y_{t-1}, \ldots, y_2, y_1).$$

The conditional expectation of y_{t+1} is conditional upon the t preceding observations of the time series process. Expressing the ARIMA (1,0,0) process again as the exponentially weighted sum of past random shocks, the conditional expectation of y_{t+1} is:

$$Ey_{t+1} = Ea_{t+1} + \phi_1 a_t + \phi_1^2 a_{t-1} + \ldots + \phi_1^n a_{t-n-1} + \ldots$$

Now, to be sure, only t random shocks of this process are known. The value of a_{t+1}, the next shock of the process, and the values of $a_0, a_{-1}, \ldots, a_{-\infty}$, which predate the start of the observed time series, are unknown. While distantly past shocks have not been observed, however, and thus are unknown, the *sum* of these shocks is known. Specifically,

$$y_t = a_t + \phi_1 a_{t-1} + \phi_1^2 a_{t-2} + \ldots + \phi_1^n a_{t-n} + \ldots$$

and this known quantity can be substituted into the expression for the conditional expectation of y_{t+1}. Thus,

$$Ey_{t+1} = Ea_{t+1} + \phi_1 y_t$$

and as the expected value of a_{t+1} is zero,

$$Ey_{t+1} = \phi_1 y_t.$$

Conditional expectation forecasts of the ARIMA (1,0,0) process are, then,

$$\begin{aligned} y_t(1) &= E(a_{t+1} + \phi_1 y_t) \\ &= \phi_1 y_t \\ &\vdots \\ y_t(n) &= E(a_{t+n} + \phi_1 a_{t+n-1} + \ldots + \phi_1^{n-1} a_{t+1} + \phi_1^n y_t) \\ &= \phi_1^n y_t. \end{aligned}$$

It should be intuitively plausible that the "best" forecast of a time series process is the *conditional* expectation of the process. What we mean by "best" in this context is that the conditional expectation forecast has the lowest possible *mean-square forecast error* (MSFE) of any expectation-based forecast.[1]

The notion of MSFE is that point estimates (as opposed to interval estimates which we will discuss shortly) of $y_t(1)$, $y_t(2)$, . . . , $y_t(n)$ will be compared with their observed values. When sufficient time has elapsed so that values of y_{t+1}, y_{t+2}, . . . , y_{t+n} are known, MSFE is computed from the formula

$$\text{MSFE} = 1/n \sqrt{\sum_{i=1}^{n} [y_{t+i} - y_t(i)]^2}.$$

The MSFE can be used to compare two forecast models of the same time series. A more important use, however, is in the estimation of *interval* forecasts.

It will ordinarily be of some use to set confidence intervals around each point estimate of $y_t(n)$. While the analyst is naturally interested in the expected value of y_{t+n}, this value is meaningless without some idea of how far away the real value of y_{t+n} is likely to be from this expected value. The point estimate forecast of the process may be generated directly from the difference equation form of the ARIMA model (though this may be computationally inefficient in some cases). To generate interval estimates, however, the model must be solved for Y_t. Solved, the ARIMA model expresses Y_t as a weighted sum of past shocks which, by convention, is expressed as

$$y_t = a_t + \psi_1 a_{t-1} + \psi_2 a_{t-2} + \ldots + \psi_k a_{t-k} + \ldots.$$

The ψ-weights in this solved ARIMA (p,d,q) $(P,D,Q)_S$ model are determined by the model structure (the values of p, d, q, P, D, Q, and S, that is) and by the values of the ϕ and Θ parameters. An ARIMA (1,0,0) $(1,0,0)_4$ model, for example,

$$(1 - \phi_1 B)(1 - \phi_4 B^4) y_t = a_t,$$

is solved as

$$\begin{aligned} y_t &= (1 - \phi_1 B)^{-1}(1 - \phi_4 B^4)^{-1} a_t \\ &= (1 + \phi_1 B + \phi_1^2 B^2 + \ldots)(1 + \phi_4 B^4 + \phi_4^2 B^8 + \ldots)\, a_t \\ &= a_t + \phi_1 a_{t-1} + \ldots + (\phi_1^4 + \phi_4)\, a_{t-4} \\ &\qquad + (\phi_1^5 + \phi_1 \phi_4)\, a_{t-5} + \ldots. \end{aligned}$$

so the ψ-weights of an ARIMA (1,0,0) $(1,0,0)_4$ model are:

$$\psi_1 = \phi_1$$
$$\psi_2 = \phi_1^2$$
$$\psi_3 = \phi_1^3$$
$$\psi_4 = \phi_1^4 + \phi_4$$
$$\psi_5 = \phi_1^5 + \phi_1\phi_4$$

and so on. Any ARIMA (p,d,q) $(P,D,Q)_S$ model can be rewritten in its ψ-weight form. To do this in the general case, the model is first solved for y_t and then the common powers of B are collected as in this example. Rewriting an ARIMA model in its ψ-weight form involves tedious arithmetic which we will avoid from this point on. Of course, in preparing forecasts, the analyst will always use a computer forecasting program to estimate the ψ-weights required for interval estimates of $y_t(n)$.

So long as the ϕ and Θ parameters of the ARIMA model lie within the bounds of stationarity-invertibility, the infinite sequence of ψ-weights converges to zero. The value of the kth weight, ψ_k, is thus approximately zero and this is a fortunate (though inevitable) consequence. As there are only t values of a time series available for computation, the ψ-weight form would not always be useful in preparing interval estimate forecasts. The sequence of ψ-weights is infinite but a_0 and all preceding shocks are unknown. Because the sequence of ψ-weights converges to zero, however, we may usually take advantage of the fact that

$$\psi_k a_{t-k} \simeq Ea_{t-k} = 0.$$

So the expected values (zero) of distantly remote shocks may be substituted into the ψ-weight form without appreciably affecting the precision of interval estimate forecasts.[2]

To derive interval forecasts, we note that y_{t+1} can be expressed as the series

$$y_{t+1} = a_{t+1} + \psi_1 a_t + \psi_2 a_{t-1} + \ldots + \psi_k a_{t-k+1} + \ldots$$

As the value of the future random shock, a_{t+1}, is unknown, we must substitute its expected value (zero) into this expression. Then assuming that all ψ-weights and past shocks are known (or, alternatively, assuming that $\psi_k a_{t-k+1}$ is approximately zero), the conditional expectation of y_{t+1} is:

$$\begin{aligned} Ey_{t+1} &= Ea_{t+1} + \psi_1 a_t + \ldots + \psi_k a_{t-k+1} + \ldots \\ &= \psi_1 a_t + \ldots + \psi_k a_{t-k+1} + \ldots . \end{aligned}$$

Using this conditional expectation as a forecast of y_{t+1}, the *error in forecasting* is:

$$e_{t+1} = y_{t+1} - Ey_{t+1}$$
$$= y_{t+1} - y_t(1) = a_{t+1}.$$

The forecast error in other words is equal to y_{t+1} minus its forecasted value. This error will always be equal to the random shock, a_{t+1}, and the forecast *variance* is thus

$$VAR(1) = Ee_{t+1}^2 = \sigma_a^2,$$

which is the variance of the white noise process and, also, a function of the expected MSFE. An interval forecast of y_{t+1} is thus

$$-1.96\sqrt{VAR(1)} < y_t(1) < +1.96\sqrt{VAR(1)}.$$

We expect y_{t+1} to lie in this interval 95% of the time.

If we now wish to forecast the next value of the process, we begin with the ψ-weight expression for y_{t+2},

$$y_{t+2} = a_{t+2} + \psi_1 a_{t+1} + \psi_2 a_t + \ldots + \psi_k a_{t-k+2} + \ldots.$$

The first two shocks are unknown. Substituting their expected value, we obtain the conditional expectation

$$Ey_{t+2} = \psi_2 a_t + \ldots + \psi_k a_{t-k+2} + \ldots.$$

Using this conditional expectation as a forecast, the error is:

$$e_{t+2} = y_{t+2} - y_t(2)$$
$$= a_{t+2} + \psi_1 a_{t+1}.$$

And variance of this forecast is

$$VAR(2) = Ee_{t+2}^2 = E[(a_{t+2} + \psi_1 a_{t+1})^2]$$
$$= E(a_{t+2}^2 + 2a_{t+2}\psi_1 a_{t+1} + \psi_1^2 a_{t+1}^2)$$
$$= Ea_{t+2}^2 + 2\psi_1 Ea_{t+2}a_{t+1} + \psi_1^2 Ea_{t+1}^2$$
$$= \sigma_a^2 + \psi_1^2\sigma_a^2 = (1 + \psi_1^2)\,\sigma_a^2.$$

Interval estimates of $y_t(2)$ are thus

$$- 1.96\sqrt{VAR(2)} < y_t(2) < +1.96\sqrt{VAR(2)}$$

with the same 95% interpretation. We note finally that VAR(2) will always be larger than VAR(1) except when $\psi_1^2 = 0$.

To forecast the next value of the process, we begin with the expression for y_{t+3},

$$y_{t+3} = a_{t+3} + \psi_1 a_{t+2} + \psi_2 a_{t+1} + \psi_3 a_t + \ldots.$$

The first three shocks of this expression are unknown, so substituting their expected value, the conditional expectation of y_{t+3} is:

$$Ey_{t+3} = \psi_3 a_t + \ldots + \psi_k a_{t-k+3} + \ldots.$$

Using this conditional expectation as our forcast, the error is:

$$\begin{aligned} e_{t+3} &= y_{t+3} - y_t(3) \\ &= a_{t+3} + \psi_1 a_{t+2} + \psi_2 a_{t+1}. \end{aligned}$$

Forecast variance is thus

$$\begin{aligned} VAR(3) = Ee_{t+3}^2 &= E[(a_{t+3} + \psi_1 a_{t+2} + \psi_2 a_{t+1})^2] \\ &= (1 + \psi_1^2 + \psi_2^2)\,\sigma_a^2. \end{aligned}$$

Continuing this procedure, we can demonstrate that the forecast variance for $y_t(n)$ is:

$$\begin{aligned} VAR(n) = Ee_{t+n}^2 &= E[(a_{t+n} + \psi_1 a_{t+n-1} + \ldots + \psi_{n-1} a_{t+1})^2] \\ &= (1 + \psi_1^2 + \ldots + \psi_{n-1}^2)\,\sigma_a^2. \end{aligned}$$

As lead time increases, forecast variance increases and the width of interval estimates of $y_t(n)$ increase. This is an intuitively satisfying result. The farther out into the future we predict, the farther out on a limb we climb.

In the next section, we will examine forecast variances of several ARIMA processes. First, however, we must note that *our discussion of VAR(n) has assumed that the true values of the* ψ-weights are known. *As these weights are determined by the ARIMA structure and by the values of* ϕ *and* Θ *parameters, this amounts to an assumption that we know the true structure of the process.* This assumption is never satisfied. In practice, ambiguity in identification and errors in estimation always leave some doubt as to the true ψ-weights. When we are dealing with only an approximation of the true ARIMA structure (albeit a close approximation), the true value of VAR(n) may be larger than the expected values we have given here.

This understimation of VAR(n) is especially a problem when forecasting relatively short time series. In general, identification and estimation of an ARIMA model becomes easier and more definite as the length of the time series increases. For a sufficiently long series, due to the approximate equivalence of the various ARIMA models, any model selected through the iterative identification/estimation/diagnosis strategy outlined in Section 2.11 will give ψ-weights quite close to their true values. When a time series is relatively short, however, we will have less confidence in the model selected and, as a consequence, we will expect our estimates of VAR(n) to be understated.

Denoting an estimate of the kth ψ-weight by ψ_k^*, the estimation error due only to a poor ARIMA model is:

$$u_k = \psi_k - \psi_k^*.$$

The true error associated with $y_t(n)$ is:

$$e_{t+n} = a_{t+n} + (\psi_1^* + u_1)a_{t+n-1} + \ldots + (\psi_{n-1}^* + u_{n-1})a_{t+1}$$

The true forecast variance is:

$$\text{VAR (n)} = [1 + (\psi_1^* + u_1)^2 + \ldots + (\psi_{n-1}^* + u_{n-1})^2]\,\sigma_a^2.$$

The estimated forecast variance will understate this true VAR(n) by the term

$$(u_1^2 + 2u_1\psi_1^* + \ldots + u_{n-1}^2 + 2u_{n-1}\psi_{n-1}^*)\sigma_a^2.$$

Underestimation of VAR(n) due to a poor model is thus a function both of the size of each u_k error and the size of each true ψ_k. This points out a salient difference between ARIMA modeling generally and ARIMA modeling for forecasts. When a model is to be used strictly for forecasting, the analyst may conclude that the "best" model is one in which the standard errors of ϕ and Θ parameters are smallest. Other things equal, such models will have the smallest u_k terms.

4.2 ARIMA Forecast Profiles

We have demonstrated that the origin-t forecast with lead time of n observations for an ARIMA (p,d,q) $(P,D,Q)_S$ process is given by the conditional expectation of y_{t+n}

$$y_t(n) = \psi_n a_t + \psi_{n+1} a_{t-1} + \ldots + \psi_{n+k} a_{t-k} + \ldots.$$

The forecast variance of $y_t(n)$ is:

$$VAR(n) = (1 + \psi_1^2 + \psi_2^2 + \ldots + \psi_{n-1}^2)\sigma_a^2.$$

The characteristic behavior of an ARIMA forecast, the "forecast profile," is thus determined solely by the ψ-weights of the process. Since the ψ-weights are determined by ARIMA structures, each class of ARIMA models has a characteristic profile.

White Noise

An ARIMA (0,0,0) process written as

$$y_t = a_t$$

has uniformly zero ψ-weights:

$$\psi_1 = \psi_2 = \ldots = \psi_k = 0.$$

Point forecasts of an ARIMA(0,0,0) process are thus

$$\begin{aligned} y_t(1) &= Ea_{t+1} = 0 \\ y_t(2) &= Ea_{t+2} = 0 \\ &\vdots \\ y_t(n) &= Ea_{t+n} = 0. \end{aligned}$$

Because all ψ-weights are zero, variance about these point estimate forecasts is constant for all lead times:

$$\begin{aligned} VAR(1) &= \sigma_a^2 \\ VAR(2) &= \sigma_a^2 \\ &\vdots \\ VAR(n) &= \sigma_a^2. \end{aligned}$$

For white noise processes, the conditional and unconditional expectations are identical. The best forecast is thus the process mean and the history of the process yields no information which can be used to improve upon this prediction.

Integrated Processes

An ARIMA (0,1,0) process, or random walk, written as

$$Y_t = Y_{t-1} + a_t$$

has unit ψ-weights

$$\psi_1 = \psi_2 = \ldots = \psi_k = 1.$$

Point forecasts of an ARIMA (0,1,0) process are thus

$$Y_t(1) = E(Y_t + a_{t+1}) = Y_t$$
$$Y_t(2) = E(Y_t + a_{t+1} + a_{t+2}) = Y_t$$
$$\vdots$$
$$Y_t(n) = E(Y_t + a_{t+1} + \ldots + a_{t+n}) = Y_t.$$

The best forecast of a random walk is the last observation. If the random walk is differenced, the forecast profile of the z_t process is a white noise profile. In terms of the Y_t process, however, forecast variance increases at a linear rate with respect to lead time:

$$\text{VAR}(1) = Ea_t^2 = \sigma_a^2$$
$$\text{VAR}(2) = E[(a_{t+2} + a_{t+1})^2] = 2\sigma_a^2$$
$$\vdots$$
$$\text{VAR}(n) = E[(a_{t+n} + a_{t+n-1} + \ldots + a_{t+1})^2] = n\sigma_a^2.$$

For each observation increase in lead time, VAR(n) increases by one unit of white noise variance.

In Figure 4.2(a), we show forecasts of "Series B," a time series of IBM stock prices introduced in Section 2.1. This series follows a random walk with a forecast profile typical of all integrated processes. After two or three steps into the future, the confidence intervals (set at 95% in this figure) become so large as to render the interval forecast meaningless. A nonstationary process in fact is defined as one with no finite variance. The limit of VAR(n) for an integrated process as lead time increases to infinity is infinity.

Autoregression

An ARIMA (1,0,0) process written as

$$y_t = \phi_1 y_{t-1} + a_t$$

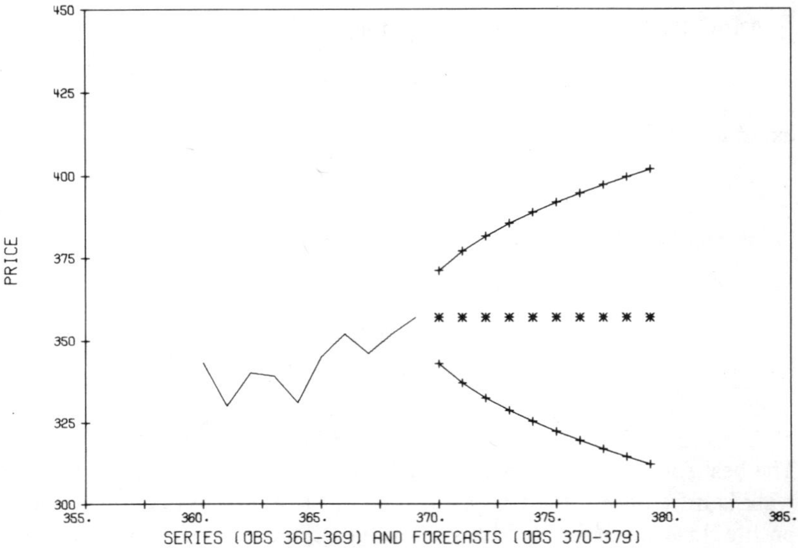

FIGURE 4.2(a) *Forecast Profile:* Series B

has exponentially decaying ψ-weights:

$$\psi_k = \phi_1^k.$$

Point forecasts of the ARIMA (1,0,0) process are thus

$$\begin{aligned}
y_t(1) &= E(a_{t+1} + \phi_1 a_t + \phi_1^2 a_{t-1} + \ldots + \phi_1^k a_{t-k+1} + \ldots) \\
&= \phi_1 y_t \\
y_t(2) &= E(a_{t+2} + \phi_1 a_{t+1} + \phi_1^2 a_t + \ldots + \phi_1^k a_{t-k+2} + \ldots) \\
&= \phi_1^2 y_t \\
&\vdots \\
y_t(n) &= E(a_{t+n} + \phi_1 a_{t+n-1} + \ldots + \phi_1^n a_t + \ldots) \\
&= \phi_1^n y_t.
\end{aligned}$$

Forecast variance about these point estimates is a function of the exponentially decaying ψ-weights:

$$\text{VAR}(1) = \sigma_a^2$$
$$\text{VAR}(2) = (1 + \phi_1^2)\,\sigma_a^2$$
$$\vdots$$
$$\text{VAR}(n) = (1 + \phi_1^2 + \ldots + \phi_1^{2n-2})\,\sigma_a^2.$$

Confidence intervals about successive forecasts increase at a rate determined by the value of ϕ_1. When ϕ_1 is small, the increase in VAR(n) for an increase in lead time is small. When ϕ_1 is large, the increase in VAR(n) for an increase in lead time is large. In any event, it is clear that successive lead time increments produce smaller and smaller increments in forecast variance. Noting that the expression for VAR(n) is a geometric progression, forecast variance approaches a limit of

$$\lim_{n\to\infty} \text{VAR}(n) = \frac{\sigma_a^2}{1 - \phi_1^2}$$

which is the variance of the y_t autoregressive process. In fact, the limit of VAR(n) as n approaches infinity will always be the variance of the ARIMA (p,d,q) $(P,D,Q)_S$ process. For the ARIMA (0,0,0) white noise process, the

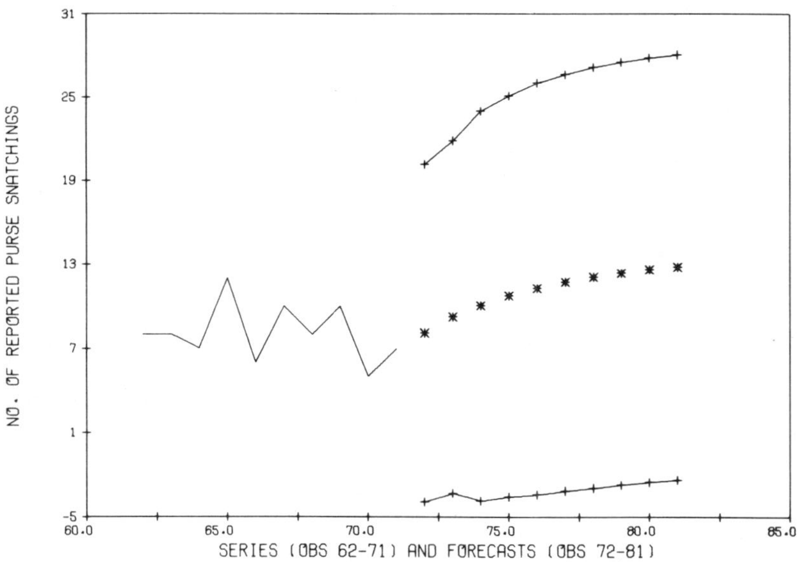

FIGURE 4.2(b) *Forecast Profile:* Hyde Park Purse Snatchings

limit of VAR(n) is σ_a^2 and for the ARIMA (0,1,0) process, VAR(n) increases without bound.

In Figure 4.2(b), we show forecasts of the Hyde Park Purse Snatchings time series. The noise component of this model is ARIMA (2,0,0) with small values of ϕ_1 and ϕ_2. The forecast profile is typical of autoregressive profiles. As lead time increases, forecasts regress to the process mean; confidence intervals about each point forecast increase with increases in lead time.

Moving Averages

An ARIMA (0,0,1) process written as

$$y_t = a_t - \Theta_1 a_{t-1}$$

has only one nonzero ψ-weight:

$$\psi_1 = -\Theta_1$$
$$\psi_2 = \psi_3 = \ldots = \psi_k = 0\,.$$

Point forecasts for the ARIMA (0,0,1) process are thus

$$y_t(1) = E(a_{t+1} - \Theta_1 a_t) = -\Theta_1 a_t$$
$$y_t(2) = E(a_{t+2} - \Theta_1 a_{t+1}) = 0$$
$$\vdots$$
$$y_t(n) = E(a_{t+n} - \Theta_1 a_{t+n-1}) = 0\,.$$

Forecast variance is determined by the single nonzero ψ-weight:

$$\text{VAR}\,(1) = Ea_{t+1}^2 = \sigma_a^2$$
$$\text{VAR}\,(2) = E\,[(a_{t+2} - \Theta_1 a_{t+1})^2)] = (1 + \Theta_1^2)\,\sigma_a^2$$
$$\vdots$$
$$\text{VAR}\,(n) = E\,[(a_{t+n} - \Theta_1 a_{t+n-1})^2] = (1 + \Theta_1^2)\,\sigma_a^2.$$

After the second step into the future, forecast variance is constant. The limit of VAR(n) is thus

$$\lim_{n\to\infty} \text{VAR}\,(n) = (1 + \Theta_1^2)\,\sigma_a^2,$$

which is the variance of the ARIMA (0,0,1) process.

In Figure 4.2(c), we show forecasts of the Swedish Harvest Index time series which we analyzed in Section 2.12.3. Here we see the distinctive

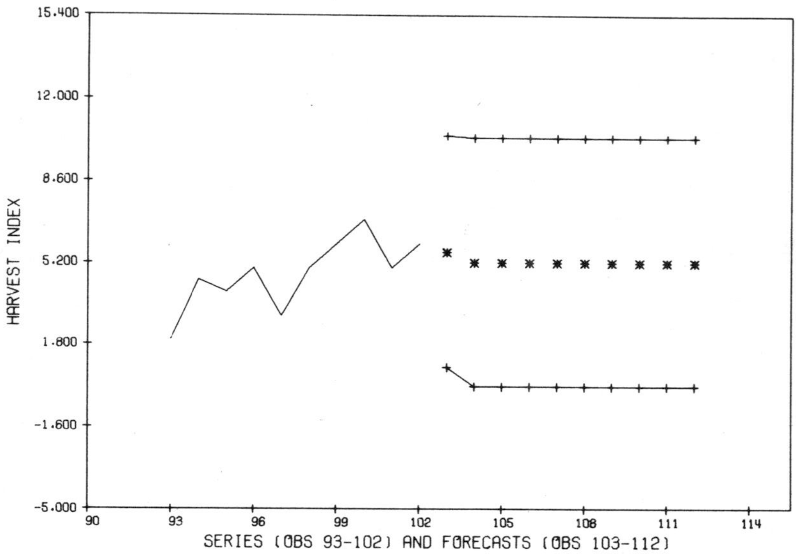

FIGURE 4.2(c) *Forecast Profile:* Swedish Harvest Index

forecast profile of a moving average process. With lead times greater than one observation, $y_t(n)$ is the process mean; VAR(n) remains constant.

As the reader may suspect by now, univariate forecasts of simple ARIMA models (as well as all univariate forecasts) tend to be statistically trivial. The best point forecast is often the process mean; and for substantial lead times, interval forecasts often approach infinity. Forecasts of complicated ARIMA seasonal models are somewhat more useful. The forecasts track the seasonal pattern of the process quite well for at least one seasonal period and thus may be used to assess "turning points" in the series. In Figure 4.2(d), we show forecasts of the Sutter County workforce series which we modeled in Section 2.12.1 as an ARIMA (0,1,1) $(0,1,1)_{12}$ process. Both moving average parameters are relatively large. The forecasts appear to track the pattern of seasonal variation quite well, although because the model is nonstationary, confidence intervals about the point forecasts grow large rapidly

4.3 Conclusion: The Uses of Forecasting

The reader who now understands how interval estimate forecasts are generated has no doubt gained a deeper insight into the nature of ARIMA

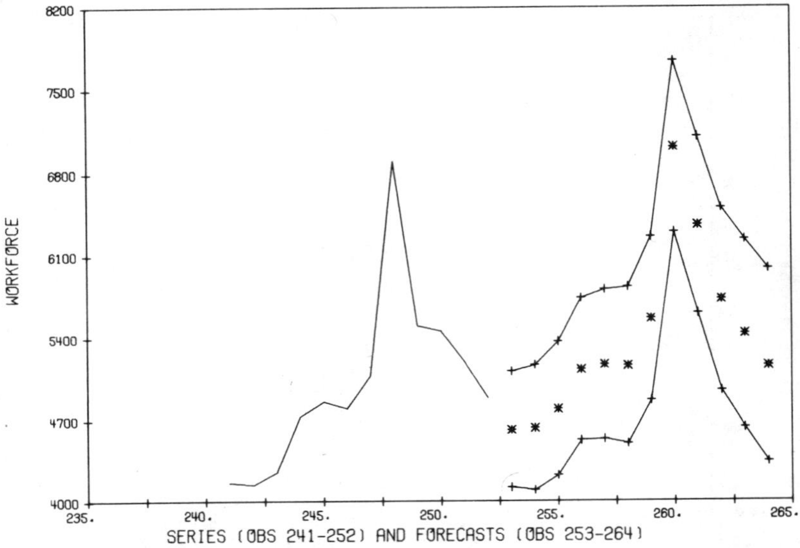

FIGURE 4.2(d) *Forecast Profile:* Sutter County Workforce

models. Beyond this not insubstantial value, however, the reader who intends to apply the principles of time series analysis to social science problems will likely see no use for univariate forecasts. In fact, however, univariate forecasts can be extremely useful as a tool of model diagnosis. We will conclude this chapter with a description of this use and with a comment on another use of forecasting which, in our opinion, is improper.

Forecasting as a Diagnostic Tool

It will often happen that two roughly identical ARIMA models produce radically different forecasts of the same time series. When this is true, the analyst will be justified in selecting the model with the "better" forecasting ability.

An example of this is seen in our analysis of the U.S. Suicide Rate time series (Section 3.5). The analysis first lead to an ARIMA (1,0,0) model for the N_t component:

$$y_t = f(I_t) + \frac{a_t}{1 - .87B}.$$

Due to the relatively large estimated value of ϕ_1, however, we respecified the N_t component as ARIMA (0,1,0):

$$Y_t = f(I_t) + \frac{a_t}{1 - B}.$$

In practice, it is almost always better to work with an ARIMA (0,1,0) model rather than an ARIMA (1,0,0) model with a large autoregressive parameter. (We will discuss the problems of estimating these parameters in Chapter 6.) There is nevertheless one situation in which the ARIMA (1,0,0) model would be preferred to the ARIMA (0,1,0) model: when the *true* noise process is autoregressive. The analyst can never know the true stochastic process underlying a time series. Yet there are many sensible procedures which the analyst can use to rule out competing models and one of these is forecasting.

While there are few major differences between ARIMA (1,0,0) and ARIMA (0,1,0) models in many contexts, this is not at all true in the context of forecasting. Using 1969 as the origin, forecasts of the two models are:

	ARIMA (1,0,0)	ARIMA (0,1,0)	ACTUAL RATE[3]
1970	11.08	11.10	11.6
1971	11.06	11.10	not available
1972	11.05	11.10	not available
1973	11.03	11.10	12.0
1974	11.02	11.10	12.1
1975	11.01	11.10	12.7
1976	11.00	11.10	12.5

The forecasts of these models are remarkably close to each other but this is expected: The two models are nearly identical. Yet forecasts of the ARIMA (1,0,0) model regress gradually to the estimated process mean (approximately 10.9 suicides per 100,000 total population) while forecasts of the ARIMA (0,1,0) model remain constant for all lead times. The actual rate does not regress to the population mean, as would be expected of an autoregressive process, but rather continues to move upward. A random walk in fact is characterized by wide swings away from the process mean. Overall, the ARIMA (0,1,0) model has a lower MSFE statistic than the ARIMA (1,0,0) model; and on this basis alone, we would select ARIMA (0,1,0) model as the "best" one for this time series.

Nevertheless, the differences between the forecasts of these two models is small and other analysts might choose the ARIMA (1,0,0) model in spite of its lesser forecasting power. In the end, model selection will hinge on a great many statistical and substantive concerns. Forecasting power is only one of these concerns and it need not be more important than any other.

Forecasting as an Impact Assessment Tool

If our experiences are typical, students of time series analysis are more fascinated with univariate forecasting than with any other application of ARIMA modeling. There seems to be a fundamental (almost spiritual) human interest in "predicting the future" which is aggravated by a course in time series analysis. The fact remains that univariate forecasts are essentially trivial and often disappointing. One can predict the future only for short lead times and in limited contexts.

Students of time series analysis often "discover" a means of assessing impacts with univariate forecasts. These methods are actually quite old and widely used in industrial engineering and quality control applications. While these methods are valid and useful in manufacturing contexts, however, they are not generally suited to social science problems and data.

As an illustration of the use of these methods in quality control engineering, consider a machine that manufactures ball bearings. Each ball bearing will vary slightly in diameter and this variance from ball bearing to ball bearing is a stochastic (time series) process. An ARIMA model of the process is used to set forecast confidence intervals around the process realization. When the manufacturing process is "in control," 95% of the ball bearings produced by the machine will lie within these confidence intervals. If a run of three or four ball bearings are observed to lie outside the confidence intervals, the quality control engineer infers that the process is "out of control." The machine is turned off and repairs are made.

The strong forecast-based inference here is possible because of certain given characteristics of the manufacturing process. One, the quality control engineer knows a priori what a "bad" ball bearing is. A "bad" ball bearing is one that is a few thousandths of an inch too small or too large; one that cannot be sold to customers. Two, process realizations are relatively long. The quality control engineer may have one thousand or more observations available for building an ARIMA model of the process. Three, process inputs are tightly controlled and known. White noise inputs to the manufacturing process arise from relatively small variations in the quality of raw materials (water, oil, electricity, steel, and so on). The quality control engineer knows not only the sources of white noise inputs but also their relative magnitudes.

These characteristics of the manufacturing process are not ordinarily seen in social processes. There is no definition of a "high" or "bad" unemployment rate, for example; social science time series realizations are ordinarily short; process inputs are unknown and erratic and there are quite often seasonal inputs which are unheard of in manufacturing processes. Forecasts of social processes are thus less certain and more prone to error than forecasts of industrial processes.

In Chapter 3, we developed impact assessment models from a foundation of scientific validity. Certain "threats to validity" are controlled through design while others are controlled through analysis. This distinction is not always clear and this is particularly true when alternative patterns of impact are compared and ruled out. When an intervention component has been misspecified, Type I or Type II decision errors are a likely result. To control this threat to validity, we have recommended a conservative strategy of model building. The strategy leads generally to a more confident statement of impact but it does so at a real cost: A relatively long postintervention time series segment is required.

Impact assessments based on forecasting do not require this great cost. Deutsch (1978) has recently proposed a variation of the time series quasi-experiment which allows for an impact assessment within a few weeks or months of the intervention. At the simplest level, Deutsch proposes to build an ARIMA forecasting model from the preintervention time series. Postintervention forecasts of the model are then compared with the actual observations. If the postintervention observations fall outside the forecast confidence intervals, Deutsch concludes that the social system has gone out of control or that the intervention has had an impact on the time series.

If "early detection methods" should prove generally reliable, Deutsch's work will represent an important advance in social science methodology. In the first published use of these methods, however, Deutsch and Alt (1977) found a statistically significant drop in gun-related crime after introduction of a strict gun-control law. In our reanalysis of those data (Sections 2.12.2 and 3.6; see also, Hay and McCleary, 1979), we found no evidence of the effect claimed by Deutsch and Alt. We attribute this difference in findings to, among other things, weaknesses of the early detection methods used by Deutsch and Alt.

A major deficiency of forecast-based impact assessments is that confidence intervals about each point estimate are subject to error. As noted in Section 4.1, the ψ-weight models used to set confidence limits about each point estimate require that the analyst know the *true* ARIMA structure of the time series process. Yet in practice, this is never the case.

Compared to industrial process time series, social science time series are relatively short. A weak seasonal component, for example, may go undetected or may have statistically insignificant parameter estimates for a time series of only 100 observations. When a few more observations are added to the time series, however, the seasonal component may suddenly assume statistical significance. Nonstationary processes present an analogous problem. With a weak trend, estimates of Θ_0 are likely to be statistically insignificant unless the time series is relatively long.

When a run of postintervention observations fall outside the forecast confidence intervals, there are always two equally plausible explanations: the process may have been impacted by a social intervention and/or the forecast confidence intervals may have been underestimated. With a relatively short time series, the latter explanation is always more plausible than the former.

But a greater problem with forecast-based impact assessments is that threats to validity cannot be controlled. A confident statement of impact requires not only a statement as to whether an impact occurred or not (which forecast-based assessment may or may not adequately give) but also a statement as to the nature of the impact. In the first place, abrupt temporary impacts will almost always "fool" a forecast-based impact assessment.[4] As an exercise, the reader may wish to try a forecast-based assessment for the Sutter County Workforce time series. In Section 3.2.2, we demonstrated that the Sutter County flood had only a substantively and statistically insignificant impact on this time series. Yet if the preintervention series is used to forecast the postintervention observations, the analyst will arrive at a radically different conclusion. The "reactive intervention" threat to internal validity will also "fool" a forecast-based impact assessment. For most ARIMA models, the last observation of the series has the greatest weight in determining forecasts, and if the last preintervention observation is an extreme value, forecast-based impact assessments will indicate a statistically and substantively significant effect.

To guard against these threats, impact assessment requires a relatively long postintervention time series segment. There is a fundamental difference between *detecting* and *modeling* an impact. Even in the more "applied" social sciences (evaluation research and policy analysis, for example), impact assessment must be concerned with the dynamic structure of social change. This concern can be addressed only from a foundation of scientific validity and from the conservative impact model-building strategy we outlined in Chapter 3.

For Further Reading

Nelson (1973: Chapter 6–8) develops univariate ARIMA forecasting at an introductory level. Granger and Newbold (1977: Chapters 4–5) develop the same material at a slightly higher level. More comprehensive treatments of forecasting which consider non-ARIMA methods as well are given by Pindyck and Rubinfeld (1976) and Makridakis and Wheelwright (1978). Makridakis-Wheelwright is written for graduate students in business while Pindyck-Rubinfeld is written for graduate students in economics. While both works are outstanding, there is a clear difference in the levels of sophistication assumed of the reader. Granger and Newbold (1977: Chapter 8) compare the performance of a variety of forecasting methods. This work is absolutely essential for any reader who plans to do forecasting. Finally, Vigderhous (1978) or Land and Felson (1976) are excellent examples of forecasting in a social science context.

NOTES TO CHAPTER 4

1. The proof of this claim is obvious when one considers that the *conditional* expectation uses *all* of the available information about the process. See Pindyck and Rubinfeld (1976: 498–499) for a formal proof.

2. The assumption is that ψ_k is zero (or some infinitesimally small number) is satisfied whenever the time series is long, say 50 observations or more, or whenever the low-order ψ-weights are so small that the sequence converges to zero within a few weights. When this assumption is *not* satisfied, the minimum MSFE forecasts are generated by backcasting the series to obtain estimates of $y_0, y_{-1}, \ldots, y_{-\infty}$. For relatively long time series, of course, the conditional expectation of y_{t+n} is the same whether the expected values or the backcasted values of distant random shocks are used. See Box and Jenkins (1976: 199–200) for a detailed description of backcasting.

3. The values of this time series are taken from the U.S. Department of Commerce publication *Historical Statistics of the United States: Colonial Times to 1970*. The values after 1970 are taken from the *1978 Statistical Abstract of the United States*. The rates for 1971 and 1972 are not given in that volume.

4. Hay and McCleary (1979: 309–310) show that one of the time series analyzed by Deutsch and Alt has an abrupt, temporary impact effect. Using the early detection method, however, Deutsch and Alt conclude that the effect is a permanent reduction in gun-related crime.

5 Multivariate ARIMA Models

For many readers, this may be the most interesting chapter of the volume. Whereas we were concerned only with *uni*variate time series analysis in preceding chapters, in this chapter, we will generalize the Box-Jenkins philosophy to *multi*variate time series analysis, that is, to the modeling of relationships between two or more time series.[1] There are many ways to view multivariate time series analysis. In *Design and Analysis of Time Series Experiments,* for example, Glass et al. (1975; Chapter 8) develop multivariate ARIMA modeling under the rubric of "concomitant variation." From this perspective, independent variable time series are introduced only for the purpose of reducing background noise (or unexplained variance) in the dependent-variable time series.

Noise reduction is not an unimportant consideration. Many of the problems addressed (or sidestepped) in preceding chapters can be seen as problems of background noise. Trend, for example, is a bothersome topic which raises philosophical dilemmas of the most complex nature. In time series analysis, trend must be equated with change. If a social process changes systematically throughout a finite realization, however, can it be assumed that the process will continue to change in the same systematic manner? The exogenous forces which underlie trend may be relatively constant during a finite period of time, so during that period, the constant term of an ARIMA model may adequately represent these forces. As the time frame grows larger, however, these forces may change subtly and gradually; their representation by a constant term may weaken.

Seasonality presents a similar dilemma. A seasonal ARIMA model mimics the effects of excluded (and often unknown) periodic exogenous forces. While the model performs remarkably well in this role, the analyst must always remember that the *essence* of seasonality has not really been captured in the model. In relatively short time frames, seasonality may be adequately explained as structured, periodic noise. Over longer time frames, however, exogenous seasonal forces may change gradually and the imitative power of the model may wane.

Finally, viewing outliers as background noise, the same dilemma arises. The input to a univariate ARIMA model is white noise and, in theory, a white noise process can generate an infinitely large random shock. A one-in-a-million random shock nevertheless complicates the analysis unless the time series being modeled is a million observations long.

These three specific problems of background noise can be mitigated by incorporating an independent-variable time series into the ARIMA model. If the same set of exogenous forces (which are responsible for trend, seasonality, and outliers) underlie two time series, then a bivariate model of the relationship may incorporate these forces indirectly. As multivariate ARIMA models "solve" all of these dilemmas, we will state unequivocally that a "fair" multivariate model is always preferred to a "good" univariate model. The analyst must remember nonetheless that a "good" or even "excellent" multivariate model gives only an approximate representation of the excluded exogenous forces.

From a concomitant variation perspective (as outlined by Glass et al.), trend, seasonality, and outliers are seen as background noise which can be reduced by incorporating an independent-variable time series into the model. The concomitant variation perspective misses the most important facet of multivariate ARIMA models, however. Multivariate ARIMA models are inherently causal. Although we acknowledge the importance of noise reduction, we will develop multivariate ARIMA modeling from a *causal* modeling perspective in this chapter.

The jump from univariate to multivariate time series analysis will not be difficult. The impact assessment models developed in Chapter 3, in fact, are multivariate models with the step function I_t as an independent variable. For a set of n independent variables, $X_{1t}, \ldots, X_{nt}$, the general multivariate ARIMA model may be written as

$$Y_t = f(X_{1t}, \ldots, X_{nt}) + N_t.$$

The functions of the several independent variables in this model are transfer functions such as those described in Chapter 3. The difference here is that these transfer functions will be identified *empirically*.

Our discussion begins with the cross-correlation function which may be used to identify a transfer function relationship between two time series. We will then illustrate the multivariate model-building strategy with one forecasting and one causal modeling example. Needless to say, our development of this material leans heavily on the principles developed in Chapters 2 and 3. The reader who is unsure of this material is thus advised to review those chapters before proceeding.

5.1 The Cross-Correlation Function

It is sometimes useful to think of autocorrelation as *within-series* correlation. In the same way that the ACF is used to identify within-series correlation, the cross-correlation function (CCF) is used to identify *between-series* correlation. Patterns of between-series correlation are used to identify a transfer function relationship *between* two time series in much the same way that the ACF is used to identify an ARIMA relationship *within* the time series.

As a first principle, we note that two nonstationary time series will always be correlated due to common patterns of drift or trend. *This correlation must always be regarded as spurious*. To eliminate between-series correlations due only to drift or trend, the time series are made stationary prior to estimation of the CCF. After an appropriate differencing,

$$x_t = (1 - B)^d(1 - B^S)^D X_t$$
$$z_t = (1 - B)^d(1 - B^S)^D Y_t,$$

the CCF may be estimated. By convention, x_t is referred to as the *input* series, or causor, and z_t is referred to as the *output* series, or effector. This terminology reflects the input-output relationship

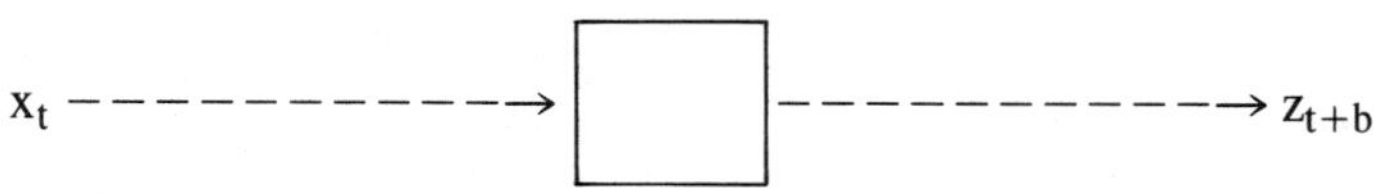

which is explicitly causal. Given two stationary time series, the CCF for lags $\pm k$ is given by the formulae

$$\mathrm{CCF}(+k) = \frac{\sum_{t=1}^{N-k} (x_t - \bar{x})(z_{t+k} - \bar{z})}{\sqrt{\sum_{t=1}^{N} (x_t - \bar{x})^2 \sum_{t=1}^{N} (z_{t+k} - \bar{z})^2}}$$

$$\mathrm{CCF}(-k) = \frac{\sum_{t=1}^{N+k} (x_{t-k} - \bar{x})(z_t - \bar{z})}{\sqrt{\sum_{t=1}^{N} (x_{t-k} - \bar{x})^2 \sum_{t=1}^{N} (z_t - \bar{z})^2}}$$

These formulae give the familiar Pearson product-moment correlation coefficient (approximately) between two time series separated by $\pm k$ observations. When $k = 0$, the formulae are identical. When $k \neq 0$, the first formula gives the *positive* half of the CCF by lagging the z_t series forward in time. The second formula gives the *negative* half of the CCF by lagging the x_t series forward in time.

A major difference between the CCF and the ACF (as noted in Section 2.8) is that the CCF need not be symmetrical about lag-zero. In other words, CCF $(+k) \neq$ CCF$(-k)$ generally. When the ACF is used to identify an ARIMA model, only one half of the ACF need be examined. ACF$(-k)$ is a mirror image of ACF$(+k)$ but this is not true of the CCF.

We are always reluctant to introduce tedious arithmetic into our argument. The relationship between CCF$(+k)$ and CCF$(-k)$ is one that cannot ordinarily be grasped without a basic demonstration, however. Apologies given, we present 10 pairs of numbers:

x_t	t	z_t
.665	1	−.160
−1.630	2	−.058
− .298	3	.333
.225	4	−.815
1.222	5	−.149
.531	6	.113
− .957	7	.611
.676	8	.266
− .723	9	−.479
.289	10	.338

These numbers were generated so that $\bar{x} = \bar{z} = 0$; the first nine values of x_t are random Normal numbers and the 10th was selected to ensure that $\bar{x} = 0$. The values of z_t were generated as

$$z_t = \frac{x_{t-2}}{2}.$$

For 8 of these 10 pairs of numbers, then, there is a perfect causal relationship:

$$x_t \dashrightarrow z_{t+2}$$

Applying the formulae for CCF(±k) to these 10 pairs of numbers, starting with k = −3,

$$\text{CCF}(-3) = \frac{(-.160)(.225) + \ldots + (.611)(.289)}{\sqrt{[(-.160)^2 + \ldots + (.338)^2][(.665)^2 + \ldots + (.289)^2]}} = .25$$

$$\text{CCF}(-2) = \frac{(-.160)(-.298) + \ldots + (.266)(.289)}{\sqrt{[(-.160)^2 + \ldots + (.338)^2][(.665)^2 + \ldots + (.289)^2]}} = -.04$$

$$\text{CCF}(-1) = \frac{(-.160)(-1.630) + \ldots + (-.479)(.289)}{\sqrt{[(-.160)^2 + \ldots + (.338)^2][(.665)^2 + \ldots + (.289)^2]}} = -.22$$

$$\text{CCF}(0) = \frac{(-.160)(.665) + \ldots + (.338)(.289)}{\sqrt{[(-.160)^2 + \ldots + (.338)^2][(.665)^2 + \ldots + (.289)^2]}} = -.11$$

$$\text{CCF}(+1) = \frac{(-.058)(.665) + \ldots + (.338)(-.723)}{\sqrt{[(-.160)^2 + \ldots + (.338)^2][(.665)^2 + \ldots + (.289)^2]}} = -.22$$

$$\text{CCF}(+2) = \frac{(.333)(.665) + \ldots + (.338)(.676)}{\sqrt{[(-.160)^2 + \ldots + (.338)^2][(.665)^2 + \ldots + (.289)^2]}} = .95$$

$$\text{CCF}(+3) = \frac{(-.815)(.665) + \ldots + (.338)(-.957)}{\sqrt{[(-.160)^2 + \ldots + (.338)^2][(.665)^2 + \ldots + (.289)^2]}} = -.13.$$

CCF(+2) is the largest of these seven numbers. The value of CCF(+2) ≈ 1 indicates the perfect causal relationship built into these numbers. There are many other "large" correlations among the seven, however. The statistical significance of any CCF(±k) estimate can be assessed by comparing it with its standard error. One unit of standard error for CCF(±k) is given by

$$\text{SE}\,[\text{CCF}\,(\pm k)] = \frac{1}{\sqrt{N-k}}.$$

For these seven estimates, then, the standard errors are:

SE [CCF (±3)]	$= \sqrt{1/7}$	= .378
SE [CCF (±2)]	$= \sqrt{1/8}$	= .354
SE [CCF (±1)]	$= \sqrt{1/9}$	= .333
SE [CCF (0)]	$= \sqrt{1/10}$	= .316.

Dividing each of these estimated CCF(±k) by its standard error, the standardized estimates are:

CCF (−3)	= .25/.378	= .66 SE
CCF (−2)	= −.04/.354	= −.11 SE
CCF (−1)	= −.22/.333	= −.66 SE
CCF (0)	= −.11/.316	= −.35 SE
CCF (+1)	= −.22/.333	= .66 SE
CCF (+2)	= .95/.354	= 2.68 SE
CCF (+3)	= −.13/.378	= −.34 SE.

As a convention, the analyst may assume that any estimate of CCF(±k) smaller in absolute value than 2 SE is zero. By this rule, only the estimate of CCF(+2) is statistically different than zero.

Among other things, this exercise illustrates the interpretation of asymmetry in the CCF. *The CCF measures not only the strength of a relationship but also the direction*. When "x_t causes z_{t+b}," evidence of the relationship is found at CCF(+b), in the positive half of the CCF, that is. When "z_t causes x_{t+b}," on the other hand, evidence of the relationship is found at CCF(−b), in the negative half of the CCF. Asymmetry in the estimated ACF is thus interpreted on the basis the causal relationship specified a priori. When one

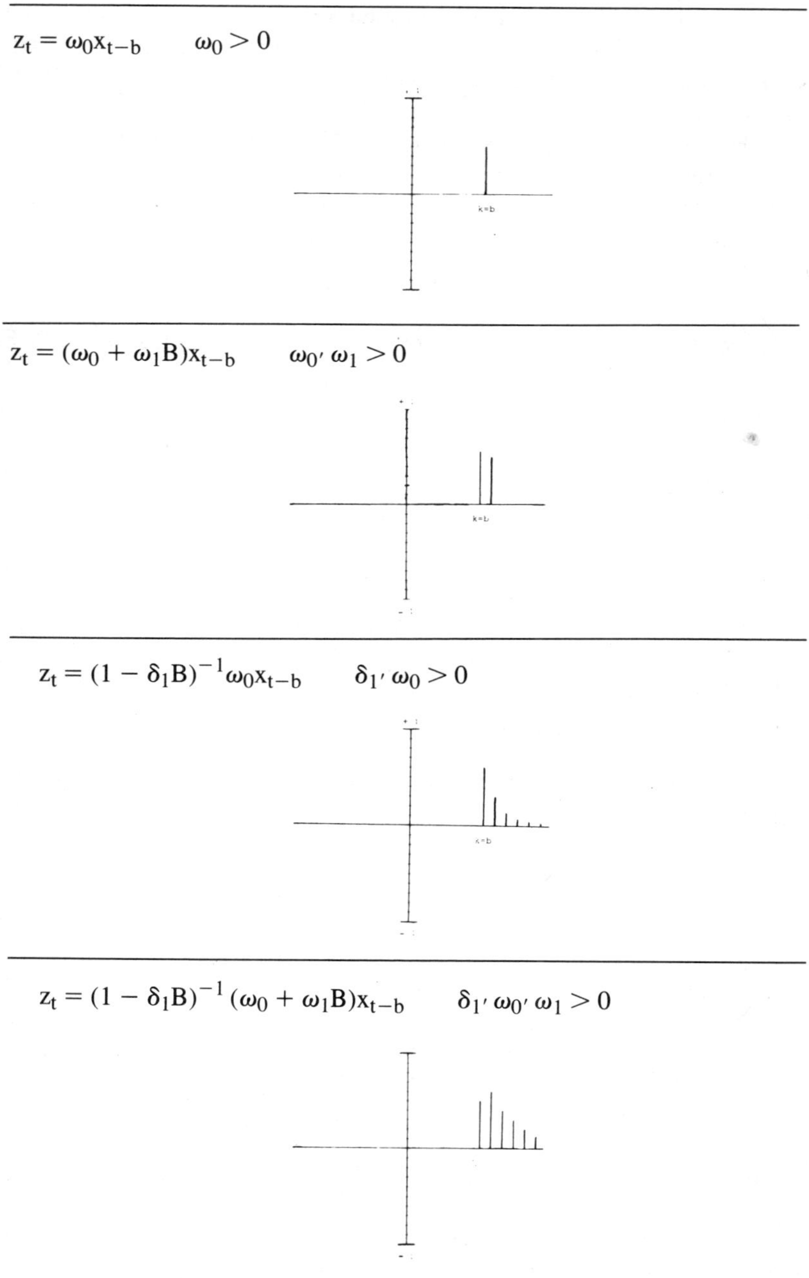

FIGURE 5.1 Expected CCFs for Several Bivariate Relationships

of the two time series has been specified as the causor, evidence of the relationship is expected in the positive CCF. But more important, the CCF(+b) estimated under the assumption that x_t is the causor will be identical to the CCF(−b) estimated under the assumption that x_t is the effector.

What has been demonstrated in the specific case for 10 pairs of numbers must now be demonstrated in the general case. Figure 5.1 shows the *expected* CCFs for several common transfer function relationships. First, the zero-order relationship

$$z_t = \omega_0 x_{t-b} + N_t$$

is expected to have a nonzero value of CCF (+b). All other lags of the CCF are expected to be zero. To demonstrate this, we define the expected CCF (+k) as

$$\text{CCF}\,(+k) = \frac{\text{COV}\,(x_{t-k}z_t)}{\sqrt{\text{VAR}\,(x_t)\,\text{VAR}\,(z_t)}}.$$

Then,

$$\begin{aligned}\text{COV}\,(x_{t-k}z_t) &= \text{E}\,[(x_{t-k})\,(\omega_0 x_{t-b} + N_t)]\\ &= \text{E}\,(\omega_0 x_{t-k}x_{t-b} + x_{t-k}N_t).\end{aligned}$$

Now in all cases, $\text{E}x_{t-k}N_t = 0$, so

$$\text{COV}(x_{t-k}z_t) = \omega_0 \text{E}x_{t-k}x_{t-b}.$$

Then assuming that x_t is a white noise process (we will cover the case in which x_t is not white noise in Section 5.3),

$$\begin{aligned}\text{COV}\,(x_{t-k}z_t) &= \omega_0\sigma_x^2 \text{ whenever } b = k\\ &= 0 \text{ otherwise.}\end{aligned}$$

Dividing this term by $\sigma_x\sigma_z$, the expected CCF is

$$\begin{aligned}\text{CCF}\,(k) &= \omega_0\frac{\sigma_x}{\sigma_z} \text{ whenever } b = k\\ &= 0 \text{ otherwise.}\end{aligned}$$

Following this same procedure, the reader may demonstrate that the converse relationship

$$x_t = \omega_0 z_{t-b} + N_t$$

is expected to have a nonzero value of CCF $(-b)$. All other lags of the CCF are expected to be zero.

The first-order transfer function relationship

$$z_t = \frac{\omega_0}{1 - \delta_1 B} x_{t-b} + N_t$$

describes a *dynamic* causal relationship between the two time series. Rewriting the relationship as the infinite series,

$$z_t = \omega_0 \sum_{i=0}^{\infty} \delta_1^i x_{t-b-i} + N_t.$$

The covariance between x_{t-k} and z_t is:

$$\begin{aligned} \text{COV}\,(x_{t-k}z_t) &= E\,[(x_{t-k})\,(\omega_0 \sum_{i=0}^{\infty} \delta_1^i x_{t-b-i} + N_t)] \\ &= E\,(\omega_0 \sum_{i=0}^{\infty} \delta_1^i x_{t-k}x_{t-b-i}) \\ &= \omega_0 E x_{t-k}x_{t-b} + \omega_0\delta_1 E x_{t-k}x_{t-b-1} \\ &\quad + \omega_0\,\delta_1^2 E x_{t-k}x_{t-b-2} + \ldots + \omega_0\delta_1^n E x_{t-k}x_{t-b-n} + \ldots. \end{aligned}$$

When $b < k$, all terms of this expression are zero. When $b = k$, however, the first term of the infinite series is nonzero, so

$$\text{COV}(x_{t-b}z_t) = \omega_0\sigma_x^2.$$

When $k = b + 1$, the second term of the infinite series is nonzero:

$$\text{COV}(x_{t-b-1}z_t) = \omega_0\delta_1\sigma_x^2.$$

And when $k = b + n$, the $n + 1$st term of the infinite series is nonzero:

$$\text{COV}(x_{t-b-n}z_t) = \omega_0\delta_1^n\sigma_x^2.$$

Dividing these covariances by $\sigma_x\sigma_z$, the expected CCF is:

$$\text{CCF}\,(k) = 0 \text{ for } k < b$$

$$\text{CCF (b)} = \omega_0 \frac{\sigma_x}{\sigma_z}$$

$$\text{CCF (b+n)} = \omega_0 \delta_1^n \frac{\sigma_x}{\sigma_z}.$$

So the CCF for a dynamic first-order transfer function relationship is expected to be zero until CCF (b). Successive positive lags, CCF (b+1), CCF (b+2), . . . , CCF (b+n), decay exponentially back to zero. The expected CCF is thus identical with the ACF expected of an ARIMA (1,0,0) process.

The CCFs shown in Figure 5.1 all suggest a causal relationship between x_t and z_{t+b}.[2] A single spike in the positive CCF is interpreted as an ω parameter. Decay from a spike is interpreted as a δ parameter. These are *expected* CCFs, of course, and, in practice, identification of a transfer function relationship may be complicated by ambiguous identification statistics. Nevertheless, this first step in multivariate model building must produce some evidence of relationship before the next step in the procedure can begin. We will now demonstrate the multivariate model-building procedure with an example.

5.2 A Forecasting Example

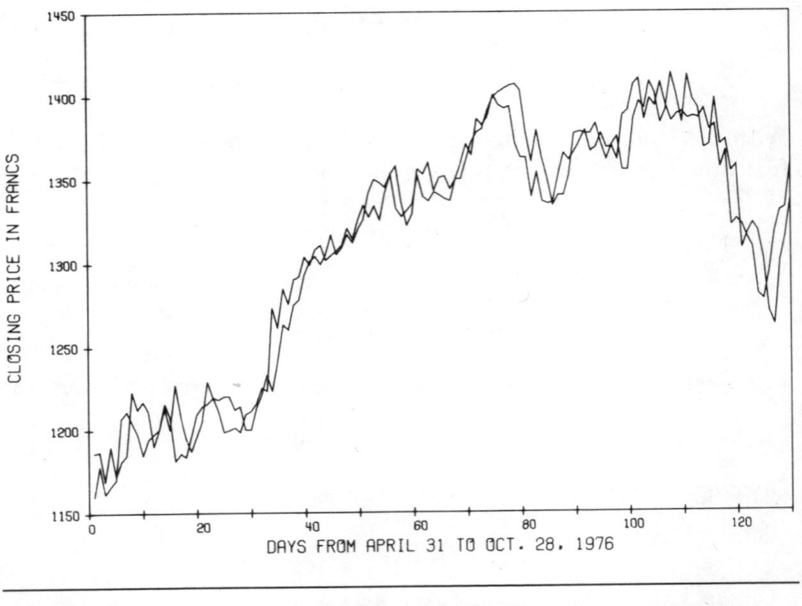

FIGURE 5.2(a) Paris and New York IBM Common Stock Prices

In Chapter 4, we noted that univariate forecasts often tend to be trivial. This is not true of multivariate forecasts. The two time series shown in Figure 5.2(a) are daily closing prices (in francs) of IBM common stock on the New York and Paris exchanges (see, Makridakis and Wheelwright, 1978: 487–488). Both series drift upward during the 130 days presented here. While one might conclude from this common pattern of drift that there is a causal relationship between these two series, it would be unwise to leap to this conclusion. Many stock price time series (perhaps even the majority) follow similar patterns of drift during this same period. Common patterns of drift or trend by themselves say nothing about the causal relationship among time series.

As a first step in building a bivariate ARIMA model, univariate models are built for both series. Univariate analysis shows that both time series are well represented by ARIMA (0,1,0) models. For the New York series,

$$(1 - B)X_t = a_t$$

```
CROSS-CORRELATIONS OF LAGS   -15 TO  15.
NO. OF VALID OBSERVATIONS      =     129.

INPUT SERIES..  NYIBM      NEW YORK IBM STOCK PRICE (DIFFERENCED)
OUTPUT SERIES.. PARISIBM   PARIS IBM STOCK PRICE (DIFFERENCED)

 LAG    CORR     SE   -1    -.8   -.6   -.4   -.2    0     .2    .4    .6    .8   +1
                       +----+----+----+----+----+----+----+----+----+----+
 -15   -.013   .094                             (    I    )
 -14    .002   .093                             (    I    )
 -13   -.066   .093                             (  XXI    )
 -12    .130   .092                             (    IXXX )
 -11   -.034   .092                             (   XI    )
 -10    .007   .092                             (    I    )
  -9   -.067   .091                             (  XXI    )
  -8   -.114   .091                             ( XXXI    )
  -7    .070   .091                             (    IXX  )
  -6   -.153   .090                             (XXXXI    )
  -5    .036   .090                              (   IX  )
  -4   -.004   .089                              (   I   )
  -3    .053   .089                              (   IX  )
  -2   -.017   .089                              (   I   )
  -1   -.040   .088                              (  XI   )
   0    .151   .088                              (   IXXX*
   1    .064   .088                              (   IXX )
   2    .659   .089                              (   IXXX)XXXXXXXXXXXX
   3   -.004   .089                              (   I   )
   4    .041   .089                              (   IX  )
   5   -.041   .090                              (  XI   )
   6    .043   .090                             (    IX   )
   7    .097   .091                             (    IXX  )
   8   -.087   .091                             (  XXI    )
   9   -.025   .091                             (   XI    )
  10    .068   .092                             (    IXX  )
  11   -.110   .092                             ( XXXI    )
  12   -.056   .092                             (   XI    )
  13   -.129   .093                             ( XXXI    )
  14    .151   .093                             (    IXXXX)
  15   -.055   .094                             (   XI    )
                                              -2SE      +2SE
```

FIGURE 5.2(b) *Identification:* CCF Estimated from the Differenced Series

```
SERIES.. RESIDUAL (NOBS= 127)  ESTIMATED MODEL RESIDUALS
NO. OF VALID OBSERVATIONS   =   127.

AUTOCORRELATIONS OF LAGS 1 - 30.
Q( 30, 127) =   41.847       SIG =     .074

 LAG   CORR    SE  -1   -.8  -.6  -.4  -.2    0   .2   .4   .6   .8  +1
                    +----+----+----+----+----+----+----+----+----+----+
   1  -.310  .090                    XXX(XXXXXI    )
   2  -.140  .100                       (XXXXXI    )
   3  -.080  .100                        (  XXI    )
   4   .160  .100                        (    IXXXX)
   5  -.120  .100                        ( XXXI    )
   6   .000  .100                        (    I    )
   7   .040  .100                        (    IX   )
   8   .090  .100                        (    IXX  )
   9  -.100  .100                        ( XXXI    )
  10  -.060  .100                        (  XXI    )
  11   .080  .100                        (    IXX  )
  12   .000  .100                        (    I    )
  13  -.070  .100                        (  XXI    )
  14   .070  .100                        (    IXX  )
  15   .100  .100                        (    IXXX )
  16  -.230  .110                       *XXXXXI     )
  17   .040  .110                       (     IX    )
  18   .040  .110                       (     IX    )
  19   .070  .110                       (     IXX   )
  20  -.160  .110                       ( XXXXI     )
  21   .130  .110                       (     IXXX  )
  22   .030  .110                       (     IX    )
  23  -.060  .110                       (   XXI     )
  24  -.020  .110                       (    XI     )
  25  -.020  .110                       (    XI     )
  26  -.030  .110                       (    XI     )
  27   .040  .110                       (     IX    )
  28   .010  .110                       (     I     )
  29   .070  .110                       (     IXX   )
  30   .040  .110                       (     IX    )
                                      -2SE        +2SE
```

FIGURE 5.2(c) *Identification:* ACF for the Residuals of the Model

$$Y_t = .767\,X_{t-2} + \frac{a_t}{1 - B}$$

and for the Paris series,

$$(1 + B)Y_t = a_t.$$

The major purpose of a preliminary univariate analysis is to make sure that both time series are stationary. As noted, a CCF estimated when one or both series are nonstationary will be overwhelmed by spurious correlations. Bivariate identification will thus require that both of these time series be differenced.

In this case, we have no a priori theory about the relationship between these series. The possibilities include the case in which a change in the New York series causes a change in the Paris series,

$$X_t \text{-----------}\!\!\rightarrow Y_{t+b},$$

```
SERIES.. RESIDUAL (NOBS= 127)  ESTIMATED MODEL RESIDUALS
NO. OF VALID OBSERVATIONS  =   127.

AUTOCORRELATIONS OF LAGS 1 - 30.
Q( 29, 127) =  25.997       SIG =      .675

 LAG   CORR    SE  -1   -.8  -.6  -.4  -.2   0    .2   .4   .6   .8  +1
                    +----+----+----+----+----+----+----+----+----+----+
  1    .120  .090                       (    IXXX )
  2   -.070  .090                       (  XXI    )
  3   -.110  .090                       ( XXXI    )
  4    .040  .090                       (    IX   )
  5   -.060  .090                       (  XXI    )
  6    .010  .090                       (    I    )
  7    .050  .090                       (    IX   )
  8    .070  .090                       (    IXX  )
  9   -.090  .090                       (  XXI    )
 10   -.100  .090                       ( XXXI    )
 11    .030  .090                       (    IX   )
 12   -.040  .090                       (   XI    )
 13   -.020  .090                       (   XI    )
 14    .070  .090                       (    IXX  )
 15    .020  .090                       (    IX   )
 16   -.170  .090                       (XXXXI    )
 17    .010  .100                       (    I    )
 18    .040  .100                       (    IX   )
 19    .130  .100                       (    IXXX )
 20   -.030  .100                       (   XI    )
 21    .200  .100                       (    IXXXX*
 22    .090  .100                       (    IXX  )
 23   -.030  .100                       (   XI    )
 24   -.090  .100                       (  XXI    )
 25   -.090  .100                       (  XXI    )
 26   -.080  .100                       (  XXI    )
 27    .050  .100                       (    IX   )
 28    .060  .100                       (    IXX  )
 29    .090  .100                       (    IXX  )
 30   -.010  .110                      (     I     )
                                     -2SE         +2SE
```

FIGURE 5.2(d) *Diagnosis:* ACF for the Residuals of the Model

$$Y_t = .987\,X_{t-2} + \frac{1 - .88B}{1 - B}\,a_t$$

the case in which a change in the Paris series causes a change in the New York series,

$$Y_t \text{-----------}\!\!\rightarrow X_{t+b},$$

and the noncausal case in which, perhaps as a result of some underlying common variable, the two series appear to be causing each other. We will operate under the assumption that the New York series is the causor but this assumption is arbitrary.

Figure 5.2(b) shows the CCF estimated from the differenced time series. The lone spike at CCF (+2) suggests that the New York series leads the Paris series by exactly two days. Had the spike instead appeared at CCF (−2), the opposite inference would have been supported. Although the assumption that the New York series was the causor was arbitrary, it is supported

```
CROSS-CORRELATIONS OF LAGS  -15 TO  15.
NO. OF VALID OBSERVATIONS     =    129.

INPUT SERIES..  NYIBM       PREWHITENED NEW YORK IBM STOCK PRICE
OUTPUT SERIES.. RESIDUAL    ESTIMATED MODEL RESIDUALS

  LAG    CORR     SE  -1    -.8   -.6   -.4   -.2    0     .2    .4    .6    .8   +1
                       +----+----+----+----+----+----+----+----+----+----+
  -15   -.070   .094                           (  XXI    )
  -14   -.067   .093                           (  XXI    )
  -13   -.062   .093                           (  XXI    )
  -12   -.053   .092                           (   XI    )
  -11   -.041   .092                           (   XI    )
  -10   -.032   .092                           (   XI    )
   -9   -.024   .091                           (   XI    )
   -8   -.017   .091                           (    I    )
   -7   -.013   .091                           (    I    )
   -6   -.035   .090                           (   XI    )
   -5   -.012   .090                            (   I   )
   -4   -.012   .089                            (   I   )
   -3   -.030   .089                            (  XI   )
   -2   -.022   .089                            (  XI   )
   -1   -.014   .088                            (   I   )
    0   -.012   .088                            (   I   )
    1    .003   .088                            (   I   )
    2   -.046   .089                            (  XI   )
    3   -.021   .089                            (  XI   )
    4   -.036   .089                            (  XI   )
    5   -.005   .090                            (   I   )
    6   -.021   .090                           (   XI    )
    7   -.027   .091                           (   XI    )
    8   -.031   .091                           (   XI    )
    9   -.030   .091                           (   XI    )
   10   -.025   .092                           (   XI    )
   11   -.024   .092                           (   XI    )
   12   -.028   .092                           (   XI    )
   13   -.033   .093                           (   XI    )
   14   -.032   .093                           (   XI    )
   15   -.048   .094                           (   XI    )
                                              -2SE      +2SE
```

FIGURE 5.2(e) *Diagnosis:* CCF for the Differenced X_t Series and the Model Residuals

empirically by this CCF. In general, the analyst need not specify which series is the causor, but instead may make the specification empirically. If the positive CCF is statistically significant, then the X_t series is the causor; and if the negative CCF is statistically significant, then the Y_t series is the causor.

To be sure, there are many nonzero values in both the positive and negative halves of the CCF. The only statistically significant value is at CCF (+2), however, and this suggests the model

$$(1 - B)Y_t = (1 - B)\omega_0 X_{t-2} + N_t.$$

A change in the price of IBM common stock on the New York exchange is followed by an analogous change in price on the Paris exchange two days later. Finally, the CCF suggests that the effect is not distributed over successive days, that is, there is no dynamic transfer of effect which would be indicated by a pattern of decay from CCF (+2) to CCF (+3) to CCF (+4) and so forth.

The next step in the bivariate model-building procedure is to identify an ARIMA noise model for the N_t component. There are a number of ways in which this identification can be made. In our experience, however, the most satisfactory way is the straightforward one: Identify the N_t component from the transfer function residuals. To do this, the analyst first assumes that N_t is white noise. The tentative bivariate model is thus

$$(1 - B)Y_t = (1 - B)\omega_0 X_{t-2} + a_t$$

$$Y_t = \omega_0 X_{t-2} + \frac{a_t}{1 - B}.$$

This tentative model has only one parameter, whose value is estimated as

$$\hat{\omega}_0 = .7674 \text{ with t statistic} = 13.16.$$

The residual ACF for this model, shown in Figure 5.2(c), suggests that an ARIMA (0,0,1) model will adequately reflect the structure of autocorrelation in these residuals. This leads to the tentative model

$$(1 - B)\, Y_t = (1 - B)\, \omega_0 X_{t-2} + (1 - \Theta_1 B)\, a_t$$

$$Y_t = \omega_0 X_{t-2} + \frac{1 - \Theta_1 B}{1 - B}\, a_t.$$

Parameter estimates for this model are:

$$\hat{\omega}_0 = .987 \text{ with t statistic} = 32.12$$

$$\hat{\Theta}_1 = .88 \text{ with t statistic} = 20.17.$$

Both estimates are statistically significant and otherwise acceptable.

Bivariate ARIMA models of the sort we have tentatively selected for these two time series may be diagramed as

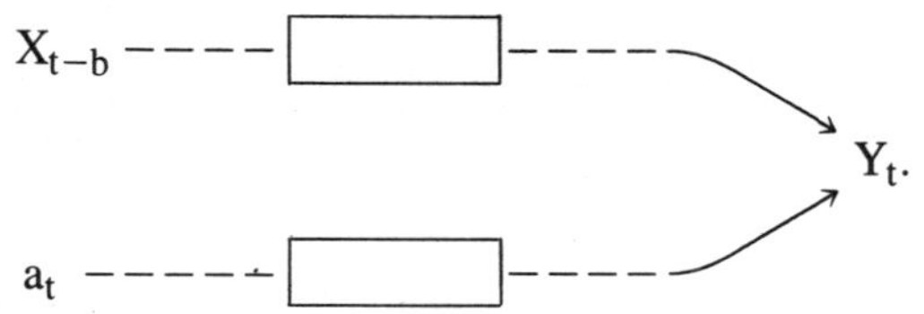

The sense of this diagram is that two *distinct* input processes (the time series observation, X_{t-b}, and the random shock, a_t) pass through two *distinct* filters (a transfer function and an ARIMA structure) and are then combined additively into an output process (the time series observation, Y_t). The difference between this diagram and the input-output diagrams we drew for univariate processes in Chapter 2 is that there are two inputs here and this hints at the special problems of diagnosing a bivariate ARIMA model. In effect, the statistical adequacy of both the transfer function component *and* the noise component must be diagnosed.

First, as one might suspect, the model residuals must not be different than white noise. The residual ACF, shown in Figure 5.2(d), indicates that the noise component of this tentative model is adequate. If model residuals are different than white noise, a new noise component must be identified.

Second, the model residuals must be independent of the causor time series. To test the hypothesis of independence, a CCF is estimated from the model residuals and the input time series (the differenced New York IBM series in this case). This CCF, shown in Figure 5.2(e), has no significant values, indicating that the causor series and the model residuals are uncorrelated. As the hypothesis of independence stands, the tentative model is acceptable. Had this CCF indicated that the input series and the model residuals were not independent, a new transfer function component would have to be identified.

Because this model satisfies both diagnostic criteria, it is accepted. There nevertheless may be other acceptable bivariate models which through meta-diagnosis, could be compared with this one. The reader is invited to explore this possibility as an exercise.

Before commenting on the relative utility of this analysis, we must point out that our only purpose was to illustrate the procedures of bivariate modeling. In particular, we do not intend to endorse international stock speculations. While we have no experience here, we have been advised (by an investment analyst who wishes to remain anonymous) that it is not an easy matter to turn a profit on foreign stock exchanges, especially on blue chip shares. The hidden costs of buying and selling stock (not to mention buying francs with dollars) make speculation a generally risky enterprise.

Still, a foreign investor will realize the incremental utility of a bivariate forecasting model over a univariate forecasting model. The Paris IBM series is well represented by an ARIMA (0,1,0) model, so forecasts of the series are:

$$Y_t(n) = Y_t.$$

Because this series follows a random walk, the best univariate forecast of a future price is the current price. Using the New York series as a lead indicator, however, a bivariate forecast is possible. To compare the two models, we use the first 125 observations to forecast the 126th, the first 126 observations to forecast the 127th, and so forth. The results of this exercise are:

Day	Observation	Univariate Forecast	Bivariate Forecast
126	1270.000	1302.000	1277.735
127	1262.000	1270.000	1273.791
128	1300.000	1262.000	1288.419
129	1315.000	1300.000	1313.757
130	1337.000	1315.000	1326.185

Bivariate forecasts are clearly superior to the univariate forecasts. The MSFEs for these two models based on only these five forecasts are 648.20 and 87.65. This should not be surprising. One would also expect a bivariate model to have a lower residual variance than a univariate model. In this case, the univariate model has an RMS = 212.89 while the bivariate model has an RMS = 81.30.

In closing, a comment on the interpretation of the bivariate model is called for. The interpretation is explicitly causal. A rise or drop in the New York series is followed by an analogous rise or drop in the Paris series two days later. While a time-lagged correlation does not imply causation, causation does imply a time-lagged correlation. The null hypothesis is thus

$$H_0\colon X_t \;---\!\not\!\!-\!\not\!\!-----\!\!\rightarrow Y_{t+2};\ \hat{\omega}_0 = 0.$$

Due to the relatively large t statistic for the estimate of ω_0, the null hypothesis must be rejected.

5.3 Prewhitening

The Paris-New York IBM model was identified with relatively little trouble because both series were well represented by ARIMA (0,1,0) models. After differencing, both series were white noise. Our concern now is with the problem of modeling bivariate relationships when the series are not white noise. When we derived the expected CCFs for zero- and first-order transfer functions in Section 5.1, we assumed that the x_t series was white noise. When this assumption is unsatisfied, as is usually the case with social science time series, the estimated CCF is uninterpretable. If the x_t and z_t series are *prewhitened*, however, an interpretable CCF can be estimated.

Figure 5.3(a) shows a time series of annual Swedish population increases (increase per thousand population) for the 1750–1849 century as reported by Thomas (1940). The rate increase in a given year is defined as

$$p_t = \text{birth rate} - \text{death rate in the } t^{\text{th}} \text{ year.}$$

The relationship between this time series and a time series of total population is a straightforward one. The total population of Sweden in 1749 was 1,760,000. Starting with this value, the total population (in thousands) at the end of 1750 is given by the expression

$$\begin{aligned} P_{1750} &= P_{1749} + P_{1749}\,(p_{1750}) \\ &= 1{,}760 + 1{,}760\,(p_{1750}) \\ &= 1{,}760\,(1 + p_{1750}). \end{aligned}$$

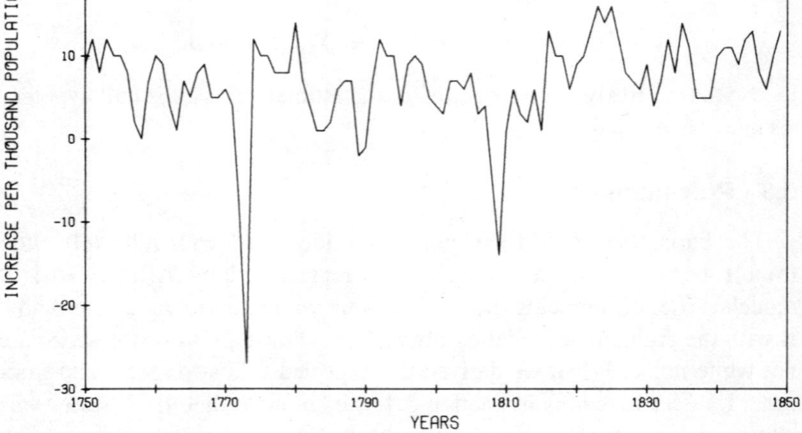

FIGURE 5.3(a) Swedish Population Rates, 1750–1849

The total population at the end of 1750, then, is given by the product of the total population at the end of 1749 and the 1750 rate increase. In the general case, total population at the end of the tth year is:

$$P_t = P_{t-1}(1 + p_t)$$
$$\frac{P_t}{P_{t-1}} = 1 + p_t.$$

Taking the natural logarithm of this expression,

$$\text{Ln}\left[\frac{P_t}{P_{t-1}}\right] = \text{Ln}(1 + p_t)$$
$$\text{Ln}(P_t) - \text{Ln}(P_{t-1}) = \text{Ln}(1 + p_t).$$

The log-transformed total population series is thus the integration (though not necessarily a random walk) of the log-transformed rate increase series.

In Section 2.12.3, we analyzed the Swedish Harvest Index time series for the years 1749–1850. The Harvest Index is a crude measure of food production wherein a value of zero indicated a total crop failure and a value of nine indicated a superabundant crop. Our analysis demonstrated that the series could be well represented by an ARIMA (0,0,1) model:

$$h_t = (1 + .39B)a_t.$$

According to Gustav Sundbärg, an early demographer quoted by Thomas, the growth of Swedish population during the 1750–1849 century could be explained almost entirely as a function of agricultural production:

> Irrespective of which party had gained control, or whether the King himself was on the throne, if the harvest was good, marriage and birth rates were high and death rates comparatively low, that is, the bulk of the population flourished. On the contrary, when the harvest failed, marriage and birth rates declined and death devastated the land, bearing witness to need and privation and at times even to starvation. Whether the factories fared well or badly or whether the bank-rate rose or fell—all these things *at this time,* were scarcely more than ripples on the surface [1940: 82].

This hypothesis (which we will now call Sundbärg's hypothesis) has not been adequately tested. When Dorothy S. Thomas, Gunnar Myrdal, and others investigated the relationships claimed by Sundbärg, the methods we have described here were unavailable; social scientists did not yet have access to computers. Early analyses of these data consisted largely of visual analyses which tend to mislead. The only satisfactory method for testing a causal hypothesis is to identify, estimate, and diagnose a bivariate time series model as we will do now.

As a first step in the analysis, we make sure that both series are stationary. The Harvest Index series, well represented by an ARIMA (0,0,1) model, is already stationary. An analysis of the population rate increase series leads also to an ARIMA (0,0,1) model:

$$p_t = (1 + .44B)a_t.$$

As both series are stationary, the CCF can be estimated from the undifferenced series.

The estimated CCF, shown in Figure 5.3(b), is not as clear as the CCF estimated in the previous example. There is no evidence of a strong, unambiguous causal relationship. Some analysts might see a two-way relationship here while others might see no relationship whatsoever. The argument is moot because, as we have noted, *the CCF can be interpreted only when the causor variable is a white noise process*. Even if this CCF did indicate a strong, unambiguous causal relationship, the evidence could not be accepted because the Harvest Index series is not white noise.

The problem here is that the CCF estimate is contaminated by within-series correlations, or autocorrelation. The hypothetical relationship between these two series is:

$$h_t \dashrightarrow p_{t+b}.$$

Any change in the harvest causes a change in the birth and death rates b years later. There are other causal factors, however, namely,

$$h_{t-j} \dashrightarrow h_t \dashrightarrow p_{t+b}.$$

When the causor series is not white noise, the CCF will reflect both between-series dependencies and within-series dependencies.

To illustrate the confounding of within- and between-series correlation, we represent the relationship between two time series as

$$z_t = v_0 x_t + v_1 x_{t-1} + \ldots + v_k x_{t-k} + N_t.$$

```
CROSS-CORRELATIONS OF LAGS  -15 TO  15.
NO. OF VALID OBSERVATIONS      =     100.

INPUT SERIES..  HARVEST    SWEDISH HARVEST INDEX 1750-1849
OUTPUT SERIES.. POP        SWEDISH POPULATION CHANGE 1750-1849

 LAG    CORR     SE  -1    -.8  -.6  -.4  -.2   0    .2   .4   .6   .8  +1
                     +----+----+----+----+----+----+----+----+----+----+
 -15  -.076  .108                        (  XXI     )
 -14  -.073  .108                        (  XXI     )
 -13   .077  .107                        (     IXX  )
 -12   .064  .107                        (     IXX  )
 -11  -.075  .106                        (  XXI     )
 -10   .128  .105                        (     IXXX )
  -9   .165  .105                        (     IXXXX)
  -8   .116  .104                        (     IXXX )
  -7   .038  .104                        (     IX   )
  -6  -.212  .103                        *XXXXI     )
  -5  -.120  .103                        ( XXXI     )
  -4  -.044  .102                        (    XI    )
  -3   .114  .102                        (     IXXX )
  -2   .084  .101                        (     IXX  )
  -1  -.119  .101                        ( XXXI     )
   0   .057  .100                        (     IX   )
   1   .391  .101                        (     IXXXX)XXXXX
   2   .265  .101                        (     IXXXX)XX
   3   .006  .102                        (     I    )
   4  -.134  .102                        ( XXXI     )
   5  -.144  .103                        (XXXXI     )
   6  -.022  .103                        (    XI    )
   7   .026  .104                        (     IX   )
   8   .108  .104                        (     IXXX )
   9   .245  .105                        (     IXXXX)X
  10   .215  .105                        (     IXXXX*
  11   .143  .106                        (     IXXXX)
  12   .010  .107                        (     I    )
  13  -.087  .107                        (  XXI     )
  14  -.105  .108                        ( XXXI     )
  15   .005  .108                        (     I    )
                                       -2SE       +2SE
```

FIGURE 5.3(b) *Identification:* CCF for the h_t and p_t Series

Now in the general case, only a finite number of the v-weights will be nonzero. To derive the expected CCF, the v-weight model is multiplied by $x_t, x_{t-1}, \ldots, x_{t-k}$. The result is a set of k equations:

$$\begin{aligned}
x_t z_t &= v_0 x_t x_t + \ldots + v_k x_t x_{t-k} + x_t N_t \\
x_{t-1} z_t &= v_0 x_{t-1} x_t + \ldots + v_k x_{t-1} x_{t-k} + x_{t-1} N_t \\
&\vdots \\
x_{t-k} z_t &= v_0 x_{t-k} x_t + \ldots + v_k x_{t-k} x_{t-k} + x_{t-k} N_t.
\end{aligned}$$

Taking the expectation of this equation system,

$$
\begin{aligned}
\mathrm{COV}\,(x_t z_t) &= v_0\sigma_x^2 + \ldots + v_k \mathrm{COV}\,(x_t x_{t-k}) \\
\mathrm{COV}\,(x_{t-1} z_t) &= v_0 \mathrm{COV}\,(x_{t-1} x_t) + \ldots + v_k \mathrm{COV}\,(x_{t-1} x_{t-k}) \\
&\vdots \\
\mathrm{COV}\,(x_{t-k} z_t) &= v_0 \mathrm{COV}\,(x_{t-k} x_t) + \ldots + v_k \sigma_x^2 .
\end{aligned}
$$

Finally, dividing the system by $\sigma_x \sigma_z$,

$$
\begin{aligned}
\mathrm{CCF}\,(0) &= v_0 \frac{\sigma_x}{\sigma_z} + \ldots + v_k \frac{\sigma_x}{\sigma_z} \mathrm{ACF}\,(k) \\
\mathrm{CCF}\,(+1) &= v_0 \frac{\sigma_x}{\sigma_z} \mathrm{ACF}\,(1) + \ldots + v_k \frac{\sigma_x}{\sigma_z} \mathrm{ACF}\,(k-1) \\
&\vdots \\
\mathrm{CCF}\,(+k) &= v_0 \frac{\sigma_x}{\sigma_z} \mathrm{ACF}\,(k) + \ldots + v_k \frac{\sigma_x}{\sigma_z} .
\end{aligned}
$$

The positive CCF thus is determined by the v-weight relationship between the two time series, by the variance of the two time series, *and by the ACF of the causor time series*.

Now in the New York-Paris IBM example of the previous section, both series were well represented by ARIMA (0,1,0) models. After differencing, both series were white noise. As the ACF of a white process is uniformly zero, the CCF is expected to be:

$$
\begin{aligned}
\mathrm{CCF}\,(0) &= v_0 \frac{\sigma_x}{\sigma_z} \\
\mathrm{CCF}\,(+1) &= v_1 \frac{\sigma_x}{\sigma_z} \\
&\vdots \\
\mathrm{CCF}\,(+k) &= v_k \frac{\sigma_x}{\sigma_z}
\end{aligned}
$$

that is, the CCF will be uncontaminated by within-series correlation. In the Swedish population example, however, the Harvest Index series is *not* white noise and, as a result, no bivariate transfer function relationship can be identified from the CCF shown in Figure 5.3(b).

In theory, within-series correlation can be removed from the CCF by solving the k-equation system directly. In practice, however, this is nearly impossible. The components of the k equations, especially the ACF of the x_t series, must be estimated for the direct solution. Rounding error alone would make a direct solution inefficient and inaccurate. A more efficient and practical method of removing within-series correlation from the CCF is to *prewhiten* both series. Noting that the causor time series is well represented by an ARIMA model,

$$x_t = (1 - \phi_1 B - \ldots - \phi_p B^p)^{-1} (1 - \Theta_1 B - \ldots - \Theta_q B^q) a_t,$$

the x_t series can be prewhitened, *turned into white noise,* that is, by inverting the model.

$$a_t = (1 - \phi_1 B - \ldots - \phi_p B^p)(1 - \Theta_1 B - \ldots - \Theta_q B^q)^{-1} x_t.$$

Starting again with the v-weight relationship between the two time series,

$$z_t = v_0 x_t + v_1 x_{t-1} + \ldots + v_k x_{t-k} + N_t.$$

Applying the inverted ARIMA (p,d,q) $(P,D,Q)_S$ model to both sides of the equation,

$$z^*_t = v_0 a_t + v_1 a_{t-1} + \ldots + v_k a_{t-k} + N^*_t,$$

where

$$z^*_t = (1 - \phi_1 B - \ldots - \phi_p B^p)(1 - \Theta_1 B - \ldots - \Theta_q B^q)^{-1} z_t$$
$$N^*_t = (1 - \phi_1 B - \ldots - \phi_p B^p)(1 - \Theta_1 B - \ldots - \Theta_q B^q)^{-1} N_t.$$

To derive the expected CCF between z^*_t and a_t, the v-weight model is multiplied through by $a_t, a_{t-1}, \ldots a_{t-k}$. The k-equation system obtained by this procedure is:

$$\begin{aligned}
a_t z^*_t &= v_0 a_t a_t + \ldots + v_k a_t a_{t-k} + a_t N^*_t \\
a_{t-1} z^*_t &= v_0 a_{t-1} a_t + \ldots + v_k a_{t-1} a_{t-k} + a_{t-1} N^*_t \\
&\vdots \\
a_{t-k} z^*_t &= v_0 a_{t-k} a_t + \ldots + v_k a_{t-k} a_{t-k} + a_{t-k} N^*_t,
\end{aligned}$$

whose expectation is:

$$\begin{aligned}
\mathrm{COV}\,(a_t z^*_t) &= v_0\sigma^2_a \\
\mathrm{COV}\,(a_{t-1} z^*_t) &= v_1\sigma^2_a \\
&\vdots \\
\mathrm{COV}\,(a_{t-k} z^*_t) &= v_k\sigma^2_a.
\end{aligned}$$

Finally, dividing these equations by $\sigma_a\sigma_{z*}$

$$\begin{aligned}
\mathrm{CCF}\,(0) &= v_0\frac{\sigma_a}{\sigma_{z*}} \\
\mathrm{CCF}\,(+1) &= v_1\frac{\sigma_a}{\sigma_{z*}} \\
&\vdots \\
\mathrm{CCF}\,(+k) &= v_k\frac{\sigma_a}{\sigma_{z*}}
\end{aligned}$$

which may be a surprising result. *The CCF between the a_t and z^*_t series is proportional to the v-weights which define the bivariate relationship between x_t and z_t.* By prewhitening the time series prior to analysis, the effects of within series correlation (autocorrelation in the causor series) can be removed from the CCF.

In Figure 5.3(c), we show a bivariate model-building strategy. Like the univariate modeling strategy outlined in Section 2.11, the bivariate strategy is an iterative procedure whereby a parsimonious but statistically adequate ARIMA model is constructed. Because the strategy deals with two time series, of course, it has many more steps than its univariate analogue. The logic nonetheless is identical with the logic of the univariate model-building strategy.

As a first step, univariate models are constructed for both series. The results of these analyses will indicate whether either series must be differenced or transformed.

The univariate ARIMA model for x_t is inverted and applied to both series: *prewhitening*. A CCF is then estimated from the a_t and z^*_t series and used to identify a transfer function model for the relationship between the x_t and z_t time series.

The parameters of the transfer function are estimated. The residuals of

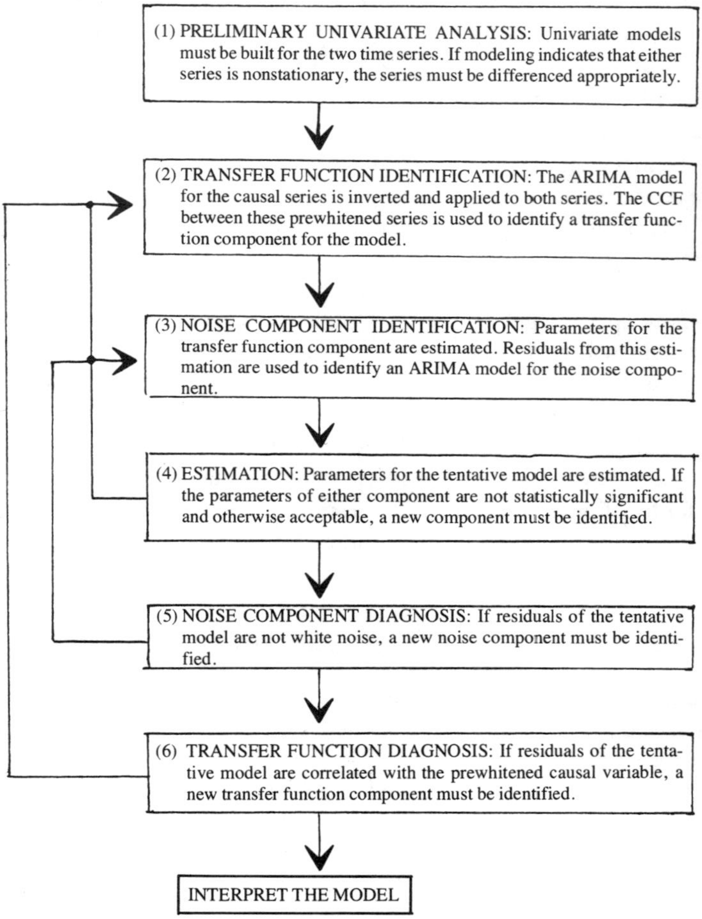

FIGURE 5.3(c) The Bivariate ARIMA Model-Building Strategy

this estimation are used to identify an ARIMA model for the N_t component.

Parameters of the fully identified model are estimated. All estimates must be statistically significant and otherwise acceptable. By "otherwise acceptable," we mean that noise component parameters must lie within the bounds of stationarity-invertibility; transfer function parameters must lie within the bounds of system stability. If the parameter estimates of either component are unacceptable, a new component must be identified.

The tentative model has two components,

$$z_t = f(x_{t-b}) + N_t,$$

and both components must pass diagnostic checks. The statistical adequacy of the noise component is diagnosed in the same way that a univariate ARIMA model is diagnosed: The model residuals must not be different than white noise.

The transfer function component has been specified so as to account for all process variance common to the x_t and z_t series. If the transfer function component is statistically inadequate, a portion of this common variance will show up as model residuals. A statistically adequate transfer function component will be independent of the noise component. To test the null hypothesis of independence, a CCF is estimated from the prewhitened x_t series and the model residuals. If the transfer function and noise components are not independent, there will be spikes at the low-order lags of the CCF. If the transfer function proves statistically inadequate by this criterion, a new transfer function must be identified, estimated, and diagnosed.

We can now apply this model-building strategy to theSwedish population growth example. Stated simply, Sundbärg's hypothesis is:

harvest in year t ———————→ population growth in year t+b.

The analysis begins with prewhitening. As the Harvest Index series is well represented by the ARIMA (0,0,1) model

$$h_t = (1 + .39B)a_t.$$

It is prewhitened as

$$a_t = (1 + .39B)^{-1}h_t.$$

The same inverse operator is used to prewhiten the population rate increase series

$$z_t^* = (1 + .39B)^{-1}p_t.$$

The $a_t z_t^*$ CCF, shown in Figure 5.3(d), gives a clear picture of the relationship between the Harvest Index and population rate increase time series. The large spike at CCF (+1) suggests the transfer function model

$$p_t = \omega_0 h_{t-1} + N_t,$$

```
CROSS-CORRELATIONS OF LAGS   -15 TO  15.
NO. OF VALID OBSERVATIONS     =     100.

INPUT SERIES..  HARVEST    PREWHITENED SWEDISH HARVEST INDEX
OUTPUT SERIES.. POP        PREWHITENED SWEDISH POPULATION CHANGE

 LAG    CORR     SE  -1    -.8   -.6   -.4   -.2    0    .2    .4    .6    .8   +1
                      +-----+-----+-----+-----+-----+-----+-----+-----+-----+-----+
 -15  -.053   .108                           (    XI     )
 -14  -.082   .108                           (   XXI     )
 -13   .078   .107                           (      IXX  )
 -12   .095   .107                           (      IXX  )
 -11  -.162   .106                           (XXXXI      )
 -10   .152   .105                           (      IXXXX)
  -9   .105   .105                           (      IXXX )
  -8   .050   .104                           (      IX   )
  -7   .111   .104                           (      IXXX )
  -6  -.254   .103                          X(XXXXI      )
  -5  -.011   .103                           (      I    )
  -4  -.076   .102                           (   XXI     )
  -3   .111   .102                           (      IXXX )
  -2   .103   .101                           (      IXXX )
  -1  -.156   .101                           (XXXXI      )
   0  -.006   .100                           (      I    )
   1   .355   .101                           (      IXXXX)XXXX
   2   .165   .101                           (      IXXXX)
   3  -.018   .102                           (      I    )
   4  -.090   .102                           (   XXI     )
   5  -.130   .103                           ( XXXI      )
   6   .019   .103                           (      I    )
   7   .004   .104                           (      I    )
   8   .044   .104                           (      IX   )
   9   .195   .105                           (      IXXXX*
  10   .123   .105                           (      IXXX )
  11   .111   .106                           (      IXXX )
  12  -.005   .107                           (      I    )
  13  -.056   .107                           (    XI     )
  14  -.105   .108                           ( XXXI      )
  15   .043   .108                           (      IX   )
                                           -2SE       +2SE
```

FIGURE 5.3(d) *Identification:* CCF for the Prewhitened p_t and h_t Series

which implies that the current year's harvest determines the next year's population growth. Our estimate of ω_0 for this relationship is:

$$\hat{\omega}_0 = .87.$$

The residual ACF for this estimate, shown in Figure 5.3(e), suggests an ARIMA (1,0,0) model for the N_t component. This leads to the full model

$$p_t = \omega_0 h_{t-1} + \frac{a_t}{1 - \phi_1 B}.$$

```
SERIES.. RESIDUAL (NOBS= 100) ESTIMATED MODEL RESIDUALS
NO. OF VALID OBSERVATIONS   =  100.

AUTOCORRELATIONS OF LAGS 1 - 30.
Q( 29, 100) =  43.730       SIG =      .039

 LAG   CORR    SE  -1   -.8  -.6  -.4  -.2   0    .2   .4   .6   .8  +1
                    +----+----+----+----+----+----+----+----+----+----+
   1   .450  .100                       (    IXXXX)XXXXXX
   2   .200  .120                      (     IXXXXX)
   3   .050  .120                      (     IX    )
   4   .080  .120                      (     IXX   )
   5   .130  .120                      (     IXXX  )
   6   .090  .120                      (     IXX   )
   7  -.050  .120                      (    XI     )
   8  -.170  .130                     (  XXXXI      )
   9  -.130  .130                     (   XXXI      )
  10  -.020  .130                     (     XI      )
  11  -.010  .130                     (      I      )
  12   .020  .130                     (      IX     )
  13  -.070  .130                     (    XXI      )
  14  -.090  .130                     (    XXI      )
  15  -.050  .130                     (     XI      )
  16   .040  .130                     (      IX     )
  17   .090  .130                     (      IXX    )
  18   .110  .130                     (      IXXX   )
  19   .160  .130                     (      IXXXX  )
  20   .100  .130                     (      IXXX   )
  21   .080  .130                     (      IXX    )
  22   .080  .130                     (      IXX    )
  23   .000  .140                     (      I      )
  24   .060  .140                     (      IXX    )
  25   .050  .140                     (      IX     )
  26   .020  .140                     (      IX     )
  27  -.070  .140                     (    XXI      )
  28  -.090  .140                     (    XXI      )
  29  -.080  .140                     (    XXI      )
  30  -.010  .140                     (      I      )
                                    -2SE          +2SE
```

FIGURE 5.3(e) *Identification:* ACF for the Residuals of the Model $p_t = .866\, h_{t-1} + a_t$

Parameter estimates for this full model are:

$$\hat{\omega}_0 = .826 \text{ with t statistic} = 4.18$$

$$\hat{\phi}_1 = .460 \text{ with t statistic} = 4.94.$$

Both parameter estimates are statistically significant and otherwise acceptable.

To diagnose the tentative model, we first require that the model residuals are not different than white noise. The residual ACF, shown in Figure 5.3(f), has no statistically significant spikes and the Q-statistic is not signifi-

```
SERIES.. RESIDUAL (NOBS=  99) ESTIMATED MODEL RESIDUALS
NO. OF VALID OBSERVATIONS   =    99.

AUTOCORRELATIONS OF LAGS 1 - 30.
Q( 28,    99) =  12.088        SIG =      .996

 LAG   CORR    SE  -1   -.8  -.6  -.4  -.2   0   .2   .4   .6   .8   +1
                    +----+----+----+----+----+----+----+----+----+----+
   1   .000  .100                       (    I    )
   2   .010  .100                       (    I    )
   3  -.100  .100                       ( XXXI    )
   4   .020  .100                       (    IX   )
   5   .100  .100                       (    IXXX )
   6   .090  .100                       (    IXX  )
   7  -.030  .100                       (   XI    )
   8  -.140  .100                       (XXXXI    )
   9  -.090  .110                      (   XXI     )
  10   .060  .110                      (     IXX   )
  11  -.010  .110                      (     I     )
  12   .070  .110                      (     IXX   )
  13  -.070  .110                      (   XXI     )
  14  -.080  .110                      (   XXI     )
  15  -.040  .110                      (    XI     )
  16   .040  .110                      (     IX    )
  17   .060  .110                      (     IXX   )
  18   .030  .110                      (     IX    )
  19   .120  .110                      (     IXXX  )
  20   .030  .110                      (     IX    )
  21   .010  .110                      (     I     )
  22   .090  .110                      (     IXX   )
  23  -.020  .110                      (    XI     )
  24   .020  .110                      (     IX    )
  25   .040  .110                      (     IX    )
  26   .030  .110                      (     IX    )
  27  -.060  .110                      (   XXI     )
  28  -.040  .110                      (    XI     )
  29  -.070  .110                      (   XXI     )
  30   .020  .110                      (     IX    )
                                     -2SE       +2SE
```

FIGURE 5.3(f) *Diagnosis:* ACF for the Residuals of the Model

$$p_t = .828\, h_{t-1} + \frac{a_t}{1 - .459B}$$

cant. Second, the model residuals and the prewhitened Harvest Index series must be uncorrelated. The CCF for the residuals and the prewhitened h_t series, shown in Figure 5.3(g), has one statistically significant spike. As this tentative model satisfies both diagnostic criteria, we accept it.

The analysis leads to a conclusion that population growth and harvests are related by the equation

$$p_t = .826h_{t-1} + \frac{a_t}{1 - .46B}.$$

In years following a crop failure (a value of Harvest Index = 0 indicates crop failure), Swedish population was expected to increase by only .661% (the

```
CROSS-CORRELATIONS OF LAGS   -15 TO  15.
NO. OF VALID OBSERVATIONS       =     99.

INPUT SERIES..  HARVEST   PREWHITENED SWEDISH HARVEST INDEX
OUTPUT SERIES.. RESIDUAL  ESTIMATED MODEL RESIDUALS

 LAG    CORR    SE  -1   -.8  -.6  -.4  -.2  0    .2   .4   .6   .8   +1
                    +----+----+----+----+----+----+----+----+----+----+
 -15  -.084  .109                       (  XXI    )
 -14   .092  .108                       (    IXX  )
 -13   .136  .108                       (    IXXX )
 -12  -.110  .107                       ( XXXI    )
 -11   .163  .107                       (    IXXXX)
 -10   .104  .106                       (    IXXX )
  -9   .006  .105                       (    I    )
  -8   .090  .105                       (    IXX  )
  -7  -.251  .104                      X(XXXXI    )
  -6   .045  .104                       (    IX   )
  -5  -.043  .103                       (   XI    )
  -4   .132  .103                       (    IXXX )
  -3   .165  .102                       (    IXXXX)
  -2  -.201  .102                       *XXXXI    )
  -1  -.049  .101                       (   XI    )
   0   .008  .101                       (    I    )
   1   .185  .101                       (    IXXXX*
   2  -.022  .102                       (   XI    )
   3  -.067  .102                       (  XXI    )
   4  -.117  .103                       ( XXXI    )
   5   .042  .103                       (    IX   )
   6   .084  .104                       (    IXX  )
   7   .063  .104                       (    IXX  )
   8   .170  .105                       (    IXXXX)
   9   .108  .105                       (    IXXX )
  10   .073  .106                       (    IXX  )
  11  -.028  .107                       (   XI    )
  12  -.009  .107                       (    I    )
  13  -.121  .108                       ( XXXI    )
  14   .051  .108                       (    IX   )
  15   .010  .109                       (    I    )
                                       -2SE      +2SE
```

FIGURE 5.3(g) *Diagnosis:* CCF for Prewhitened h_t Series and the Model Residuals

mean of the p_t series). In years following a superabundant crop (Harvest Index = 9), population was expected to increase by more than 1.4%.

Overall, the construction *and* interpretation of this model are simplified because it is a *bi*variate model. *In all cases,* bivariate ARIMA models (such as the Paris-New York IBM model of the previous section and population-harvest model of this section) are constructed by the routine outlined in Figure 5.3(c). In our experiences, students will have little trouble with bivariate analysis if this strategy is followed mechanically. The interpretation of a bivariate model is also straightforward: "x causes y." In both of the bivariate example analyses of this chapter, changes in one variable are *followed* by changes in the other. While this finding does not "prove" causation, the causal null hypothesis is rejected and the causal hypothesis is consequently more plausible than it was prior to the analysis. Model construction and interpretation are not so simple in the *multi*variate case, however. When a model has more than one input (or independent-variable) time series, the model-building strategy becomes less mechanical and interpretations of the model become more complicated. Nevertheless, it is only in the multivariate case that ARIMA models and methods achieve their full potential as tools of social research.

5.4 Multivariate Population Growth Model[3]

The bivariate ARIMA analysis supports Sundbärg's hypothesis. No matter how crucial agricultural production may have been during that century, however, we cannot conclude that the harvest was the sole determinant of population growth. A crop failure no doubt effected an increase in the death rate but the affect on the birth rate must have been less substantial. To better explain Swedish population growth, then, we can add other independent variables to the model.

A likely predictor of birth rates is shown in Figure 5.4(a). These are annual fertility rates (per thousand) for the 1750–1849 century. The fertility rate in the tth year is defined as

$$f_t = \text{births per 1000 female population.}$$

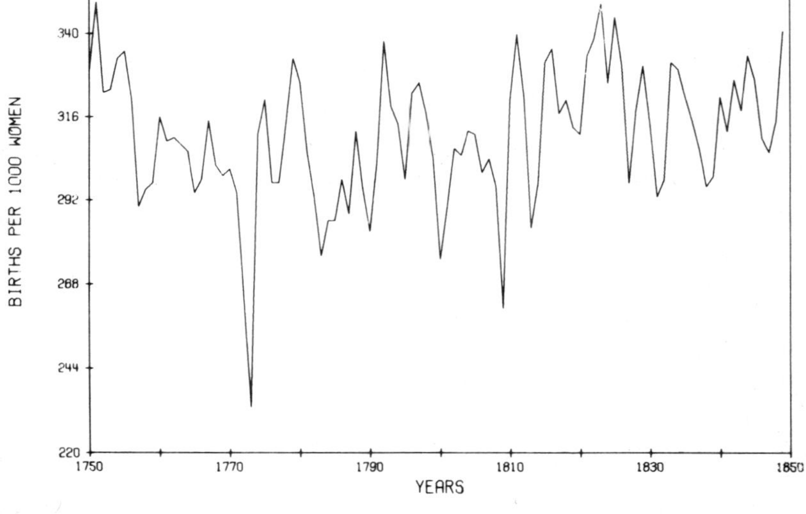

FIGURE 5.4(a) Swedish Fertility Rates, 1750–1849

```
CROSS-CORRELATIONS OF LAGS  -15 TO  15.
NO. OF VALID OBSERVATIONS      =     98.

INPUT SERIES..  FERTILE     PREWHITENED SWEDISH FERTILITY RATE
OUTPUT SERIES.. POP         PREWHITENED SWEDISH POPULATION CHANGE

LAG   CORR    SE      -1  -.8  -.6  -.4  -.2   0   .2   .4   .6   .8   +1
                      +----+----+----+----+----+----+----+----+----+----+
-15  -.070  .110                          (  XXI    )
-14   .129  .109                          (    IXXX )
-13  -.123  .108                          ( XXXI    )
-12   .059  .108                          (    IX   )
-11   .002  .107                          (    I    )
-10   .107  .107                          (    IXXX )
 -9  -.053  .106                          (   XI    )
 -8   .039  .105                          (    IX   )
 -7  -.075  .105                          (  XXI    )
 -6  -.132  .104                          ( XXXI    )
 -5   .160  .104                          (    IXXXX)
 -4   .098  .103                          (    IXX  )
 -3   .027  .103                          (    IX   )
 -2  -.092  .102                          (  XXI    )
 -1   .093  .102                          (    IXX  )
  0   .712  .101                          (    IXXXX)XXXXXXXXXXXXX
  1   .028  .102                          (    IX   )
  2   .095  .102                          (    IXX  )
  3  -.009  .103                          (    I    )
  4   .085  .103                          (    IXX  )
  5   .037  .104                          (    IX   )
  6   .141  .104                          (    IXXXX)
  7   .010  .105                          (    I    )
  8   .082  .105                          (    IXX  )
  9   .005  .106                          (    I    )
 10   .048  .107                          (    IX   )
 11   .011  .107                          (    I    )
 12   .087  .108                          (    IXX  )
 13  -.010  .108                          (    I    )
 14   .097  .109                          (    IXX  )
 15  -.011  .110                          (    I    )
                                        -2SE       +2SE
```

FIGURE 5.4(b) *Identification:* CCF for the Prewhitened p_t and h_t Series

The fertility rate is thus a type of birth rate. Assuming no effective birth control methods, fertility rate is a measure of the number of females of child-bearing age (15 to 45 years old) in the population. An analysis of the f_t series shows that it is well represented by the ARIMA (2,0,0) model

$$(1 - .62B + .23B^2)f_t = a_t.$$

To prewhiten this series, the ARIMA (2,0,0) model is applied:

$$\begin{aligned} a_t &= (1 - .62B + .23B^2)f_t \\ &= f_t - .62f_{t-1} + .23f_{t-2}. \end{aligned}$$

```
SERIES.. RESIDUAL (NOBS= 100) ESTIMATED MODEL RESIDUALS
NO. OF VALID OBSERVATIONS  =  100.

AUTOCORRELATIONS OF LAGS 1 - 30.
Q( 29, 100) =   26.740        SIG =      .586

 LAG    CORR     SE  -1    -.8   -.6   -.4   -.2    0     .2    .4    .6    .8   +1
                     +----+----+----+----+----+----+----+----+----+----+
   1    .250   .100                          (      IXXXX)X
   2    .030   .110                         (       IX     )
   3    .050   .110                         (       IX     )
   4    .050   .110                         (       IX     )
   5    .050   .110                         (       IX     )
   6    .150   .110                         (       IXXXX  )
   7    .020   .110                         (       IX     )
   8   -.090   .110                         (     XXI      )
   9   -.080   .110                         (     XXI      )
  10    .050   .110                         (       IX     )
  11    .050   .110                         (       IX     )
  12   -.140   .110                         (  XXXXI       )
  13   -.040   .110                         (      XI      )
  14   -.110   .110                         (    XXXI      )
  15   -.100   .110                         (    XXXI      )
  16    .010   .110                         (       I      )
  17    .070   .110                         (       IXX    )
  18    .030   .120                         (       IX     )
  19    .070   .120                         (       IXX    )
  20   -.030   .120                         (      XI      )
  21   -.030   .120                         (      XI      )
  22    .010   .120                         (       I      )
  23   -.030   .120                         (      XI      )
  24   -.110   .120                         (    XXXI      )
  25   -.110   .120                         (    XXXI      )
  26   -.040   .120                         (      XI      )
  27   -.100   .120                         (    XXXI      )
  28   -.180   .120                         (XXXXXXI       )
  29   -.130   .120                         (   XXXXI      )
  30   -.110   .120                         (    XXXI      )
                                          -2SE          +2SE
```

FIGURE 5.4(c) *Identification:* ACF for the Residuals of the Model $p_t = .239\, f_t + a_t$

The same operator is then applied to the population rate increase time series

$$z_t^* = (1 - .62B + .23B^2)p_t$$
$$= p_t - .62p_{t-1} + .23p_{t-2}.$$

The CCF for a_t and z_t^*, shown in Figure 5.4(b), suggests a zero-order transfer function for $f_t$and p_t with no time lag.

$$p_t = \omega_0 f_t + N_t.$$

```
SERIES.. RESIDUAL (NOBS= 100) ESTIMATED MODEL RESIDUALS
NO. OF VALID OBSERVATIONS  =  100.

AUTOCORRELATIONS OF LAGS 1 - 30.
Q( 29, 100) =  27.040        SIG =      .570

 LAG    CORR     SE  -1    -.8  -.6  -.4  -.2   0    .2   .4   .6   .8  +1
                      +----+----+----+----+----+----+----+----+----+----+
   1    .250  .100                      (       IXXXX)X
   2    .030  .110                     (        IX      )
   3    .050  .110                     (        IX      )
   4    .050  .110                     (        IX      )
   5    .050  .110                     (        IX      )
   6    .150  .110                     (        IXXXX   )
   7    .020  .110                     (        IX      )
   8   -.090  .110                     (      XXI       )
   9   -.080  .110                     (      XXI       )
  10    .050  .110                     (        IX      )
  11    .050  .110                     (        IX      )
  12   -.140  .110                     (  XXXXI         )
  13   -.040  .110                     (       XI       )
  14   -.110  .110                     (    XXXI        )
  15   -.110  .110                     (    XXXI        )
  16    .010  .110                     (        I       )
  17    .070  .110                     (        IXX     )
  18    .030  .120                     (        IX      )
  19    .060  .120                     (        IXX     )
  20   -.030  .120                     (       XI       )
  21   -.030  .120                     (       XI       )
  22    .000  .120                     (        I       )
  23   -.030  .120                     (       XI       )
  24   -.120  .120                     (    XXXI        )
  25   -.110  .120                     (    XXXI        )
  26   -.040  .120                     (       XI       )
  27   -.100  .120                     (    XXXI        )
  28   -.180  .120                     (XXXXXI          )
  29   -.130  .120                     (    XXXI        )
  30   -.110  .120                     (    XXXI        )
                                     -2SE            +2SE
```

FIGURE 5.4(d) *Identification:* ACF for the Residuals of the Model $p_t = .239\,f_t - .01\,h_{t-1} + a_t$

Population growth is determined by fertility rates in the same year. As the fertility rate is a type of birth rate, the zero time lag, indicated by a spike at CCF (0), makes good sense. The estimate of ω_0 is:

$$\hat{\omega}_0 = .239.$$

The residual ACF for this estimate, shown in Figure 5.4(c), suggests an ARIMA (1,0,0) model for N_t. The fully specified bivariate model is thus

$$p_t = \omega_0 f_t + \frac{a_t}{1 - \phi_1 B}.$$

Parameter estimates for this model are:

$$\hat{\omega}_0 = .234 \text{ with t statistic} = 11.44$$
$$\hat{\phi}_1 = .26 \text{ with t statistic} = 2.55.$$

Both parameter estimates are statistically significant and otherwise acceptable. Diagnosis of the model residuals indicates that the noise and transfer function components are independent and that the model residuals are white noise. We thus accept this tentative model.

So far, the analysis of population rate increases leads us to conclude that the growth of Swedish population during the 1750–1849 century was due to (or caused by) the effects of *two* exogenous variables: agricultural production and fertility. The bivariate models for these two relationships are:

$$p_t = .826h_{t-1} + \frac{a_t}{1 - .46B}$$

and

$$p_t = .234f_t + \frac{a_t}{1 - .26B}.$$

A logical next step would be to incorporate both exogenous variables into a single multivariate ARIMA model. On the basis of the bivariate models, the multivariate model is specified as

$$p_t = \omega_0 f_t + \omega_0^* h_{t-1} + N_t.$$

Parameter estimates for the transfer function component are:

$$\hat{\omega}_0 = .239$$
$$\hat{\omega}_0^* = -.010.$$

Note that the ω-parameter for the effect of the Harvest Index series has dropped substantially in absolute value. This estimate could change dramatically when noise parameters are estimated, however. The residual ACF for these estimates, shown in Figure 5.4(d), suggests an ARIMA (1,0,0) model for the N_t component. The full model is thus

$$p_t = \omega_0 f_t + \omega_0^* h_{t-1} + \frac{a_t}{1 - \phi_1 B}$$

Parameter estimates for this tentative model are:

$$\hat{\omega}_0 = .231 \text{ with t statistic} = 2.64$$

$$\hat{\omega}_0^* = .046 \text{ with t statistic} = .28$$

$$\hat{\phi}_1 = .26 \text{ with t statistic} = 2.64.$$

There is no need to diagnose this tentative model. The estimate of ω_0^* is not statistically significant and must be dropped from the model.

This analysis would seem to disconfirm Sundbärg's hypothesis. When fertility rates are considered, the Harvest Index time series accounts for only a statistically insignificant proportion of the variance in the population change time series.

There is a simpler explanation for the finding of this analysis, however. A multivariate ARIMA model of the sort

$$y_t = f(x_{1t}) + \ldots + f(x_{nt}) + N_t$$

is generally nonlinear but is linear in terms of its components. To estimate parameters for the full model, *all components must be independent*.

As it turns out (and as a diagnosis of this model would have indicated), the fertility rate and harvest time series are highly correlated. The harvest in effect determines the values of future fertility rates.

$$h_t \text{ - - - - - - - - - - - -}\rightarrow f_{t+b}.$$

Thomas (1940) cites a number of plausible mechanisms for this relationship. First, in years following a crop failure, marriage rates (and hence, fertility rates) drop. Second, and more important, in years following a crop failure, young women who might otherwise bear children in Sweden are likely to emigrate (primarily to Finland and the United States during this period). As a result of emigration, the average age of the female population rises dramatically in years following a crop failure and fertility drops accordingly

If there is indeed a causal relationship between harvests and fertility rates, the two transfer function components of the multivariate ARIMA model are not independent. The model we built implies that

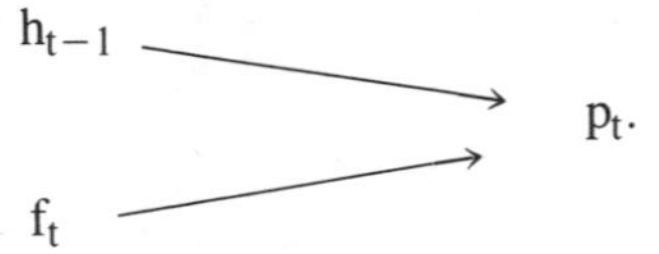

If there is a causal relationship between the harvest and fertility rates, however, the true model is:

$$h_{t-1} \text{-------}\!\!\rightarrow f_t \text{-------}\!\!\rightarrow p_t.$$

Our finding of no effect for the Harvest Index time series could thus be due only to a misspecification of the model.

As a first step in building a multivariate population growth model, the hypothesized causal relationship between harvests and fertility rates must be tested. The time series are prewhitened as

$$a_t = (1 + .39B)^{-1}h_t$$
$$z_t^* = (1 + .39B)^{-1}f_t.$$

```
CROSS-CORRELATIONS OF LAGS  -15 TO  15.
NO. OF VALID OBSERVATIONS     =    100.

INPUT SERIES..  HARVEST   PREWHITENED SWEDISH HARVEST INDEX
OUTPUT SERIES.. FERTILE   PREWHITENED SWEDISH FERTILITY RATE

 LAG   CORR    SE  -1   -.8  -.6  -.4  -.2   0    .2   .4   .6   .8  +1
                    +----+----+----+----+----+----+----+----+----+----+
 -15  -.006  .108                       (    I    )
 -14  -.024  .108                       (   XI    )
 -13   .121  .107                       (    IXXX )
 -12   .093  .107                       (    IXX  )
 -11  -.110  .106                       ( XXXI    )
 -10   .039  .105                       (    IX   )
  -9   .146  .105                       (    IXXXX)
  -8  -.045  .104                       (   XI    )
  -7   .053  .104                       (    IX   )
  -6  -.220  .103                      X(XXXXI    )
  -5   .029  .103                       (    IX   )
  -4   .005  .102                       (    I    )
  -3   .024  .102                       (    IX   )
  -2   .096  .101                       (    IXX  )
  -1  -.269  .101                     XX(XXXXI    )
   0   .017  .100                       (    I    )
   1   .417  .101                       (    IXXXX)XXXXX
   2   .347  .101                       (    IXXXX)XXXX
   3  -.079  .102                       (  XXI    )
   4  -.086  .102                       (  XXI    )
   5   .013  .103                       (    I    )
   6   .024  .103                       (    IX   )
   7  -.091  .104                       (  XXI    )
   8  0.000  .104                       (    I    )
   9   .185  .105                       (    IXXXX*
  10   .099  .105                       (    IXX  )
  11   .017  .106                       (    I    )
  12  -.031  .107                       (   XI    )
  13  -.050  .107                       (   XI    )
  14  -.051  .108                       (   XI    )
  15   .118  .108                       (    IXXX )
                                      -2SE     +2SE
```

FIGURE 5.4(e) *Identification:* CCF for the Prewhitened f_t and h_t Series

```
SERIES.. RESIDUAL (NOBS=  99)  ESTIMATED MODEL RESIDUALS
NO. OF VALID OBSERVATIONS  =     99.

AUTOCORRELATIONS OF LAGS 1 - 30.
Q( 29,   99) =   52.133        SIG =      .005

LAG   CORR    SE  -1   -.8  -.6  -.4  -.2   0   .2   .4   .6   .8   +1
                   +----+----+----+----+----+----+----+----+----+----+
      .480  .100                        (   IXXXX)XXXXXXX
  2   .260  .120                      (     IXXXXX)X
  3   .140  .130                     (      IXXXX  )
  4   .230  .130                     (      IXXXXXX)
  5   .160  .130                     (      IXXXX  )
  6   .000  .130                     (      I      )
  7   .060  .130                     (      IXX    )
  8  -.050  .140                     (     XI      )
  9  -.030  .140                     (     XI      )
 10   .010  .140                     (      I      )
 11   .090  .140                     (      IXX    )
 12   .110  .140                     (      IXXX   )
 13   .060  .140                     (      IXX    )
 14   .030  .140                     (      IX     )
 15  -.010  .140                     (      I      )
 16   .000  .140                     (      I      )
 17   .100  .140                     (      IXXX   )
 18   .180  .140                     (      IXXXXX )
 19   .160  .140                     (      IXXXX  )
 20   .050  .140                     (      IX     )
 21   .020  .140                     (      IX     )
 22   .060  .140                     (      IXX    )
 23   .090  .140                     (      IXX    )
 24   .080  .140                     (      IXX    )
 25  -.020  .140                     (     XI      )
 26  -.010  .140                     (      I      )
 27  -.050  .140                     (     XI      )
 28  -.070  .140                     (    XXI      )
 29  -.040  .140                     (     XI      )
 30   .010  .140                     (      I      )
                                   -2SE           +2SE
```

FIGURE 5.4(f) *Identification:* ACF for the Residuals of the Model

$$f_t = \frac{3.498}{1 - .372B} h_{t-1} + a_t$$

The CCF for a_t and z_t^*, shown in Figure 5.4(e), shows a strong and unambiguous casual relationship between the Harvest Index and fertility rates time series. Statistically significant spikes at CCF (+1) and CCF (+2) imply the model

$$f_t = \frac{\omega_0}{1 - \delta_1 B} h_{t-1} + N_t.$$

Parameter estimates for this model are:

$$\hat{\omega}_0 = 3.498$$
$$\hat{\delta}_1 = .372.$$

```
CROSS-CORRELATIONS OF LAGS  -15 TO  15.
NO. OF VALID OBSERVATIONS     =     99.

INPUT SERIES..  POP          SWEDISH POPULATITION CHANGE
OUTPUT SERIES.. RESIDUAL   ESTIMATED MODEL RESIDUALS

 LAG    CORR    SE   -1  -.8  -.6  -.4  -.2   Ø    .2   .4   .6   .8  +1
                     +----+----+----+----+----+----+----+----+----+----+
 -15   -.Ø36  .1Ø9                       (   XI    )
 -14   -.ØØ6  .1Ø8                       (    I    )
 -13   -.13Ø  .1Ø8                       ( XXXI    )
 -12    .Ø45  .1Ø7                       (    IX   )
 -11   -.Ø43  .1Ø7                       (   XI    )
 -1Ø    .Ø34  .1Ø6                       (    IX   )
  -9   -.Ø23  .1Ø5                       (   XI    )
  -8   -.ØØ7  .1Ø5                       (    I    )
  -7    .Ø11  .1Ø4                       (    I    )
  -6   -.141  .1Ø4                       (XXXXI    )
  -5    .116  .1Ø3                       (    IXXX )
  -4    .121  .1Ø3                       (    IXXX )
  -3   -.Ø34  .1Ø2                       (   XI    )
  -2   -.145  .1Ø2                       (XXXXI    )
  -1   -.ØØ6  .1Ø1                       (    I    )
   Ø    .528  .1Ø1                       (    IXXXX)XXXXXXXX
   1    .3Ø8  .1Ø1                       (    IXXXX)XXX
   2    .196  .1Ø2                       (    IXXXX*
   3    .Ø75  .1Ø2                       (    IXX  )
   4    .122  .1Ø3                       (    IXXX )
   5    .Ø74  .1Ø3                       (    IXX  )
   6    .134  .1Ø4                       (    IXXX )
   7    .Ø33  .1Ø4                       (    IX   )
   8   -.Ø53  .1Ø5                       (   XI    )
   9   -.Ø95  .1Ø5                       (  XXI    )
  1Ø   -.Ø7Ø  .1Ø6                       (  XXI    )
  11    .ØØ4  .1Ø7                       (    I    )
  12    .136  .1Ø7                       (    IXXX )
  13    .127  .1Ø8                       (    IXXX )
  14    .1ØØ  .1Ø8                       (    IXXX )
  15    .Ø35  .1Ø9                       (    IX   )
                                       -2SE      +2SE
```

FIGURE 5.4(g) *Identification:* CCF for the f_t^* Series

The residual ACF for these estimates, shown in Figure 5.4(f), suggest an ARIMA (1,0,0) model for the N_t component. The full model is thus

$$f_t = \frac{\omega_0}{1 - \delta_1 B} h_{t-1} + \frac{a_t}{1 - \phi_1 B}.$$

Parameter estimates for the full model are:

$$\hat{\omega}_0 = 3.572 \text{ with t statistic} = 6.17$$

$$\hat{\delta}_1 = .439 \text{ with t statistic} = 3.34$$
$$\hat{\phi}_1 = .50 \text{ with t statistic} = 5.45.$$

All parameter estimates are statistically significant and otherwise acceptable. Diagnosis of this model indicates that the noise and transfer function components are independent and that the model residuals are white noise.

The analysis confirms the hypothesis that there is a strong causal relationship between harvests and fertility rates and the implications of this finding for a multivariate population growth model are clear. The multivariate model must reflect the structure

$$h_{t-1} \dashrightarrow f_t \dashrightarrow p_t.$$

To build this structure, we require a fertility rate time series that has been purged of the harvest effect. The residuals of the harvest-fertility bivariate model will prove adequate for this purpose. As this bivariate model is:

$$f_t = \frac{3.6}{1 - .44B} h_{t-1} + \frac{a_t}{1 - .50B},$$

we define a new fertility rate time series, f_t^*, as

$$f_t^* = a_t = (1 - .50B)\, f_t - \frac{1 - .50B}{1 - .44B} 3.6\, h_{t-1}.$$

The f_t^* series is white noise and, by definition, uncorrelated with the h_t series. These two time series may thus be incorporated directly into the multivariate population growth model.

The transfer function relationship between the h_t and p_t series has already been determined. The transfer function relationship between f_t^* and p_t must be identified, however. We would ordinarily begin by prewhitening the series. But f_t^* is hypothesized to be the causor and, as f_t^* is already white noise, prewhitening is not required. The CCF estimated from f_t^* and p_t, shown in Figure 5.4(g), has significant spikes at CCF (0) and CCF (+1), suggesting the transfer function relationship

$$p_t = \frac{\omega_0}{1 - \delta_1 B} f_t^* + N_t.$$

Adding this structure to the previously identified relationship between p_t and

h_t, the full model is:

$$p_t = \frac{\omega_0}{1 - \delta_1 B} f_t^* + \omega_0^* h_{t-1} + N_t.$$

Parameter estimates for this transfer function are:

$$\hat{\omega}_0 = .234$$
$$\hat{\omega}_0^* = 1.017$$
$$\hat{\delta}_1 = .616.$$

The residual ACF for these estimates, shown in Figure 5.4(h), suggests an ARIMA (1,0,0) model for the N_t component. The fully specified model is thus

$$p_t = \frac{\omega_0}{1 - \delta_1 B} f_t^* + \omega_0^* h_{t-1} + \frac{a_t}{1 - \phi_1 B}.$$

Parameter estimates for this full model are:

$$\hat{\omega}_0 = .234 \text{ with t statistic} = 8.94$$
$$\hat{\omega}_0^* = .952 \text{ with t statistic} = 6.44$$
$$\hat{\delta}_1 = .620 \text{ with t statistic} = 9.70$$
$$\hat{\phi}_1 = .255 \text{ with t statistic} = 2.48.$$

All parameter estimates are statistically significant and otherwise acceptable. Diagnosis indicates that all components are independent of each other and that the model residuals are white noise. We accept this tentative model.

At this point, we must review our progress by comparing three models of Swedish population growth. First, the univariate model

$$p_t = (1 + .44B)a_t$$

has an RMS = 27.8. Second, the bivariate model

$$p_t = .826h_{t-1} + \frac{a_t}{1 - .46B}$$

has an RMS = 23.6. By incorporating the effects of harvests, the predictive power of the model increases substantially. Finally, the multivariate model

$$p_t = \frac{.23}{1 - .62B} f_t^* + .952h_{t-1} + \frac{a_t}{1 - .26B}$$

has an RMS = 12.3. Again, by incorporating fertility rates, the predictive power of the model increases substantially.

The bivariate and multivariate models have increased our understanding of Swedish population growth in another, more important sense. While the univariate model does an adequate job of *predicting,* it does nothing to *explain* the substantive phenomenon. The bivariate and multivariate models, on the other hand, explain population growth in *causal* terms.

Figure 5.4(h) diagrams the multivariate population growth model as an input-output system. According to this diagram, population changes are due to the effects of three exogenous forces. Two of the three exogenous forces are white noise processes, a_t and f_t^*. The third exogenous force is h_{t-1}, the Harvest Index time series. Fertility rates, denoted by f_t in the diagram, play

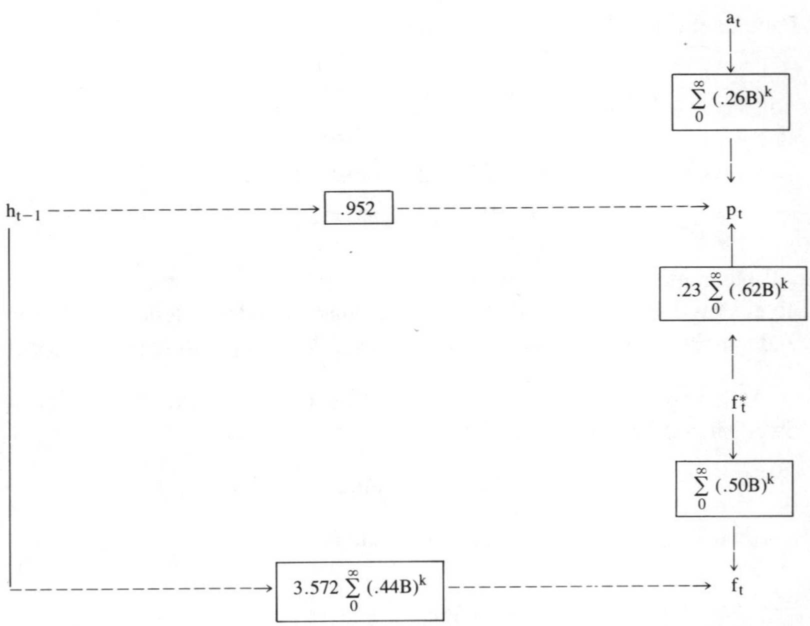

FIGURE 5.4(h) An Input-Output Diagram of the Multivariate Model

no causal role in this system. The f_t series is included in the model only because it shares two sources of variance with the p_t series. Causal arrows in the diagram lead *to* the f_t variable but not *from* it.

The reader who is familiar with structural equations models (see, e.g., Blalock, 1971; Goldberger and Duncan, 1973; Heise, 1975) or econometric models (see, e.g., Kmenta, 1971; Johnston, 1972) will immediately recognize Figure 5.4(h) for what it is: a structural model of Swedish population growth. The reader may surmise from this that the structural equations or econometric approaches to time series analysis and the ARIMA approach are substantially the same. In fact, this multivariate ARIMA model can be used for any purpose that a structural equations or econometric model would be used for; and of course, the ARIMA model is subject to any and all criticisms that could be made of a structural equations or econometric models.

The major difference between econometric or structural equations approaches to time series analysis and the ARIMA approach is that econometric models are ordinarily identified theoretically. ARIMA models are identified empirically, of course, and because of this, ARIMA models require relatively long time series. Beyond this practical point, there is no difference.

A structural equations model ordinarily begins with a set of "structural equations" deduced from theory. These elementary equations are then manipulated until a "reduced form equation" emerges. The reduced form equation includes all variables of the system and, under ideal conditions, parameters can be estimated directly from the reduced form. This is also true of the multivariate ARIMA model. The empirically identified ARIMA model for Swedish population growth is:

$$p_t = \frac{.23}{1 - .62B} f_t^* + .95h_{t-1} + \frac{a_t}{1 - .26B}.$$

However, noting that f_t^* is related to f_t as

$$f_t = \frac{3.6}{1 - .44B} h_{t-1} + \frac{f_t^*}{1 - .50B}$$

so

$$f_t^* = (1 - .50) f_t - \frac{(3.6)(1 - .50B)}{1 - .44B} h_{t-1}.$$

This expression for f_t^* may be substituted into the multivariate model to obtain

$$p_t = \frac{(.23)(1 - .50B)}{1 - .62B} f_t$$

$$- \left[\frac{(.23)(3.6)(1 - .50B)}{(1 - .62B)(1 - .44B)} - .952\right] h_{t-1}$$

$$+ \frac{a_t}{1 - .26B} .$$

This formidable model is the reduced form equation that would have resulted from the elementary structural equations. In this case, and in the general case, structural equations or econometric approaches to time series analysis and the ARIMA approach lead to the same model. The reader will best understand the ARIMA approach to time series analysis by viewing it as a special case of the structural equations or econometric approach. Multivariate ARIMA models are structural equations or econometric models in which the relationships among variables have been identified empirically. Empirical identification implies that relatively long time series are available, of course.

The final step in multivariate ARIMA analysis is model interpretation. As this relates to Sundbärg's hypothesis, the model shows clearly that, during the 1750–1849 century, *the harvest had a profound influence on population growth*.

5.5 Conclusion and Recommendations

So far, there has been little published research in the social sciences using multivariate ARIMA methods. One reason often cited for this situation is that the computer software required for multivariate ARIMA analysis is not widely available. We will address this point directly in the next chapter. For the time being, we note only that while this has been true in the past, there are now many suitable computer programs readily available in academic computing centers.

Another reason often cited for this situation is that the time series data required for ARIMA models are not available. An ARIMA model often requires a time series of 100 observations or more and data of this quantity are seldom found in the social sciences. Data availability varies from substantive area to substantive area, of course. While long time series may indeed be rare, we suspect that they are not so rare as people think. In the course of writing this volume, we encountered many data sets long enough for ARIMA analysis. Moreover, as the use of computers spreads, we suspect that long time series will become increasingly common in the social sciences.

Finally, there is a popular misconception about the nature of ARIMA methods which might explain why these methods are not widely used. Regarding structural equations or econometric approaches to time series analysis, for example, the popular notion is that ARIMA models are empirical and *a*theoretical (or mindless) while econometric models are sometimes empirical but always theoretical. On this point, Hibbs writes.

> Box-Tiao or Box-Jenkins methods are essentially models for "ignorance" that are not based on theory and, in this sense, are void of explanatory power. Although these models are in many situations likely to yield good estimates of endogenous reponses to external interventions, they provide no insight into the causal structure underlying the transmission of exogenous impulses through a dynamic system of interdependent social, economic, or political relationships [1977: 172].

In this oft-quoted passage, however, Hibbs refers only to *univariate* ARIMA models. When multivariate models are considered, the differences between these two approaches to time series analysis are largely practical differences which have nothing to do with the quality of the models.

When ARIMA models were introduced to the social sciences a decade ago, an unproductive debate ensued over the relative merits of ARIMA and econometric approaches to time series analysis. More recently, these two approaches have converged. An eclectic approach to time series analysis recognizes that multivariate ARIMA models and econometric models are identical in every substantial respect.

The advantages of ARIMA models over econometric or structural equations models are obvious. Lag structures among variables are identified more precisely; seasonal variance is accounted for in a systematic manner; model parameters are estimated with a high degree of reliability; and so forth. But all of these advantages follow from the quantity of data required for the ARIMA model. ARIMA models require relatively long time series and, in this sense, multivariate ARIMA models are "better" than econometric models only because long time series are "better" than short time series.

The advantages of econometric or structural equations models over ARIMA models are also a consequence of this data requirement. In macroeconomic applications, an econometric model may incorporate *hundreds* of time series variables. An ARIMA model based on this same number of time series would have *thousands* of parameters to be estimated and, thus, would require time series of a thousand or more observations. Assuming that time series of more than 200 or 300 observations are unavailable, multivariate ARIMA models are ordinarily limited to a dozen or fewer time series variables. Thus, when a rich body of theory points to many variables and many structural relationships, the econometric or structural equations approach to time series analysis will be superior to the ARIMA approach.

In our opinion, multivariate ARIMA models will become an important research tool for the social sciences. ARIMA models will never *replace* structural models as a research tool, however. These two approaches to time series analysis each have strengths and weaknesses. The particular approach taken should be determined by the quality and quantity of data available and by the problem itself.

For Further Reading

There is a considerable "transfer of learning" between the ARIMA and regression approaches to time series analysis. Hibbs (1974) and Ostrom (1978) both develop several useful regression models for time series analysis. The Hibbs work has become a classic in its field and should not be missed. Structural equations or econometric approaches to time series analysis differ from the more general regression approaches in that a model based on *several* regression equations is posited. Kmenta (1971) develops much of econometric theory without linear algebra. Johnston (1972) develops the same material but from a linear algebra basis; in addition, Johnston (1972: Chapter 4) includes an excellent introduction to linear algebra. The Kmenta book is widely used as an undergraduate text while the Johnston book is widely used as a graduate text. Dhrymes (1974), Malinvaud (1970), and Theil (1971) are generally thought of as "advanced" econometrics textbooks and are not suitable introductions. Goldberger and Duncan (1973) or Heise (1975) present this material from a more eclectic basis and thus may be more suitable than an econometric text. Hibbs (1977) outlines the relationship between structural equations and ARIMA approaches. While Hibbs's argument concerns only impact assessment models, it is general to any multivariate time series model. New advances in this field generally appear in such journals as *Econometrica, Journal of the American Statistical Association, Political Methodology,* and *Sociological Methods and Research.* The reader who wishes to keep abreast of new developments is advised to watch these journals.

NOTES TO CHAPTER 5

1. We use "multivariate" to mean time series models in which one output (dependent) series is explained as a function of several input (independent) series, that is,

$$Y_t = f(X_{1t}, X_{2t}, \ldots X_{nt}) + N_t.$$

Other authors may use this term to mean a time series model that has several output series, each a distinct measure of a single underlying concept.

2. The modeling strategy we develop in this chapter is only for the simplest case of *uni*directional cause, that is, for systems in which "x causes y":

$$x_t \dashrightarrow y_{t+b}.$$

*Bi*directional causal structures

$$x_t \longleftarrow\!\!-\!-\!-\!-\!-\!-\!-\!\!\longrightarrow y_t$$

must be modeled through a more complicated strategy. The more complicated causal systems are beyond the scope of this introductory volume. The interested reader is directed to Granger and Newbold (1977: Chapter 7) for a discussion of advanced multivariate modeling topics.

3. We use the Thomas (1940) data only to demonstrate a strategy for multivariate ARIMA modeling. We do not generally endorse these methods for demographic accounting. Demography may be the only social science field in which sophisticated, proven mathematical models are routinely available. Keyfitz (1977: Chapter 8) gives a readable introduction to these models; see also Land (1980) for a comprehensive system of models.

6 Nonlinear Estimation

In this final chapter of the volume, we describe the set of numerical procedures by which ARIMA model parameters are estimated. These procedures are relatively more complicated than the procedures used to estimate the parameters of many statistical models used by social scientists. We have left this material until last because some readers may not be interested in the details of estimation. Models and parameter estimates can be competently interpreted without reference to the mechanical routines of estimation. Those readers who wish only to interpret the results of time series research may thus not wish to invest the time and effort required to master this material. Those readers who plan to estimate ARIMA models in the course of instruction or research, on the other hand, will find that this material is absolutely essential.

The statistical models most commonly used by social scientists are *linear* models. For a finite set of variables, $\{Y, X_1 \ldots, X_n, u\}$, a linear model is defined as any linear combination of those variables. The familiar linear prediction equation, for example,

$$Y = b_0 + b_1X_1 + \ldots + b_nX_n + u,$$

is a linear combination of a finite set of variables and, thus, is a linear model. In contrast, the conventional log-linear version of this same prediction equation,

$$Y = b_0(X_1)^{b_1} \ldots (X_n)^{b_n}(u),$$

is *not* a linear combination and, thus, is *not* a linear model. If the model is log-transformed, however:

$$Ln(Y) = Ln(b_0) + b_1 Ln(X_1) + \ldots + b_n Ln(X_n) + Ln(u),$$

a linear model is defined, though not in terms of the original set of variables. Finally, the model

$$Y = Ln(b_0)\,[Ln(X_1)]^{b_1} \ldots [Ln(X_n)]^{b_n}[Ln(u)],$$

is *intrinsically* nonlinear. This model is not a linear combination of the variables nor is there any transformation which will result in a linear combination.

When a model is linear (or when it can be made linear with a transformation), maximum likelihood estimates of its parameters can be obtained as analytical solutions to a set of equations. This procedure is a straightforward one which could be executed (though tediously) with paper and pencil. When a model is intrinsically nonlinear, however, as the general ARIMA model is, analytical solutions for parameter estimates are ordinarily undefined. Numerical solutions must be obtained, usually by means of an iterative algorithm wherein parameter estimates are successively refined to a desired degree of precision. While these numerical algorithms could theoretically be executed with paper and pencil, the sheer number of iterative computations required is so large as to render this strategy practically impossible. Estimation of ARIMA model parameters is hence, by definition, dependent on sophisticated computer software.

Our discussion of ARIMA parameter estimation will begin with the logic of maximum likelihood estimation (MLE). After developing this logic for a simple problem, we demonstrate that the MLEs of ARIMA parameters are identical with the least-squares estimates. Due to this identity, MLEs of ARIMA parameters can be derived through a numerical minimization of the sum of squares function. We illustrate the general principles of least-squares estimation by plotting the sum of square functions for several time series analyzed in earlier chapters. The most commonly used numerical algorithms for minimizing the sum of squares function are discussed, and at the end of the chapter we describe the characteristics of several ARIMA software packages.

6.1 Maximum Likelihood Estimation

Consider a coin-flip experiment in which the probability of realizing k heads in N flips is given by the probability density function:

$$P(\text{k heads in N flips}) = \frac{N!}{k!\,(N-k)!}\,p^k(1-p)^{N-k}$$

for p = probability of heads on any flip. If a fair coin is flipped 10 times with the result of 6 heads, the probability of this result is:

$$P(\text{6 heads in 10 flips}) = \frac{10!}{6!\,4!}\,(.5)^6(.5)^4$$
$$= .205065.$$

This probability may now be used to test the null hypothesis that the coin is fair, that is, that p = .5.

Statistical inferences of this sort are called *classical* inferences. In contrast, *Bayesian* inferences, of which MLE is a special case, use the empirical outcome of an experiment to derive a set of model parameters (p in this case) which is most likely to have generated those outcomes. Again, if a coin is flipped 10 times with the result of 6 heads, a *likelihood function,* L, is defined as:

$$L = (p)\,(p)\,(p)\,(p)\,(p)\,(p)\,(1-p)\,(1-p)\,(1-p)\,(1-p)$$
$$= p^6\,(1-p)^4.$$

This likelihood function is identical with the classical probability density function except for a multiplicative constant which in this case is the term

$$\frac{N!}{k!\,(N-k)!}.$$

The likelihood function describes a relationship between the model parameter, p, and the likelihood that a particular value of the parameter has generated the set of outcomes. Substituting successively larger values of p into the likelihood function, starting with p = 0.0, the relationship emerges as:

p	likelihood
0.0	0.000000
.1	.000006
.2	.000026
.3	.000175
.4	.000531
.5	.000977
.6	.001194
.7	.000953
.8	.000491

.9	.000053
1.0	0.000000

a parabola whose maximum value occurs when p = .6. *This is the MLE of p, that is, the value of p which is most likely to generate 6 heads in 10 flips.*

But there is no guarantee that the likelihood function attains its maximum value at exactly p = .6. It is possible, for example, that the maximum occurs at p = .59 or .61. To derive the *exact* maximum of the likelihood function, a differential calculus operation is required. In the general case, the likelihood for the result of k heads in N flips is:

$$L = p^k(1-p)^{N-k}.$$

The calculus operation will be simplified, however, if the likelihood function is log-transformed:

$$Ln(L) = kLn(p) + (N-k)Ln(1-p).$$

The log-likelihood function is linear and, thus, will always be easier to differentiate than the likelihood function.

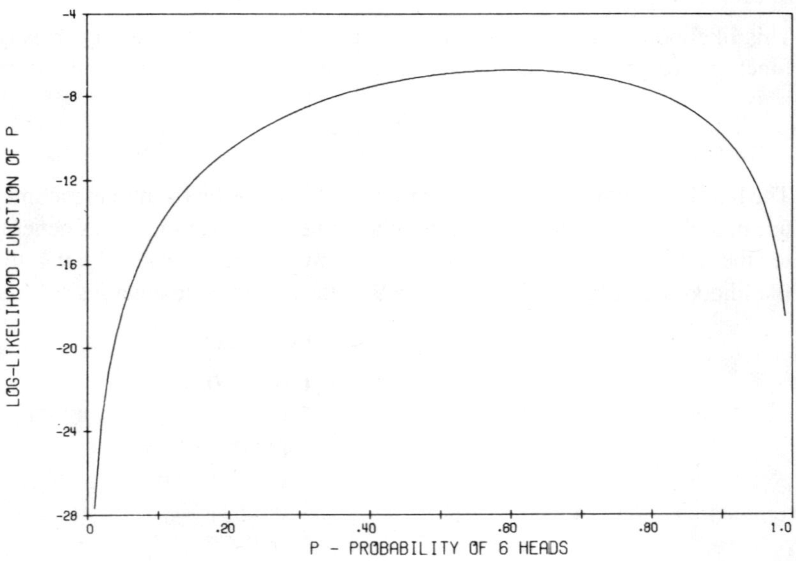

FIGURE 6.1(a) Log-Likelihood Function for a Coin-Flip Experiment: 6 Heads in 10 Flips

As the log-transformation is a monotonic linear transformation, this simplification will have no bearing on the outcome. Figure 6.1(a) shows the log-likelihood function for the result of 6 heads in 10 flips. The log-likelihood function shown here is unimodal with a clearly defined maximum point. While we will not do so here, it is easily demonstrated that the likelihood and log-likelihood functions are maximized by the same value of p.

To obtain the MLE of p, the log-likelihood function is differentiated with respect to p. This operation is:

$$\frac{\partial\,[\mathrm{Ln}(L)]}{\partial\,p} = k/p - (N-k)\,/\,(1-p).$$

The solution to this derivative

$$\frac{\partial\,[\mathrm{Ln}(L)]}{\partial\,p} = 0$$

$$k/p - (N-k)\,/\,(1-p) = 0$$

$$k - kp = pN - pk$$

$$p = k/N$$

is the MLE of p. Substituting $k = 6$ and $N = 10$ into this formula, results in $p = .6$.

We do not require a calculus background of the reader but the principles of calculus maximization should be clear from the log-likelihood function shown in Figure 6.1(a). The log-likelihood function (and the likelihood function itself) increases until $p = k/N$ and then decreases. Calculus maximization is a simple method of discovering the exact point at which the log-likelihood function attains its maximum value.

The logic of MLE may now be generalized to the more difficult problem of obtaining ARIMA parameter estimates. In the coin-flip experiment, binomial outcomes, heads or tails, were observed and MLE procedures were used to obtain a value of the parameter p which was most likely to have generated those outcomes. In a time series analysis, white noise outcomes (random shocks) are observed and MLE procedures are used to obtain values of ARIMA parameters which are most likely to have generated those shocks. Of course, the random shocks are not really observed but we will assume for the moment that they are.

Denoting the first random shock of a time series as a_1, the classical probability density function is:

$$P(a_1) = \frac{e^{-(a_1^2/2\sigma_a^2)}}{\sigma_a\sqrt{2\pi}}.$$

This formidable appearing expression is merely the "Normal curve," the probability density function associated with all Normally distributed variables. As white noise has the statistical property

$$a_t \sim NID(0,\sigma_a^2),$$

the Normal (or Gaussian) function is implied.

If one were interested only in classical inference (testing a null hypothesis that all model parameters were zero, for example), inferences could follow directly from this function. To obtain MLEs of the ARIMA parameters, however, a likelihood function is required. The likelihood function for a_1 is:

$$L(a_1) = \frac{e^{-(a_1^2/2\sigma_a^2)}}{\sigma_a}.$$

This likelihood function differs from the probability density function only by the multiplicative constant, $1/\sqrt{2\pi}$.

The second shock of the series, a_2, has an identical likelihood function. Because each shock of a white noise process is independent of every other shock, the likelihood function for a_1 *and* a_2 is simply the *product* of their individual likelihoods. That is,

$$L(a_1,a_2) = [L(a_1)\,L(a_2)] = \frac{e^{-(a_1^2/2\sigma_a^2)}}{\sigma_a} \times \frac{e^{-(a_2^2/2\sigma_a^2)}}{\sigma_a}$$

$$= \frac{e^{-(a_1^2 + a_2^2)/2\sigma_a^2}}{\sigma_a^2}.$$

In the same way, and for the same reasons, the likelihood function for all N shocks of a time series is:

$$L(a_1, \ldots, a_N) = \frac{e^{-(\sum_{i=1}^{N} a_i^2)/2\sigma_a^2}}{\sigma_a^N}.$$

The log-likelihood function is thus:

$$\mathrm{Ln}\,(\mathrm{L}) = -\sum_{i=1}^{N} a_i^2/2\sigma_a^2 - \mathrm{NLn}\,(\sigma_a).$$

The utility of log-transforming the likelihood function is apparent in this case. The log-likelihood function is the simple sum of two terms. Moreover, the second term of the log-likelihood function,

$$-\mathrm{NLn}(\sigma_a),$$

will be negative whenever $\sigma_a > 1$; will be zero whenever $\sigma_a = 1$; and will positive whenever $\sigma_a < 1$. But the first term of the log-likelihood function,

$$-\sum_{i=1}^{N} a_i^2/2\sigma_a^2,$$

will always be negative. The maximum of the log-likelihood function will thus occur when the sum of squared shocks is at its minimum. In other words,

$$\mathrm{Max}\,[-\sum_{i=1}^{N} a_i^2/2\sigma_a^2 - \mathrm{NLn}\,(\sigma_a)] = \mathrm{Min}\,[\sum_{i=1}^{N} a_i^2].$$

This important result means that *the MLEs of ARIMA parameters are the "least-squares" estimates*. To obtain MLEs of ARIMA parameters, then, a set of parameters must be selected so as to minimize the sum of squared random shocks.

The same calculus procedure used to maximize the log-likelihood function can be used to minimize the sum of squares function. First, the sum of squares function must be written in a way that explicitly represents each parameter. Noting that the general ARIMA (p,0,q) model (for a stationary time series or a differenced nonstationary series) is:

$$(1 - \phi_1 B - \ldots - \phi_p B^p) z_t = (1 - \Theta_1 B - \ldots - \Theta_q B^q) a_t,$$

then the random shock, a_t, is defined as:

$$a_t = \frac{1 - \phi_1 B - \ldots - \phi_p B^p}{1 - \Theta_1 B - \ldots - \Theta_q B^q} z_t.$$

So the sum of squares function may be expressed as:

$$\sum_{i=1}^{N} a_i^2 = \sum_{i=1}^{N} [\frac{1 - \phi_1 B - \ldots - \phi_p B^p}{1 - \Theta_1 B - \ldots - \Theta_q B^q} z_i]^2.$$

We will call this expression the *conditional* sum of squares function. Denoting a vector of ARIMA parameter estimates as

$$\hat{\beta} = \{\hat{\phi}_p, \hat{\phi}_p, \hat{\Theta}_q, \hat{\Theta}_Q, \hat{\delta}_r, \hat{\omega}_s\},$$

we will denote the conditional sum of squares function as $S(\hat{\beta})$. This conditional sum of square function is conditional upon a specific set of parameter estimates.[1] The function will change its value as the parameter estimates change. The *absolute minimum* of this function, its smallest possible value, that is, will occur when solved derivatives,

$$0 = \frac{\partial S(\hat{\beta})}{\partial \hat{\phi}_p} = \frac{\partial S(\hat{\beta})}{\partial \hat{\phi}_p} = \frac{\partial S(\hat{\beta})}{\partial \hat{\Theta}_q} = \frac{\partial S(\hat{\beta})}{\partial \hat{\Theta}_Q} = \frac{\partial S(\hat{\beta})}{\partial \hat{\delta}_r} = \frac{\partial S(\hat{\beta})}{\partial \hat{\omega}_s},$$

are taken as the parameter estimates. Because these solved derivatives minimize the sum of squares function, they are the MLEs of the ARIMA parameters.

When a model is nonlinear, however, these derivatives are not easily solved. To illustrate this point, consider the simple *linear* model with

$$Y_i = b_0 + b_1 X_i + u_i$$

for the ith case. The sum of the squared disturbance terms for this model is:

$$\sum_{i=1}^{N} u_i^2 = \sum_{i=1}^{N} (Y_i - b_0 - b_1 X_i)^2.$$

The sum of squares is minimized when its solved derivatives are taken as the parameter estimates. These derivatives are:

$$\frac{\partial S(\hat{\beta})}{\partial \hat{b}_0} = -2 \sum_{i=1}^{N} (Y_i - \hat{b}_0 - \hat{b}_1 X_i)$$

$$\frac{\partial S(\hat{\beta})}{\partial \hat{b}_1} = -2X_i \sum_{i=1}^{N} (Y_i - \hat{b}_0 + \hat{b}_1 X_i).$$

These solved derivatives are called the "normal equations" of a model. Whenever a model is linear, its normal equations will be linear. In this particular case, there are two linear normal equations in two unknowns, $\hat{b}_0$ and $\hat{b}_1$. The equations may thus be solved for the unknowns to give MLEs of the parameters.

But now consider the simple ARIMA (0,0,1) model,

$$y_t = (1 - \Theta_1 B)a_t$$

or

$$a_t = \sum_{k=0}^{\infty} \Theta_1^k y_{t-k}.$$

The sum of squared shocks for this model is:

$$\sum_{t=1}^{N} a_t^2 = \sum_{t=1}^{N} [\sum_{k=0}^{\infty} \Theta_1^k y_{t-k}]^2$$

and the solved derivative is:

$$\frac{\partial S(\hat{\beta})}{\partial \hat{\Theta}} = 2 \sum_{t=1}^{N} [\sum_{k=0}^{\infty} \hat{\Theta}_1^k y_{t-k}] [k \sum_{k=0}^{\infty} \hat{\Theta}_1^{k-1} y_{t-k}] = 0.$$

This normal equation is nonlinear. In all cases, a nonlinear model will have nonlinear normal equations which cannot be solved with analytical methods.

In the next section, we will describe numerical (as opposed to analytical) methods for minimizing the $S(\hat{\beta})$ function. Meanwhile, Figures 6.1(b) and 6.1(c) illustrate the logic of minimization. Figure 6.1(b) is a plot of the sum of squares function for an ARIMA (0,1,1) model fit to the IBM stock price time series originally presented in Section 2.1 (Series B). This model is:

$$(1 - B)Y_t = z_t$$
$$z_t = (1 - \Theta_1 B)a_t.$$

It has only one parameter and the x-axis (the horizontal axis, that is) gives all

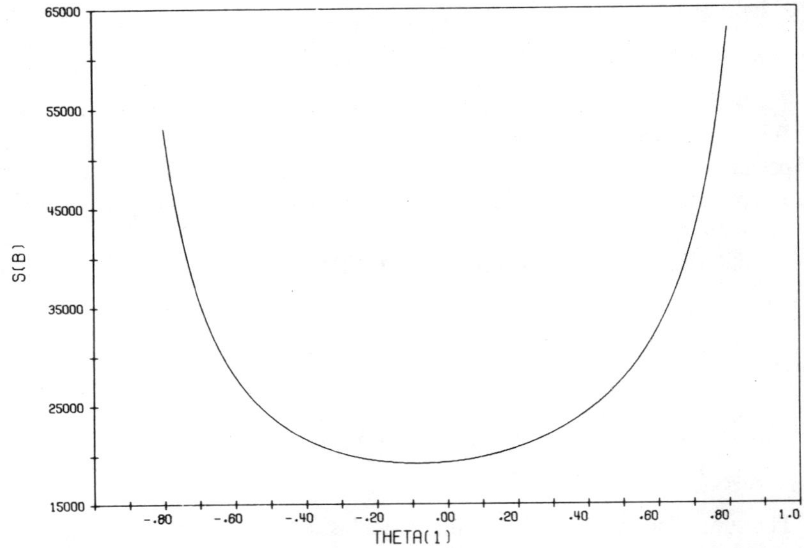

FIGURE 6.1(b) S ($\hat{\beta}$) for an ARIMA (0,1,1) Model Fit to Series B

values of $\hat{\Theta}_1$ within the bounds of invertibility. The y-axis of the plot (the vertical axis, that is) gives the corresponding value of the sum of squares function, $S(\hat{\Theta}_1)$. The minimum value of this sum of squares function appears to be at or near zero. In other words, $\hat{\Theta}_1 \simeq 0$ minimizes the $S(\hat{\Theta}_1)$ function. This approximate value will be the least-squares estimate of Θ_1. It will also be the MLE because, as has been demonstrated, the least-squares estimates and the MLEs of ARIMA parameters are identical.

Figure 6.1(c) shows a plot of the sum of squares function for the Directory Assistance time series analyzed in Section 3.1. The model used for this plot is:

$$Y_t = \frac{1 - \Theta_{12}B^{12}}{(1 - B)(1 - B^{12})} a_t + \omega_0 I_{147}.$$

It has two parameters. The sum of squares, $S(\hat{\Theta}_{12}, \hat{\omega}_0)$, is thus a three-dimensional surface. The x-axis of this plot gives a range of values for $\hat{\omega}_0$ while the y-axis gives the range of values for $\hat{\Theta}_{12}$. The contours of this plot give the height or depth of the sum of squares surface for each pair of

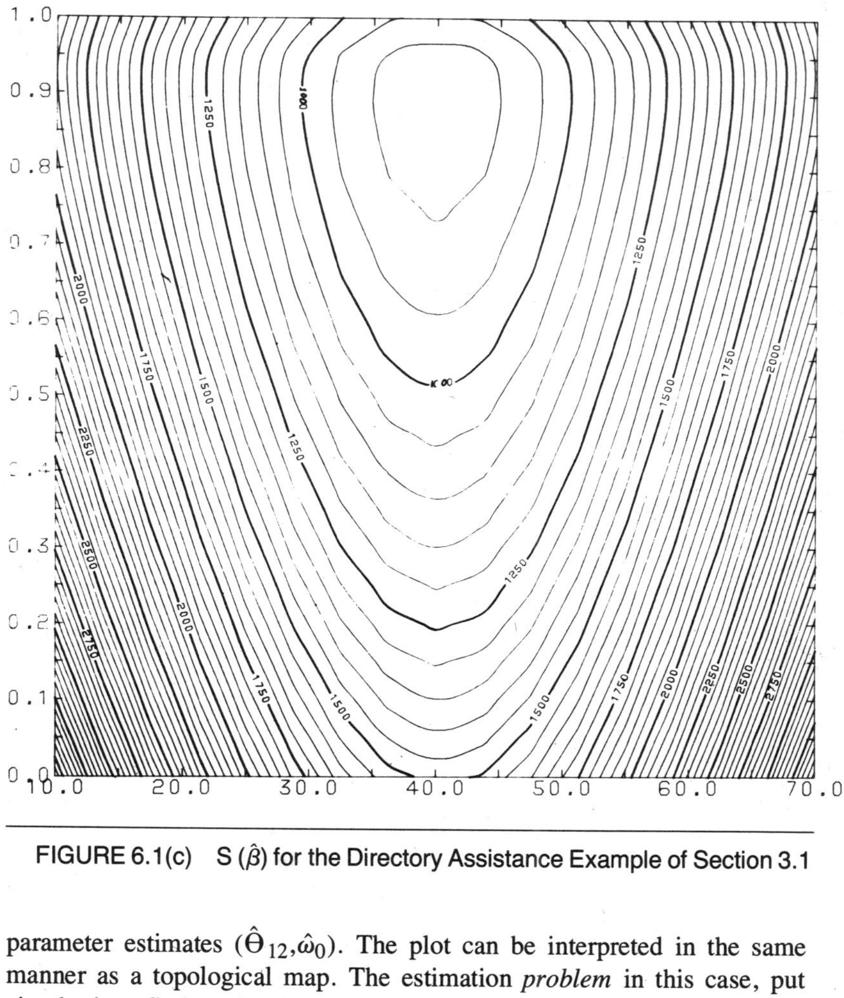

FIGURE 6.1(c) S $(\hat{\beta})$ for the Directory Assistance Example of Section 3.1

parameter estimates $(\hat{\Theta}_{12}, \hat{\omega}_0)$. The plot can be interpreted in the same manner as a topological map. The estimation *problem* in this case, put simply, is to find a pair of parameter estimates which minimize the sum of squares function. The minimum occurs in the "depression" or "valley" at the top center of the plot, corresponding approximately with $\hat{\Theta}_{12} \simeq .85$ and $\hat{\omega}_0 \simeq 40{,}000$. These parameter estimates minimize the sum of squares function.

6.2 Numerical Solutions to the S($\hat{\beta}$) Minimum

The problem of ARIMA parameter estimation is to discover a set of parameter estimates which minimizes the sum of squares function. Discovering this set of parameter estimates is no simple task, however. The general ARIMA model is nonlinear, so calculus methods may not be used. Instead, numerical methods are required.

To illustrate the basic concepts of numerical solutions, consider the problem of discovering the square root of a positive number, A. The problem is to discover a number x such that

$$x^2 = A.$$

A numerical algorithm for this problem is:

$$x_{n+1} = 1/2(x_n + A/x_n).$$

An algorithm in this context is defined as an iterative numerical procedure; subscripts in the algorithm refer to the number of the iteration. To execute this algorithm, a "guess" for the value of x_1 is plugged into the equation to obtain a value for x_2; this value of x_2 is then plugged into the equation to obtain a value for x_3, and so forth. After n iterations, the value of x_{n+1} will be very close to the square root of A. For example, let A = 20 and let x_1 = 10. Then,

$$x_2 = 1/2(10 + 20/10) = 6$$

$$x_3 = 1/2(6 + 20/6) = 4.66667$$

$$x_4 = 1/2(4.66667 + 20/4.66667) = 4.47619$$

$$x_5 = 1/2(4.47619 + 20/4.47619) = 4.47214$$

and so forth. The change in x from iteration to iteration becomes smaller and smaller as the value of x comes closer and closer to the solution. It can be demonstrated that this algorithm *converges* to $\sqrt{A}$ that is,

$$\lim_{n \to \infty} x_n = \sqrt{A}.$$

Convergence is one of the most important properties of a numerical algorithm. If an algorithm fails to converge on the proper solution, it is practically worthless. More important, some algorithms converge at a faster rate than others. An algorithm that takes millions of iterations to converge to an acceptable solution may also be practically worthless.

This particular algorithm could obviously be continued indefinitely, with successive values of x_n approaching nearer and nearer the limit of $\sqrt{A}$. If A is not a perfect square, of course, no value of x_n will ever be exactly equal to

$\sqrt{A}$. For this reason, we require a measure of *precision*. Such a measure is ϵ which is defined as:

$$\epsilon_n = \left| x_n^2 - A \right| .$$

In this example, for successive iterations of the algorithm

n	x_n	ϵ_n
1	10.00000	80.00000
2	6.00000	16.00000
3	4.66667	1.77778
4	4.47619	.03628
5	4.47214	.00004

With only four iterations, x_n is precise to the fourth decimal place. The required degree of precision, the value of ϵ_n that is, will ordinarily be determined a priori by situational needs. When extremely precise estimates are required, the value of ϵ_n will be set arbitrarily small to reflect those requirements and the algorithm will be run until the desired value of ϵ_n is realized.

What has not been made clear is that the number of iterations required to achieve a given value of ϵ_n depends also on the *initial value,* x_1. In this case, had the initial value been $x_1 = 15$ instead of $x_1 = 10$, more iterations would have been required:

n	x_n	ϵ_n
1	15.00000	205.00000
2	8.16667	46.69444
3	5.30782	8.17295
4	4.53792	.59274
5	4.47261	.00424
6	4.47214	.00001.

And in general, the farther the initial value is from the solution, the more iterations required to achieve an estimate of a given precision.

There are actually many suitable algorithms for solving the square root of A. The algorithm demonstrated here was selected only because it illustrates the basic concepts of numerical methods, including *convergence, precision,* and the relationship between precision and the *initial value* of the algorithm. These same concepts apply as well to the algorithms used for ARIMA parameter estimation. Numerical solutions to the minimization of $S(\hat{\beta})$ start with some likely guess for each ARIMA parameter. A sum of squares is evaluated conditional upon these guesses. With an appropriate algorithm, successive parameter values converge on a set of estimates which minimize the sum of squares function. The algorithm continues until a reasonable degree of precision is achieved. In most cases, parameter estimates precise to the fourth decimal place are adequate. The number of iterations required to achieve this degree of precision depends, of course, on the starting values of the algorithm, that is, on the initial guesses made for each parameter.

But unlike the simple problem of numerically solving a square root, the problem of solving the minimum $S(\hat{\beta})$ requires a relatively sophisticated algorithm. We will now describe two algorithms which have been developed for ARIMA estimation. The first, a *grid search* algorithm, was used in some of the earlier time series analysis programs such as *TMS* (Bower et al., 1974). As we will demonstrate, the grid search method is not ideally suited to ARIMA estimation. The second method we will discuss, the Marquardt algorithm has proved well suited to ARIMA estimation. Marquardt's algorithm, or variations of it, are used in most contemporary software packages. It is an extremely sophisticated method, however, so a full development would require a level of mathematical sophistication (not to mention a tolerance for esoteric detail) beyond that of most readers. Our development and description will consequently emphasize the broader conceptual and practical issues. Finally, our discussion will use examples in which the ARIMA models have only one or two parameters. The $S(\hat{\beta})$ functions for these models can be depicted graphically as two-dimensional curves or three-dimensional surfaces. Most readers will gain a better understanding of the principles of estimation when presented with graphic examples. It is a simple matter to generalize these principles to models with $n-1$ parameters and $S(\hat{\beta})$ functions in n-space, however.

6.2.1 The Grid Search Method

Returning now to the problem of finding a number x such that $x^2 = 20$, suppose that no efficient algorithm is available. A numerical solution could still be obtained (though with considerably more effort) by means of a "grid search" algorithm. To illustrate this method, let $x_1 = 6$ but now

$$x_{n+1} = x_n + \Delta.$$

In other words, the value of x increases with each iteration by a constant, Δ. Taking $\Delta = -.5$, the results of this grid search are:

n	X_n	ϵ_n
1	6.00	16.00
2	5.50	10.25
3	5.00	5.00
4	4.50	.25
5	4.00	4.00
6	3.50	7.75
7	3.00	11.00

and so forth. The smallest value of ϵ occurs when $x = 4.5$. While this estimate of $\sqrt{20}$ is relatively imprecise, it can be used as the basis for a finer search of the ϵ grid.

Figure 6.2(a) shows a plot of the precision function, ϵ, for several values of x. The precision function is a "W" with its smallest values *(minima)* at $x = \pm\sqrt{20}$. From the numerical results, it is clear that $\epsilon = 0$ somewhere in the interval

$$4.0 < x < 5.0.$$

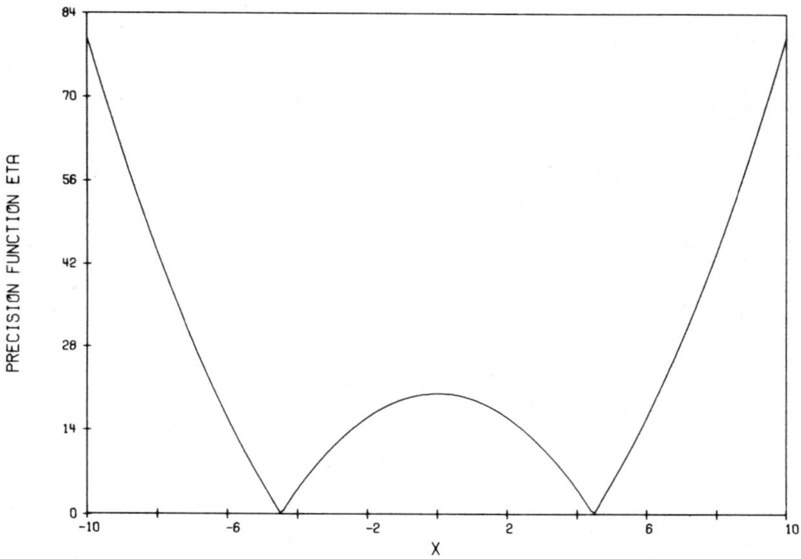

FIGURE 6.2(a) The Precision Function, ϵ

So the grid search could continue with $x_1 = 4.0$ and $\Delta = .1$:

$$x_{n+1} = x_n + .1.$$

The results of this grid search would narrow the interval around $\epsilon = 0$ considerably. By continually narrowing the interval to be searched, this crude method will yield an estimate of $\sqrt{20}$ to any desired degree of precision.

This same crude grid search method can be used to minimize a sum of squares function to obtain MLEs of ARIMA parameters. Referring to the $S(\hat{\beta})$ plot for an ARIMA (0,1,1) model of the IBM stock price time series, Figure 6.1(b), it appears that the minimum is at $\hat{\Theta}_1 \simeq 0$. We will now use a grid search to obtain a more precise estimate of Θ_1.

The grid search begins with an initial guess for the value of $\hat{\Theta}_1$. The $S(\hat{\Theta}_1)$ function is then evaluated for this guess. Say that the initial guess is $\hat{\Theta}_1 = -.75$. The conditional sum of squares function is then evaluated as:

$$S(\hat{\Theta}_1 = -.75) = \sum_{t=1}^{N} (z_t - \hat{z}_t)^2 = \sum_{t=1}^{N} \hat{a}_t^2.$$

The random shock estimates used to evaluate the $S(\hat{\Theta}_1)$ function are computed directly from the differenced time series. The first shock, for example is computed as:

$$z_1 = \hat{a}_1 + .75\hat{a}_0,$$

$$\text{thus} \qquad \hat{a}_1 = z_1 - .75\hat{a}_0.$$

The immediate problem is that the value of $\hat{a}_0$ (an estimate of the random shock which precedes the first time series observation) is unknown. A suitable estimate is often the unconditional expected value of a_0, or zero.[2] Using $\hat{a}_0 = 0$,

$$\hat{a}_1 = z_1 - .75(0) = z_1.$$

The second random shock is then estimated as:

$$\hat{a}_2 = z_2 - .75\hat{a}_1.$$

This value of $\hat{a}_2$ is then used to compute the value of $\hat{a}_3$ and so on until all N random shocks have been estimated. The conditional sum of squares function is then evaluated as the sum of all N squared random shock estimates.

The function evaluated in this way is interpreted as the sum of squares *if* $\hat{\Theta}_1 = -.75$.

As the grid search continues, the value of $\hat{\Theta}_1$ is increased by the constant Δ and the sum of squares function is reevaluated. Repeating this procedure several times, a sum of squares function is evaluated over a grid of predetermined $\hat{\Theta}_1$ values. Figure 6.1(b), in fact, was generated in this manner.

To efficiently employ the grid search method, the algorithm should be started with a value of $\hat{\Theta}_1$ as near the solution as possible. In Figure 6.1(b), for example, much effort was wasted evaluating the conditional sum of squares function for values of $\hat{\Theta}_1$ greater than .1 and less than $-.1$. In addition, the value of Δ (the increment or decrement in $\hat{\Theta}_1$) should be large enough to ensure that the solution will be discovered within a reasonable number of iterations and yet small enough to ensure that a reasonably precise estimate of the solution will be discovered.

The first of these issues is a relatively simple one to deal with in practice. The conditional sum of squares function obviously need not be evaluated for parameter values outside the bounds of stationarity-invertibility. While such values may occasionally minimize the function (usually because the ARIMA model has been incorrectly identified), they are undefined and need not be considered. More important, however, the ACF used to identify the ARIMA model will ordinarily give the analyst some information as to an appropriate set of initial values for the grid search. If nothing else, the ACF will always indicate whether parameter values are positive or negative, large or small.

Having decided upon a *range* of parameter values for the grid search, the analyst must decide upon a reasonable value of Δ. There is a trade-off here between precision and computational expense. Suppose, for example, that a solution precise to the fourth decimal place is required. For this degree of precision, searching a grid on the interval

$$-1 < \hat{\Theta}_1 < +1$$

would require (literally) hundreds of thousands of iterative evaluations. But on the other hand, if a solution precise to the second decimal place is adequate, the same grid search can be completed with only a few thousand iterative evaluations.

To balance the need for precision with the need for computional efficiency, the value of Δ can be varied. For the IBM stock price time series, we have executed a grid search across the entire bounds of invertibility with $\Delta = .1$. This grid search required only 21 evaluations of $S(\hat{\Theta}_1)$. The results,

Increment = .1		Increment = .01	
$\hat{\Theta}_1$	$S(\hat{\beta})$	$\hat{\Theta}_1$	$S(\hat{\beta})$
−1.0	1625637	−0.20	19483.3
−0.9	116298	−0.19	19437.3
−0.8	52987	−0.18	19395.8
−0.7	35519	−0.17	19358.8
−0.6	27948	−0.16	19326.3
−0.5	23929	−0.15	19298.2
−0.4	21595	−0.14	19274.2
−0.3	20222	−0.13	19254.6
−0.2	19483	−0.12	19239.0
−0.1	19220	−0.11	19227.6
0.0	19363	−0.10	19220.2
0.1	19896	−0.09	19216.8
0.2	20851	−0.08	19217.5
0.3	22315	−0.07	19222.0
0.4	24471	−0.06	19230.5
0.5	27694	−0.05	19242.9
0.6	32818	−0.04	19259.2
0.7	42017	−0.03	19279.3
0.8	62856	−0.02	19303.4
0.9	141803	−0.01	19331.3
1.0	2736045	0.00	19363.0

TABLE 6.2 Grid Search for $\hat{\Theta}_1$

listed in Table 6.2 and plotted in Figure 6.1(b), indicate that the solution lies in the interval

$$-.2 < \hat{\Theta}_1 < 0.0.$$

A grid search of this interval with $\Delta = .01$ also requires only 21 evaluations of $S(\hat{\Theta}_1)$. These results, also listed in Table 6.2, indicate that the solution lies in the interval

$$-.1 < \hat{\Theta}_1 < -.08.$$

If greater precision is required, this interval can be searched with $\Delta = .001$. As a general rule, relatively large values of Δ can be used at the start of the grid search to narrow the range of the search. Relatively small values of Δ can then be used to obtain precise estimates of the solution.

Although grid search methods are conceptually simple and easily executed, they are extremely costly. For the IBM stock price time series, a reasonably precise estimate of Θ_1 was discovered with fewer than 100 evaluations of the $S(\hat{\Theta}_1)$ function. Given access to a computer, these calculations presented no real problem. As the number of parameters in the model increases, however, the number of evaluations of $S(\hat{\beta})$ required for a grid search increases exponentially in the parameter space. If a model has only one parameter, and if 100 evaluations gives reasonably precise estimates of the parameter, a grid search is practical. If the model has two parameters, however, the grid search will require 10,000 evaluations (100^2) to obtain parameter estimates of equal precision. And if the model has four parameters, which is not at all uncommon in social science applications, the grid search will require 100,000,000 evaluations of $S(\hat{\beta})$. Executing the grid search in stages, each with a smaller value of Δ as in this example, would reduce the number of evaluations required to, say, several hundred thousand. But even this number of evaluations would be prohibitively expensive.

6.2.2 Marquardt's Algorithm

While useful as a pedagogical tool, grid search methods of estimating ARIMA parameters are virtually worthless in practice. To be sure, any method will require the evaluation of $S(\hat{\beta})$ at a number of points and, in this sense, all methods are similar to the grid search method. To be practically useful, however, a method must have some decision rule to limit the number of points requiring evaluation. An ideal method will superficially scan the $S(\hat{\beta})$ function, locate the general area of the minimum, and then will *converge* on the minimum in as few steps as possible. When the minimum is relatively far away, the method will take large steps, but as the minimum draws nearer, steps will become smaller. In this way, the ideal method will balance the need for precision with the costs of computation.

Marquardt (1963) has proposed an algorithm which, in this sense, is nearly ideal for the estimation of ARIMA parameters. While there are other algorithms which are also nearly ideal, they differ from Marquardt's algorithm only in minor details. The analyst typically encounters a nonlinear estimation algorithm as a "black box" hidden inside the time series analysis computer program. The black box usually performs adequately so long as the analyst supplies appropriate inputs to the program and so long as outputs are intelligently interpreted. To do this, the analyst need only have a general idea of the inner workings of the black box. The esoteric details can be left to computer programmers and numerical analysts. Our development and discussion of Marquardt's algorithm will thus focus on general principles and practical issues. Those readers who are interested in the detailed workings of the black box are directed to Box and Jenkins (1976: Chapter 7), Makridakis and Wheelwright (1978: Chapter 26), Draper and Smith (1966: Chapter 10) or Nelson (1973: Chapter 5.8) for treatments of these issues in varying detail, orientation, and contexts. Although our discussion will deal only with one version of the Marquardt algorithm, the terminology and principles we develop will be general to nearly all time series software.

To grasp the logic of Marquardt's algorithm, one must at least understand the basic principles of calculus minimization. We have introduced some of these principles in our earlier discussion but we must now give a more explicit introduction. The efficiency of a grid search can be increased substantially by increasing or decreasing the size of a step across the grid (increasing or decreasing the value of Δ, that is) depending upon how near or far away the solution is. An obvious decision rule for optimizing step size involves the derivatives of $S(\hat{\beta})$. If these derivatives were linear, of course, they could be solved to obtain MLEs of the ARIMA parameters. A numerical solution to the minimum $S(\hat{\beta})$ is required *because* these derivatives are either *non*linear, unavailable, or intractable.

But numerical estimates of the derivatives are always available. To illustrate just how these derivatives can be used to optimize step size, consider again the problem of estimating the value of $\sqrt{20}$ through a grid search. Denoting the *size* of the n_{th} step by Δx where

$$\Delta x = x_n - x_{n-1},$$

the corresponding change in the precision function is:

$$\Delta \epsilon = \epsilon_n - \epsilon_{n-1}.$$

Invoking the Mean Value Theorem of differential calculus, it can be demonstrated that the derivative of the precision function at some point between x_{n-1} and x_n is given by the ratio of $\Delta\epsilon$ to Δx. Of course, when the difference between x_{n-1} and x_n is made infinitesimally small, its limit *is* the derivative, that is,

$$\lim_{\Delta x \to 0} \quad \frac{\Delta \epsilon}{\Delta x} = \frac{\partial \epsilon}{\partial x}$$

Now let $x_1 = 7$ and let $\Delta x = .5$. The grid search then has the result:

n	x_n	ϵ_n	$\Delta\epsilon/\Delta x$
1	7.00	29.00	
2	6.50	22.25	13.50

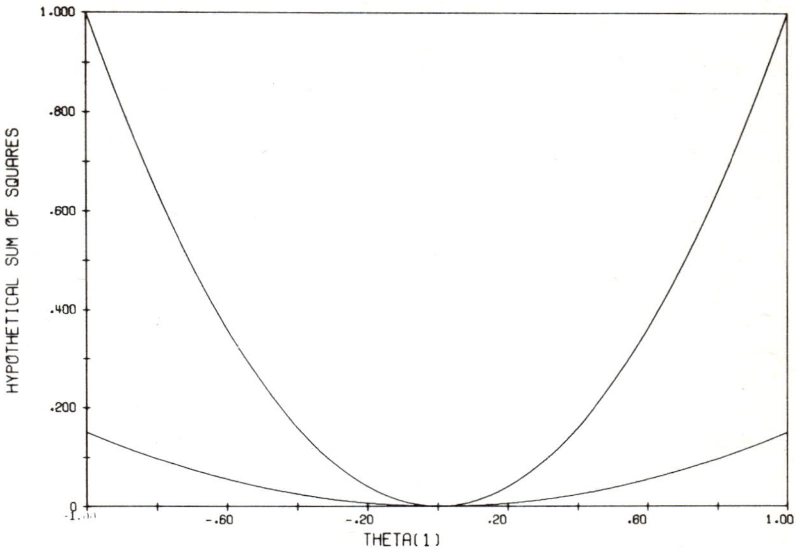

FIGURE 6.2(b) Two Hypothetical Sum of Squares Functions Differing Only in Steepness

3	6.00	16.00	12.50
4	5.50	10.25	11.50
5	5.00	5.00	10.50
6	4.50	.25	9.50
7	4.00	4.00	− 7.50 .

The numerical derivative, $\Delta\epsilon/\Delta x$, grows smaller as x approaches the minimum of the precision function; and as x *passes* the minimum, the numerical derivative changes its sign from positive to negative. The derivative is interpreted literally as the slope of a straight line drawn tangent to the precision function at a point between x_{n-1} and x. The numerical derivative thus tells one how *steep* the precision function is at that point. To optimize step size in the grid search, an algorithm can make the size of each step proportional to the steepness of the function, that is, proportional to the numerical derivative. When the function is steep, as it always will be at points far from the minimum, the numerical derivative will be large and positive. Steps should thus be large and positive. As the function becomes less steep, as it always will be at points near the solution, the numerical derivative will become small and step sizes may be made smaller. Finally, if the numerical derivative changes its sign, the search can stop for this indicates that the solution has been passed.

This method of optimizing the size of a grid search step is called (loosely) *the method of steepest descent*. An obvious shortcoming of this method is that, for the algorithm to be efficient, the $S(\hat{\beta})$ function must be relatively steep at points far away from the solution. Figure 6.2(b) shows two hypothetical functions which vary in steepness. For one of the functions, a steepest descent algorithm may or may not prove efficient, while for the other function, a steepest descent algorithm will probably be no more efficient than a simple grid search.

Strictly speaking, Marquardt's algorithm is a compromise between the steepest descent method and the *Gauss-Newton (iterative linearization)* method. The Gauss-Newton method solves a linear approximation to the nonlinear ARIMA model. Although the general ARIMA model itself is nonlinear, it can be expanded as a Taylor series around some initial guess for the parameters. Representing the general ARIMA model as

$$Y_t = a_t + f(\hat{\beta}, Y_{t-k}),$$

in which $f(\hat{\beta}, Y_{t-k})$ is a function of past observations of the time series and the model parameter estimates and where this function is nonlinear in the parameters, a Taylor series expansion is:

$$Y_t = a_t + f(\hat{\beta}_1, Y_{t-k}) + (\hat{\beta} - \hat{\beta}_1)\left[\frac{\partial f(\hat{\beta}_1, Y_{t-k})}{\partial \hat{\beta}_1}\right] + \ldots,$$

where $\hat{\beta}_1$ is an initial guess for $\hat{\beta}$. The Taylor series is an infinite expansion, of course, but only the first two terms of this expansion are required for the Gauss-Newton algorithm. The truncated Taylor series can be written as

$$Y_t - f(\hat{\beta}_1, Y_{t-k}) = (\hat{\beta} - \hat{\beta}_1)\left[\frac{\partial f(\hat{\beta}_1, Y_{t-k})}{\partial \hat{\beta}_1}\right] + a_t .$$

The quantity on the left-hand side of the operation is simply the difference between the time series observation Y_t and its predicted value given the parameter estimates $\hat{\beta}_1$. The derivative on the right-hand side of the equation is evaluated as a number for each time series observation, Y_t. An ARIMA (0,0,1) model, for example, has the derivative

$$\frac{\partial f(\hat{\beta}_1, Y_{t-k})}{\partial \hat{\beta}_1} = k \sum_{k=0}^{\infty} \hat{\beta}_1{}^{k-1} Y_{t-k},$$

where $\hat{\beta}_1$ is an initial guess for the parameter Θ_1. Although this derivative is nonlinear, and thus cannot be solved analytically, it can be *evaluated*. Denoting the evaluated derivative at Y_t by c_t, the linear approximation can be written as

$$\hat{a}_t = (\hat{\beta} - \hat{\beta}_1)c_t + a_t,$$

which is a linear model. An estimate of the quantity $(\hat{\beta} - \hat{\beta}_1)$ can be obtained with an OLS solution to the linear approximation. Once obtained, the estimate is interpreted as the *distance* between the MLE of the parameters and the initial guess at those parameters. If this distance is denoted by Δ_1 where

$$\Delta_1 = \hat{\beta} - \hat{\beta}_1,$$

then a second guess for the parameter estimates is:

$$\hat{\beta}_2 = \hat{\beta}_1 + \Delta_1.$$

The nonlinear function and the derivative are then evaluated for this set of parameter estimates, that is, the linear approximation

$$\hat{a}_t = (\hat{\beta} - \hat{\beta}_2)c_t + a_t$$

is solved to obtain a value for Δ_2. The Gauss-Newton algorithm continues iteratively until it converges, that is, until

$$\Delta_n = \epsilon,$$

where ϵ is a specified degree of precision.

Like the steepest descent method, the Gauss-Newton method is quite efficient for some types of nonlinear functions and some estimation situations. But the Gauss-Newton method, too, may be inefficient for other types of functions and other situations. For example, the Gauss-Newton algorithm may oscillate, producing positive and negative values of Δ_n before converging slowly to an acceptably precise set of parameter estimates. More important, since the Gauss-Newton algorithm solves a linear approximation to the nonlinear ARIMA model, any given iteration may reduce the linear sum of squares while increasing the nonlinear sum of squares.

In short, neither the method of steepest descent nor the Gauss-Newton method is ideal for ARIMA parameter estimation. The major practical dif-

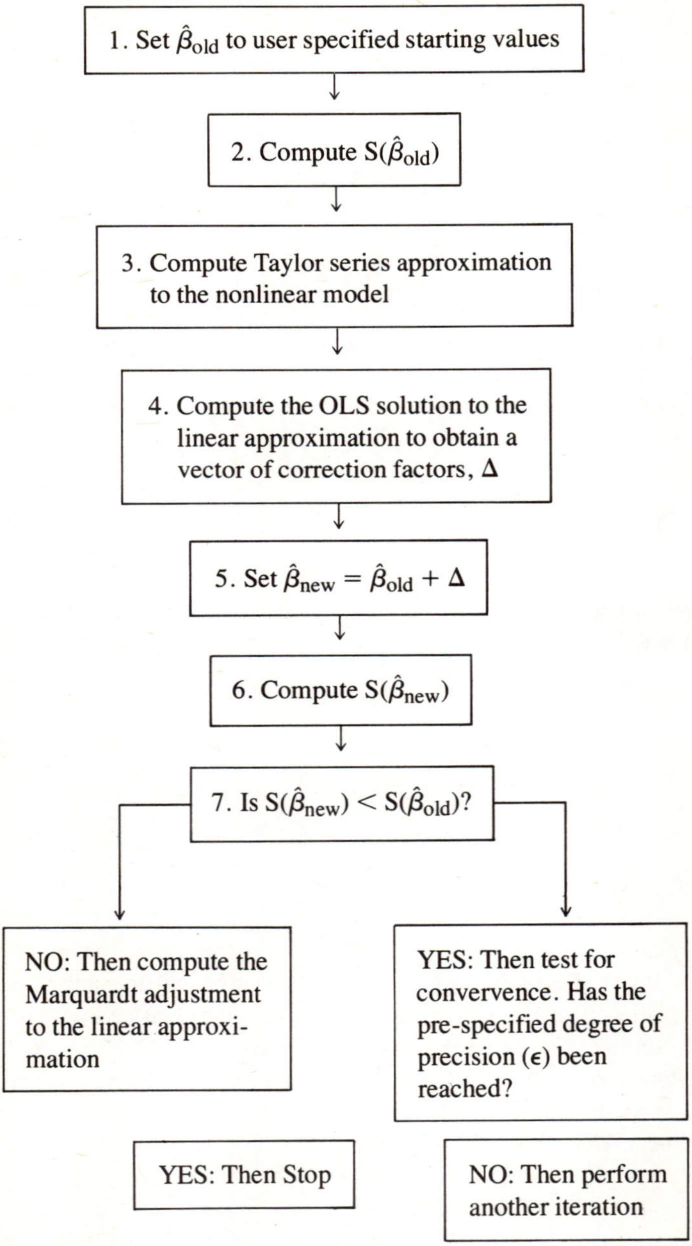

FIGURE 6.2(c) A Flowchart Diagram of Marquardt's Algorithm

ference between these two methods is the manner in which a step size (Δ) is selected for each iteration. To correct the deficiencies of the Gauss-Newton method, Marquardt incorporated several features of the steepest descent method into the algorithm. The result, often referred to as "Marquardt's compromise," combines the strongest features of each method.

Figure 6.2(c) presents the general principles of Marquardt's algorithm in a flow-chart format. The algorithm begins with a set of initial guesses, or starting values, for each parameter of the model. A truncated Taylor series expansion around these initial values is used as a linear approximation to the nonlinear model. The linear approximation is solved by OLS regression to obtain a vector of correction factors, Δ. A new set of parameter estimates is formed by adding the correction factors to the old parameters estimates, that is,

$$\hat{\beta}_{\text{new}} = \hat{\beta}_{\text{old}} + \Delta.$$

The sum of squares function is then evaluated for the new parameter estimates. If the new sum of squares is smaller than the old sum of squares, the new parameter estimates are substituted for the old. Convergence is tested, and if the specified degree of precision has not been achieved, another iteration is executed.

As described here, and as diagramed in Figure 6.2(c), the Marquardt algorithm is almost identical with the Gauss-Newton algorithm. Indeed, the two methods are identical *so long as each iteration results in a reduced sum of squares*. When an iteration does *not* result in a reduction of the sum of squares, that is, when

$$S(\hat{\beta}_{\text{old}}) < S(\hat{\beta}_{\text{new}}),$$

the iteration does not end. Instead, the linear approximation to the nonlinear function is adjusted and a new vector of correction factors is estimated. This adjustment is too complicated (and has too many variations) to be described in detail here. However, it is essentially an interpolation between the Gauss-Newton and steepest descent correction factors. This adjustment procedure is often called the "test-point" solution. Depending upon the shape of the nonlinear $S(\hat{\beta})$ function, the adjustment procedure may be repeated a number of times until a set of parameter values is found which reduces the sum of squares. Only then is the iteration completed, the convergence tests executed, and a new iteration (possibly) begun.

Figure 6.2(d) shows a contour plot of the sum of squares function for the Sutter County Workforce series analyzed in Section 2.12.1. The ARIMA

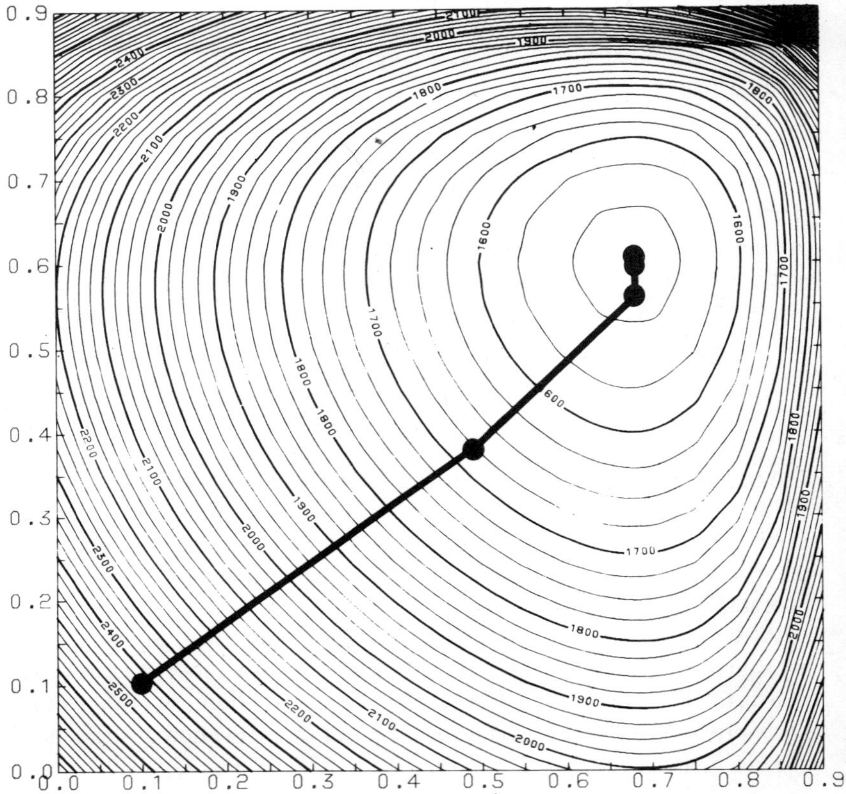

FIGURE 6.2(d) S ($\hat{\beta}$) for the Sutter County Workforce Example of Section 2.12.1: Marquardt's Algorithm "Finds" the Minimum in Four Steps

(0,1,1) $(0,1,1)_{12}$ model on which this plot is based gives ranges for the two moving average parameters along its x- and y-axes. Superimposed on this plot is the path taken by Marquardt's algorithm to obtain a solution precise to the third decimal place. Note that the steps taken by the algorithm in its first and second iterations are relatively large while the steps taken in its third and fourth iterations are relatively small. The size of each step is roughly proportional to the steepness of the function at that point. The statistics for each of the four iterations are:

Iteration	$\hat{\Theta}_1$	$\hat{\Theta}_{12}$	$S(\hat{\Theta}_1)$x 10^{-4}
Starting Values	.1000	.1000	2430.0
1	.3828	.4856	1678.0
2	.5590	.6839	1542.0
3	.5970	.6812	1539.0
4	.6031	.6800	1539.0 .

The algorithm reached this solution quickly. More complicated models and models whose sum of squares functions are not so ideally shaped (symmetrical and steep with a well-defined minimum, that is) might have required many more iterations before achieving a solution of this precision. The algorithm stopped in this case because the relative change in the sum of squares function from the third to fourth iterations was less than the specified degree of precision, $\epsilon = .001$. Had we required a greater degree of precision, say $\epsilon = .0001$ for a relative change in the sum of squares, the algorithm would not have stopped after the fourth iteration. And of course, had we specified a more realistic set of starting values, say $\hat{\Theta}_1 = \hat{\Theta}_{12} = .5$, convergence would have been achieved in fewer than four iterations.

6.3 Monitoring and Controlling the Algorithm

In practice, Marquardt's algorithm converges to a reasonably precise set of ARIMA parameter estimates within relatively few iterations, usually 5 to 15. The grid search, steepest descent, and Gauss-Newton methods, in contrast, may require hundreds or even thousands of iterations to achieve a solution of the same precision. From time to time, however, Marquardt's algorithm may iterate 50 or 100 times. Computational costs not withstanding, the parameter estimates obtained in such a case are generally suspect and certainly are of a lower quality than parameter estimates ordinarily must be. Most software packages print out the details of estimation. By monitoring these details, the analyst can often spot statistical anomalies and make adjustments to guarantee the quality of parameter estimates.

Fortunately, the rate at which Marquardt's algorithm converges (and hence, whether it will converge at all) depends on a few key factors which, for the most part, are controlled by the analyst. These factors include (1) the degree of precision specified by the analyst, (2) the quality of the starting values supplied to the algorithm by the analyst, and (3) the empirical adequacy of the ARIMA model identified by the analyst. The key here is the analyst. The parameter estimates given by the algorithm may often be no better than the information given the algorithm by the analyst.

Of these three factors, the least important is the degree of precision specified. Most computer programs stop iterating when any of four stopping criteria is satisfied: (1) a specified maximum number of iterations (say 50) have been executed; (2) the last iteration has resulted in a minimal relative change in the sum of squares evaluation, that is,

$$\left| \frac{S(\hat{\beta}_{\text{new}}) - S(\hat{\beta}_{\text{old}})}{S(\hat{\beta}_{\text{old}})} \right| \leq \epsilon$$

(3) the last iteration has resulted in a minimal relative change in the parameter estimates, that is,

$$\left| \frac{\hat{\beta}_{\text{new}} - \hat{\beta}_{\text{old}}}{\hat{\beta}_{\text{old}}} \right| \leq \epsilon$$

or (4) when the last iteration has resulted in a minimal ratio of the initial sum of squares to the final sum of squares.

Relaxing the tolerances of any stopping criterion will obviously result in a more rapid convergence to a less precise solution. Conversely, tightening any of the stopping criteria will result in a slower convergence to a more precise solution. Most computer programs provide default stopping criteria which ensure an adequate balance between precision and efficiency. *The analyst should always know exactly what these default stopping criteria are. More important, on any estimation, the analyst should always know why the algorithm stopped.* If the algorithm stops because it has converged on a solution, there is no problem. It may happen, however, that the algorithm did not converge, but rather stopped because a specified maximum number of iterations had been executed. The parameter estimates printed out in this situation would be of extremely poor quality. Beyond this, however, precision in effect is a given. It has been determined by the analyst a priori and should not be changed just to make the algorithm converge more quickly.

The starting values for parameter estimates are another matter. It should be intuitively clear that the rate of convergence will be optimal when the starting values supplied by the analyst are near the solution. What is not intuitively clear is that faulty starting values may lead to faulty parameter estimates. Although information from model identification may be used to estimate starting values, we have found that almost any naive value for the ϕ_p, ϕ_P, Θ_q, Θ_Q, and δ_r parameters of a general model will result in rapid convergence. Of course, the values supplied should always lie within the bounds of stationarity, invertibility, and stability.

Specification of a starting value for the ω_s parameter of a general model is more problematic. The ω-parameter is analogous to the unstandardized slope parameter of a linear regression model: both are expressed in the metric of the independent variable. To ensure convergence, the starting value should be as close as possible to the final estimate. At the very least, the starting value should be of the same order of magnitude as the final estimate. In the Directory Assistance analysis of Section 3.1, for example, the parameter $\hat{\omega}_0$ is approximately -40,000. The sum of squares function for this estimation is plotted in Figure 6.1(c). Examining that plot, the reader may see how crucial a realistic starting value for $\hat{\omega}_0$ is. An unrealistic starting value would result in an extremely large initial sum of squares which would cause many problems for the algorithm. On the other hand, any starting value of the same magnitude (for example, $-10{,}000$) will be adequate. When in doubt, the analyst may use OLS estimates of the ω-parameters as starting values.

Many software packages automatically compute adequate starting values for most analyses. Nonetheless, the analyst should remain sensitive to this issue and should make it a point to know the heuristic mechanisms used in the computer program. The program should display the starting values and these should be routinely inspected and compared to the final parameter estimates. An initial sum of squares, computed from the starting values, that is highly divergent from the sum of squares on the first iteration may indicate that one or more of the starting values is inappropriate.

Finally, the most crucial information supplied to the algorithm by the analyst is the ARIMA model. If the model is inappropriate, then the sum of squares function may be ill-formed with no solution clearly defined. Although a solution may be defined in a technical sense, the $S(\hat{\beta})$ function may be "flat" or "defective," in which case the algorithm may converge slowly or not at all and parameter estimates will be unstable.

The most graphic illustration of this problem is the case of parameter redundancy. As noted in Section 2.7, ARIMA (p,0,q) models may be reduced to simpler ARIMA (p,0,0), ARIMA (0,0,q), or ARIMA (0,0,0) models for certain values of the ϕ- and Θ-parameters. When parameters are redundant, the $S(\hat{\beta})$ function may not have one clearly defined minimim, but rather may have several minima, each associated with a particular configuration of the redundant parameters. When the algorithm attempts to solve a function of this sort, it may oscillate between the several minima without ever converging. The most common case of parameter redundancy may be

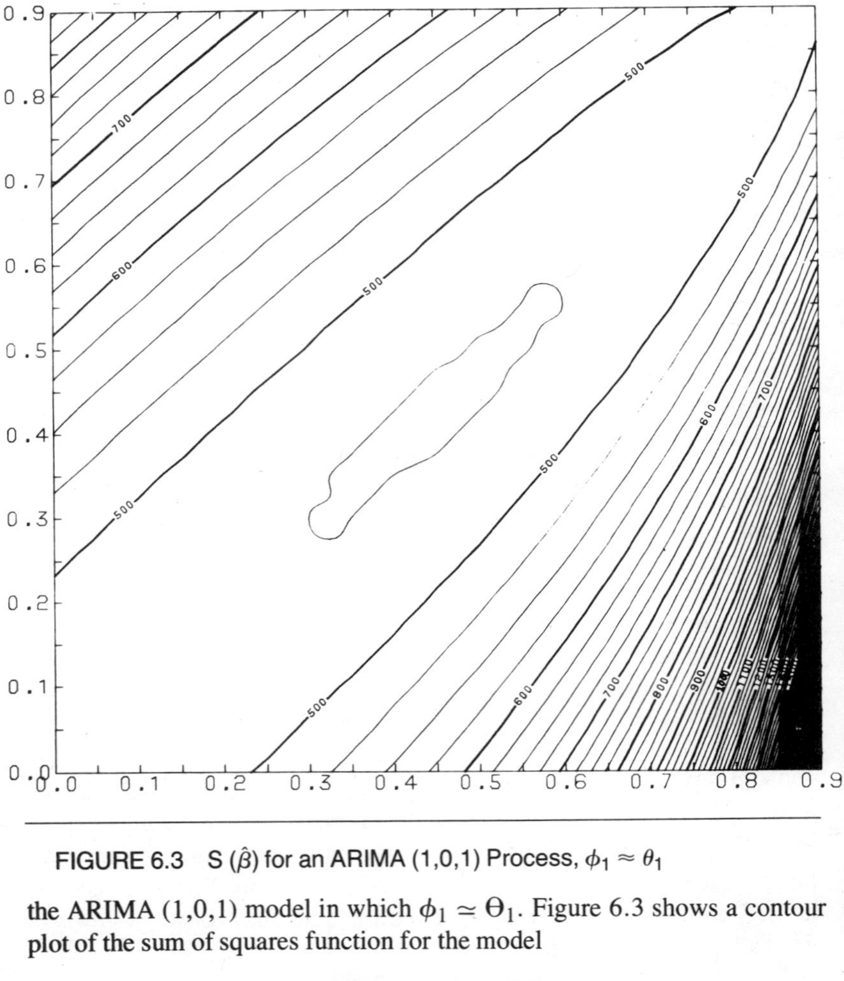

FIGURE 6.3 S ($\hat{\beta}$) for an ARIMA (1,0,1) Process, $\phi_1 \approx \theta_1$

the ARIMA (1,0,1) model in which $\phi_1 \simeq \Theta_1$. Figure 6.3 shows a contour plot of the sum of squares function for the model

$$(1 - .5B)y_t = (1 - .5B)a_t.$$

Note that the minimum of this function is not a clearly defined point, but rather is a long valley extending along the axis of $\hat{\phi}_1 = \hat{\Theta}_1$. Since the two parameters are redundant, many values of $\hat{\phi}_1$ and $\hat{\Theta}_1$ will minimize this function. A *unique* solution is undefined. Marquardt's algorithm in this case will produce a variety of solutions depending upon the starting values supplied it. If the algorithm is given $\hat{\phi}_1 \simeq \hat{\Theta}_1 \simeq .4$ for starting values, for example, it will produce different final estimates than if it is given $\hat{\phi}_1 \simeq \hat{\Theta}_1 \simeq .6$ for starting values. If this point is unclear, the reader should compare this sum of squares plot with the one shown in Figure 6.2(d). That sum of squares function is nearly ideal for the same reasons that this one is not.

Multicollinearity is a related issue. Because Marquardt's algorithm solves a linear approximation to the nonlinear function, time series programs print out many of the statistics usually associated with linear regression models. For example, most time series programs print out a correlation matrix of parameter estimates. This matrix should always be inspected as a quality control check on parameter estimates. If two parameters are highly correlated (say at the .7 level or higher), an inappropriately specified model may be indicated. When ARIMA parameters are highly correlated, the $S(\hat{\beta})$ function will have a shape similar to the one shown in Figure 6.3(a); that is, the minimum may be a long, shallow valley rather than a steep hole.

Given the wide range of models in the general ARIMA class, it is difficult to generalize about the effects of an inappropriately specified model on estimation. Fortunately, the empirical model-building strategies we developed in Chapters 2, 3, and 5 will nearly always lead to the identification of a parsimonious, statistically adequate model. If these strategies are followed routinely, and if common sense is exercised, it is unlikely that the analyst will encounter estimation problems due to model inappropriateness. The analyst nonetheless should be sensitive to this possibility.

6.4 Time Series Software

As the use of time series analysis in social research proliferates, so does the development of time series software. Computer programs for time series analysis (especially for managerial forecasting applications) have long been available commercially. It is only recently, however, that this software has become available to the academic community. When we started work on this volume in 1976, only one generalized Box-Jenkins program was widely available in academic computing centers. Only two years later, several programs (including the major statistical packages such as SPSS, BMDP, and SAS) either have Box-Jenkins capabilities or are planning to incorporate Box-Jenkins capabilities in the near future.

In this section, we will briefly review the computational requirements of Box-Jenkins time series analysis and then describe several available programs. Our intent is *not* to recommend a specific program, but rather to point out issues and discuss the manner in which these issues are handled by several programs. Our discussion of software will avoid the specific details of various programs. Development in this field is proceeding rapidly, so any description of the current state of the art will soon be outdated. Although many commercial time-sharing firms provide for ARIMA modeling, we have restricted our discussion to software currently available to academic computing facilities. In an appendix to this chapter, we give the names and addresses of several software suppliers. The reader is urged to contact these organizations directly for a description of current offerings. This list has been compiled from our personal knowledge, so the inclusion or omission of any program implies nothing about its quality or availability. Finally, the American Statistical Association regularly evaluates statistical software and we suggest that the reader consult these reports.

Speaking generally, computer programs for Box-Jenkins time series analysis are remarkably similar. Throughout this volume, we have referred to the calculation and display of certain statistics, plots, and so forth, and most Box-Jenkins programs provide all of these capabilities. The only major differences among programs tend to be in the formats of input specifications, the presentation of output, and the algorithm used for nonlinear estimation. The model-building strategies presented in earlier chapters always consisted of an identification-estimation-diagnosis sequence and, not surprisingly, most ARIMA time series programs are organized around these functions. Although these three steps are usually executed by the analyst in sequence, there are many interpretive decisions to be made at each step. We have found that flexible, interactive software which facilitates movement within and between these three steps to be optimal. Batch systems are nevertheless adequate and seem to be more widely available at this time.

In discussing time series software, it is useful to distinguish between three different levels of software organization. These are (1) a subroutine library, (2) a "stand-alone" statistical program, and (3) a statistical system. Time series software, generally written in a high-level scientific language such as FORTRAN, is available in each of these forms for computers of varying size and manufacture.

Subroutine Libraries

A subroutine library usually consists of a series of subroutines, each of which performs a discrete and distinct statistical function. One subroutine, for example, might compute the ACF and PACF while another subroutine might plot the ACF and PACF. To use a subroutine library, the user must write a "main program" which reads in the data, calls the appropriate subroutines, stores the intermediate results, and prints the desired output. Subroutine libraries offer great flexibility at the cost of detailed specification and programming.

The International Mathematical and Statistical Library (1979) is a large comprehensive subroutine library available at most research computing centers. It contains a variety of routines for the identification, estimation, and forecasting of univariate and single-input transfer function ARIMA models. It also contains a useful routine for generating ARIMA time series with known parameters to be used in testing and simulation.

Stand-Alone Statistical Programs

A stand-alone statistical program usually consists of a main program and its associated subroutines. It is designed to accept input from the user in a certain format, perform the necessary calculations, and print the results.

The earliest program designed specifically for interrupted time series analysis was TMS (Bower et al., 1974). This program implemented the methodology developed by Box and Tiao (1965) and elaborated by Glass et al., (1975). While a pioneer in its day, TMS is now generally considered to be obsolete. It is restricted to nonseasonal ARIMA models for the most part; it is restricted to a narrow subset of intervention models; furthermore, its use of a grid search estimation method makes it prohibitively expensive when compared to programs using a variation of the Marquardt algorithm.

In 1972, David J. Pack developed a subroutine library specifically designed for ARIMA analyses; and in intervening years, he has consistently expanded its scope. The most recent version of this program (Pack, 1977) contains two main programs which read input parameters and call appropriate subroutines for computation and display of results. The inclusion of these main programs places the Pack software more properly in the stand-alone program category. We have found these systems to be extremely flexible, easily transportable, and available at a modest cost. The major drawback to the Pack programs is that specification of the input parameters is in fixed fields and can be tedious and confusing to the inexperienced user. Moreover, the documentation is terse and assumes a thorough knowledge of ARIMA models and modeling procedures.

Batch Statistical Systems

A statistical system can be thought of as a series of statistical programs, each performing a certain kind of analysis. The series of programs are linked by a common set of data manipulation capabilities and a unified, "natural" user's input language. From the researcher's point of view, statistical systems are often the easiest to use; but they are also relatively costly to acquire and maintain.

The year 1979 was marked by the introduction of Box-Jenkins capabilities to three major academic and research statistical systems: SPSS (Statistical Package for the Social Sciences), BMDP (Biomedical Computer Programs), and SAS (Statistical Analysis System). The inclusion of ARIMA statistical procedures in these widely distributed and well-documented systems should greatly facilitate the use of time series analysis by social scientists. Each of these packages features extensive file definition and data management capabilities combined with a keyword syntax for input specifications. In SPSS and SAS, the Box-Jenkins routines are integrated into the unified system whereas in BMDP, they compromise a separate program (BMDP2T) which shares syntax and file structure with other BMDP series programs. Each of these packages supports univariate Box-Jenkins analyses. BMDP2T also supports intervention and transfer function analyses and can be used interactively (see below). SPSS is currently developing transfer function capabilities and SAS is considering them.

Interactive Statistical Systems

The iterative nature of Box-Jenkins time series analysis immediately suggests the desirability of interactive data analysis. In an interactive environment, the researcher ordinarily "converses" with a statistical system by means of a typewriterlike computer terminal. A command is entered from the terminal keyboard, the desired calculations are performed, and the results are immediately displayed and/or printed at the terminal console. The analyst can then enter a new command based on interpretation of the information just received. For example, in the present context, the analyst might estimate a model, perform diagnostic checking on the residuals, and then immediately estimate a revised model based on the diagnostic information just displayed. A procedure of this sort would usually take at least two separate runs using a batch program. Interactive or conversational analysis is also attractive in that input errors can be easily detected by the system and reported to the user for immediate correction. Since communication between the computer and a terminal is usually over regular phone lines, researchers may utilize interactive software, often unavailable locally, from a distant computer facility for a relatively nominal fee.

As we have mentioned, many commercial time-sharing firms offer Box-Jenkins capabilities, but the cost of these services is often prohibitive for academic research and instruction. Network linkages between university computing centers are becoming more popular and one or more members of the EDUNET network may offer ARIMA analysis programs. At present, interactive statistical systems are not distributed as widely as batch systems. Again, development is proceeding rapidly in this field and such systems will probably be more readily available in the future. BMDP2T, described above, also may be operated in an interactive mode. MINITAB is a flexible, general interactive statistical system for research and instruction using small data sets. It can be easily adapted for different sizes and models of computers and presently supports univariate ARIMA analysis. Intervention analysis capabilities are currently under development. SYBIL/RUNNER is an interactive system oriented primarily toward applied managerial forecasting. It includes ARIMA univariate and transfer function capabilities in its extensive forecasting repertoire. SCSS, the SPSS Conversational Statistical System, is a very powerful interactive system that runs on several different models of computers. SCSS does not currently support ARIMA time series analysis but these capabilities are being considered for the future. Finally, SCRUNCH, an interactive impact assessment system developed for the CDC 6000 computer by Richard A. Hay, Jr., was used to conduct many of the analyses for this volume.

For Further Reading

Almost all discussions of nonlinear estimation require a background in calculus. Draper and Smith (1966: Chapter 10) present a solid introduction which is as accessible as possible to the nonmathematician. Pindyck and Rubinfeld (1976: Chapter 15), Nelson (1973: Chapter 5.8), or Granger and Newbold (1977: Chapter 3.5) give briefer developments but at more or less the same level of sophistication. Makridakis and Wheelwright (1978: Chapter 26) work through the arithmetic of a Marquardt solution. This material is perhaps the most accessible treatment of those discussed here.

NOTES TO CHAPTER 6

1. The reader who is familiar with Box and Jenkins (1970: Chapter 7) will realize that the *conditional* sum of squares function is also conditional upon the starting shocks of the estimation, that is, on the values of a_0, a_{-1}, . . . , a_{-t}, which precede the first observation of the time series. We will return to this point later.

2. Cf. Note 1. As noted, the conditional sum of squares function is conditional upon a vector of parameter estimates, $\hat{\beta}$, and also upon a set of initial shocks, a_0, a_{-1} . . . , a_{-t}. In this case, the $S(\hat{\beta})$ function requires an estimate of only one initial shock to be evaluated: $\hat{a}_0$. An ARIMA (0,0,2) model would require values for $\hat{a}_0$ and $\hat{a}_{-1}$ and an ARIMA (0,0,q) model would require values for $\hat{a}_0$, $\hat{a}_{-1}$, . . . , $\hat{a}_{1-q}$. For *nonseasonal* series of moderate length, say 100 observations or more, the *un*conditional expectation of these shocks (zero) should prove sufficient. For *seasonal* models of any length, however, Box and Jenkins (1976: Chapter 7) recommend that the *conditional* expectation of these shocks be used. The conditional expectation is derived through a recursive procedure known as backforecasting or backcasting. In essence, backcasting is executed by "running" the ARIMA model backward in time to generate forecasts of a_0,

$a_{-1}, \ldots, a_{-t}$. These forecasts are then used to initiate an evaluation of the $S(\hat{\beta})$ function. Since backcasting is much more expensive than using the unconditional expectations, many computer programs use both methods in a two-stage approach. Parameter estimates are first computed by using the unconditional expectations. These parameter estimates are then used as starting values in a second stage to backcast the conditional expectations. This approach minimizes computational cost while providing superior parameter estimates. The reader should be aware that, *for seasonal models,* parameter estimates based on the unconditional expectations of initial shocks (zero) are *different than* and nearly always *inferior to* parameter estimates based on the conditional expectations of initial shocks (backcasted values).

Appendix to Chapter 6: ARIMA Software Distributors

Program	**Distributor**
IMSL	International Mathematical and Statistical Libraries, Inc. 7500 Bellaire Blvd. Houston, TX 77036
PACK	Automatic Forecasting Systems PO Box 563 Hatboro, PA 19040
BMDP2T	Health Sciences Computing Facility Department of Biomathematics University of California Los Angeles, CA 90024
MINITAB	Professor Thomas A. Ryan Statistics Department 215 Pond Laboratory The Pennsylvania State University University Park, PA 16802
SPSS/SCSS	SPSS, Inc. Suite 3300 444 North Michigan Ave. Chicago, IL 60611
SYBIL/RUNNER	Applied Decision Systems 33 Hayden Ave. Lexington, MA 02173
SAS	SAS Institute ,, Inc. PO Box 10066 Raleigh, NC 27605

APPENDIX A: SYMBOLS AND CONVENTIONS

(1) Acronyms

ACF: *Auto*c*orrelation* *F*unction: see Sec. 2.8
ARIMA: *Auto*r*egressive* *I*ntegrated *M*oving *A*verage
CCF: *C*ross-*C*orrelation *F*unction; see Sec. 5.1
COV: Covariance of
MLE: *M*aximum *L*ikelihood *E*stimation; see Sec. 6.1
MSFE: *M*ean *S*quare *F*orecast *E*rror; see Sec. 4.1
OLS: *O*rdinary *L*east *S*quares
PACF: *P*artial *Auto*c*orrelation* *F*unction; see Sec. 2.8
RMS: *R*esidual *M*ean *S*quare; see Sec. 2.11
SE: *S*tandard *E*rror of
VAR: Variance of

(2) Time Series Conventions

Y_t: the tth observation of a time series
y_t: the tth observation of a deviate time series
$y_t = Y_t - \Theta_0; Ey_t = 0$
z_t: the tth observation of a differenced time series
$z_t = (1 - B)Y_t$
a_t: the tth random shock of a white noise process
$a_t \sim NID(0, \sigma_a^2)$
B: the backward shift operator whose operation is
$B^n Y_t = Y_{t-n}; B^n B^m = B^{n+m}$; see Sec. 2.3
ϕ_p: the pth-order autoregressive parameter
Θ_q: the qth-order moving average parameter
p: the order of autoregression (regular)
P: the order of autoregression (seasonal)
q: the order of a moving average (regular)
Q: the order of a moving average (seasonal)
d: the order of differencing (regular)
D: the order of differencing (seasonal)
S: the length of a seasonal cycle
S = 12 for monthly data
= 4 for quarterly data
= 52 for weekly data
N_t: the noise component of a model
$f(I_t)$: the intervention component of an impact assessment model; see Chapter 3
$f(x_t)$: the causal component of a multivariate model; see Chapter 5
δ_r: the rth-order *output* parameter of a transfer function
ω_s: the sth-order *input* parameter of a transfer function
$\hat{\beta}$: a vector of parameter estimates

$\hat{\beta} = \{\hat{\phi}_p, \hat{\Phi}_P, \hat{\theta}_q, \hat{\Theta}_Q, \hat{\delta}_r, \hat{\omega}_s\}$

$S(\hat{\beta})$: the conditional sum of squares function for a vector of parameter estimates

(3) Mathematical Symbols

$\sim$: is distributed as

$\propto$: is proportional to

∞: infinity

$\sum_{i=1}^{n}$: the sum of n terms

$\hat{}$: is an estimate of; e.g., $\hat{\phi}_1$ is an estimate of the parameter ϕ_1

E: is the expected value of; see the appendix to Chapter 2; also see Sec. 4.1

$\lim_{n\to\infty}$: the limit as n approaches infinity

$=$: equals

$\simeq$: is approximately equal to

$<$: is less than

$>$: is greater than

$\leqslant$: is less than or equal to

$\geqslant$: is greater than or equal to

$\frac{\partial y}{\partial x}$: the derivative of y with respect to x

Δx: the change in x

APPENDIX B: DATA SETS

All time series are read from left to right, top to bottom.

Section 2.1, Figure 2.1(a)

German Immigration to the United States (in thousands); 45 annual observations, 1870–1914; source: Fried (1969).

118	83	141	150	87	48	32	29	29	35
85	210	251	195	180	124	84	107	110	100
92	114	119	79	54	32	32	23	17	17
19	22	28	40	46	41	38	38	32	26
31	32	28	34	36					

Section 2.1, Figure 2.1(b)

Closing Price of IBM Common Stock, "Series B"; 369 daily observations, May 17, 1961–November 2, 1962; source: Box and Jenkins (1976).

460	457	452	459	462	459	463	479	493	490
492	498	499	497	496	490	489	478	487	491
487	482	479	478	479	477	479	475	479	476
476	478	479	477	476	475	475	473	474	474
474	465	466	467	471	471	467	473	481	488
490	489	489	485	491	492	494	499	498	500
497	494	495	500	504	513	511	514	510	509
515	519	523	519	523	531	547	551	547	541
545	549	545	549	547	543	540	539	532	517
527	540	542	538	541	541	547	553	559	557
557	560	571	571	569	575	580	584	585	590
599	603	599	596	585	587	585	581	583	592
592	596	596	595	598	598	595	595	592	588
582	576	578	589	585	580	579	584	581	581
577	577	578	580	586	583	581	576	571	575
575	573	577	582	584	579	572	577	571	560
549	556	557	563	564	567	561	559	553	553
553	547	550	544	541	532	525	542	555	558
551	551	552	553	557	557	548	547	545	545
539	539	535	537	535	536	537	543	548	546
547	548	549	553	553	552	551	550	553	554
551	551	545	547	547	537	539	538	533	525
513	510	521	521	521	523	516	511	518	517
520	519	519	519	518	513	499	485	454	462
473	482	486	475	459	451	453	446	455	452
457	449	450	435	415	398	399	361	383	393

385	360	364	365	370	374	359	335	323	306
333	330	336	328	316	320	332	320	333	344
339	350	351	350	345	350	359	375	379	376
382	370	365	367	372	373	363	371	369	376
387	387	376	385	385	380	373	382	377	376
379	386	387	386	389	394	393	409	411	409
408	393	391	388	396	387	383	388	382	384
382	383	383	388	395	392	386	383	377	364
369	355	350	353	340	350	349	358	360	360
366	359	356	355	367	357	361	355	348	343
330	340	339	331	345	352	346	352	357	

Section 2.4, Figures 2.4(a-d)

Nonfatal Disabling Mine Injuries; 49 annual observations, 1930–1978; source: Dr. J. Richard Zelonka, Mine Safety and Health Administration, U.S. Department of Labor.

99981	77958	56283	59129	65559	63426	67540
66259	49636	51773	57776	61057	66774	64594
63691	57117	55350	57660	53472	35405	37264
35553	30074	24258	17718	18885	19816	18792
14160	12163	11902	11197	10944	11133	11070
11138	10446	10115	9639	9917	11552	11916
12329	11220	8545	11107	14389	14794	13554

Section 2.12.1, Figure 2.12.1(a)

Sutter County Workforce; 252 monthly observations, January 1946–December 1966; source: Friesema et al. (1979).

890	992	979	959	1110	1546	1539	3401	2092	1436	1301	1287
1488	1517	1707	1729	1788	2008	2203	3713	2946	2082	2033	1937
1711	1775	1902	1846	2083	2262	2193	3792	2343	2313	2179	1975
1880	1930	2060	1843	2052	2039	2351	3394	3581	2489	2468	2134
1903	1719	1617	1818	2067	2457	2600	3530	2693	2448	2250	1972
1682	1730	1814	1900	2051	2290	2599	3428	2262	2242	2103	1825
1670	1681	1713	1954	1976	2272	2612	3590	2496	2441	2340	2090
1812	1788	1837	1993	2021	2199	2622	3787	2914	2487	2314	2139
2124	2214	2234	2279	2423	2290	2903	4485	3085	2852	2629	2435
2227	1944	2125	2260	2299	2323	2659	3761	2779	2761	2446	2278
1879	1881	2165	2199	2308	2529	2573	3946	3200	2574	2422	2446
2828	2879	2800	2835	2585	2787	3334	4746	3613	3463	3274	2801
2488	2386	2428	2678	2744	2772	3520	3833	3377	3013	2871	2592
2375	2304	2464	2557	2739	2714	3102	3961	3772	3245	3104	2869
2513	2385	2756	2927	2940	3180	3791	4093	4309	3532	3408	2839
2792	2798	3007	3086	3201	3428	3754	4917	3760	3609	3471	3347
3333	3456	3569	3900	3909	4098	4826	5770	5108	4360	4100	3562

3284	3278	3424	3843	3614	3536	4505	5456	4881	4041	3724	3525
3437	3324	3977	4025	4016	4031	4489	5563	5709	4620	4160	4012
3987	4155	4054	4485	4558	4462	4594	6481	6345	5142	4824	4573
4158	4140	4251	4734	4858	4798	5080	6905	5504	5457	5198	4890

Section 2.12.2, Figures 2.12.2(a) and 2.12.2(e)

Boston Armed Robberies; 118 monthly observations, January 1966–October 1975; source: Deutsch and Alt (1977).

41	39	50	40	43	38	44	35	39	35	29	49
50	59	63	32	39	47	53	60	57	52	70	90
74	62	55	84	94	70	108	139	120	97	126	149
158	124	140	109	114	77	120	133	110	92	97	78
99	107	112	90	98	125	155	190	236	189	174	178
136	161	171	149	184	155	276	224	213	279	268	287
238	213	257	293	212	246	353	339	308	247	257	322
298	273	312	249	286	279	309	401	309	328	353	354
327	324	285	243	241	287	355	460	364	487	452	391
500	451	375	372	302	316	398	394	431	431		

Section 2.12.3, Figure 2.12.3(a)

Swedish Harvest Index; 102 annual observations, 1749–1850; source: Thomas (1940).

2	9	6	6	9	6	6	1	1.5	6	9	6	1.5	0
1.5	1	6	6	6	9	9	6	0	1	6	6	1	6
9	6	9	1	1	1	1	6	1	6	9	8	6	9
8	6	8	8	8	6	6	1	1	1	6	6	8	6
3.5	6	4.5	3.5	8	8	2.5	1.5	4	7	8	2	3	2.5
7	9	7	4.5	8	8	6	2	8	8	6	4.5	4	8
6	4	7	6	3.5	7	6	7	2	4.5	4	5	3	5
6	7	5	6										

Section 2.12.4, Figure 2.12.4(a)

Hyde Park Purse Snatchings; 71 28-day periods, January 1969–September 1973; source: Reed (1978).

10	15	10	10	12	10	7	17	10	14	8	17
14	18	3	9	11	10	6	12	14	10	25	29
33	33	12	19	16	19	19	12	34	15	36	29
26	21	17	19	13	20	24	12	6	14	6	12
9	11	17	12	8	14	14	12	5	8	10	3
16	8	8	7	12	6	10	8	10	5	7	

Section 3.1, Figure 3.1(a)

Average Daily Calls to Directory Assistance, Cincinnati, OH; 180 monthly observations, January 1962–December 1976; source: Dr. A. J. McSweeny, Department of Psychology, University of West Virginia; see McSweeney (1978).

350	339	351	364	369	331	331	340	346	341	357	398
381	367	383	375	353	361	375	371	373	366	382	429
406	403	429	425	427	409	402	409	419	404	429	463
428	449	444	467	474	463	432	453	462	456	474	514
489	475	492	525	527	533	527	522	526	513	564	599
572	587	599	601	611	620	579	582	592	581	630	663
638	631	645	682	601	595	521	521	516	496	538	575
537	534	542	538	547	540	526	548	555	545	594	643
625	616	640	625	637	634	621	641	654	649	662	699
672	704	700	711	715	718	652	664	695	704	733	772
716	712	732	755	761	748	748	750	744	731	782	810
777	816	840	868	872	811	810	762	634	626	649	697
657	549	162	177	175	162	161	165	170	172	178	186
178	178	189	205	202	185	193	200	196	204	206	227
225	217	219	236	253	213	205	210	216	218	235	241

Section 3.2.1, Figure 3.21(a)

Perceptual Speed Scores for a Schizophrenic Patient; 120 daily observations; source: Glass et al. (1975).

55	56	48	46	56	46	59	60	53	58	73
69	72	51	72	69	68	69	79	77	53	63
80	65	78	64	72	77	82	77	35	79	71
73	77	76	83	73	78	91	70	88	88	85
77	63	91	94	72	83	88	78	84	78	75
75	86	79	76	87	66	73	62	27	52	47
65	59	77	47	51	47	49	54	58	56	50
54	45	66	39	51	39	27	39	37	43	41
27	29	27	26	29	31	28	38	37	26	31
45	38	33	33	25	24	29	37	35	32	31
28	40	31	37	34	43	38	33	28	35	

Section 3.3, Figures 3.3(a) and 3.3(b)

Minneapolis Public Drunkenness Intakes: 151 monthly observations, January 1966–July 1978; source: Professor Michael C. Musheno, Department of Criminal Justice, Arizona State University; see Aaronson et al. (1978).

529	567	747	719	707	728	758	746	725	725	555
526	612	570	597	666	677	651	736	757	708	601

569	526	538	525	668	692	762	693	775	843	578
708	434	489	528	505	576	805	895	707	803	834
645	745	637	588	429	552	598	735	831	720	691
670	649	586	559	374	442	396	555	707	616	473
289	183	204	183	140	201	203	201	290	256	169
174	192	170	169	170	124	152	163	114	208	176
191	230	258	356	375	339	291	150	187	163	162
206	168	138	183	128	148	133	103	122	123	140
157	149	171	139	169	172	162	138	124	140	128
112	154	159	176	245	168	242	246	138	199	232
198	219	243	218	203	211	267	233	223	211	267
248	244	265	268	242	244	276	371			

Section 3.5, Figure 3.5(a)

U.S. Suicide Rate (per 100,000 total population); 50 annual observations, 1920–1969; source: U.S. Department of Commerce (1975).

10.2	12.4	11.7	11.5	11.9	12	12.6	13.2
13.5	13.9	15.6	16.8	17.4	15.9	14.9	14.3
14.3	15	15.3	14.1	14.4	12.8	10.2	10
11.2	11.5	11.5	11.2	11.4	11.4	10.4	10
10	10	10.2	10	9.8	10.7	10.6	10.6
10.6	10.4	10.9	11	10.8	11.1	10.9	10.8
10.7	11.1						

Section 5.2, Figure 5.2(a)

Paris IBM Closing Prices (in francs); 130 daily observations, April 31, 1976–October 1976; source: Makridakis and Wheelwright (1978).

1186	1187	1169	1190	1172	1181	1185	1223	1212
1217	1211	1190	1200	1213	1200	1227	1208	1195
1187	1196	1205	1229	1219	1218	1220	1220	1212
1214	1200	1200	1213	1221	1233	1223	1241	1263
1260	1278	1293	1301	1304	1299	1305	1317	1305
1309	1317	1312	1320	1326	1342	1350	1348	1345
1353	1358	1336	1322	1329	1356	1353	1360	1343
1351	1352	1344	1350	1350	1361	1372	1378	1380
1391	1399	1402	1404	1406	1407	1403	1378	1360
1379	1362	1350	1334	1340	1340	1352	1377	1378
1377	1377	1383	1371	1361	1370	1375	1355	1355
1386	1396	1392	1408	1401	1383	1391	1413	1400
1383	1412	1397	1392	1368	1370	1398	1370	1373
1354	1358	1308	1317	1323	1318	1302	1270	1262
1300	1315	1337						

Section 5.2, Figure 5.2(a)

New York IBM Closing Prices (in francs); 130 daily observations, April 31, 1976–October 28, 1976; source: Makridakis and Wheelwright (1978).

1160.25	1178.10	1161.31	1166.31	1170.09	1206.92
1210.95	1204.05	1197.34	1184.61	1194.09	1197.51
1200.28	1215.61	1207.43	1181.32	1185.79	1183.50
1196.35	1209.28	1213.94	1215.81	1219.26	1210.88
1198.59	1199.73	1201.25	1198.38	1209.28	1211.07
1215.93	1224.94	1223.03	1273.18	1260.97	1284.54
1275.22	1290.16	1291.90	1303.60	1298.83	1308.01
1310.75	1301.54	1304.46	1307.18	1310.66	1320.64
1314.15	1325.85	1334.41	1327.05	1334.02	1325.10
1341.84	1353.29	1332.12	1327.34	1331.08	1335.01
1339.20	1336.68	1342.07	1340.29	1337.97	1352.25
1350.40	1359.48	1370.75	1363.92	1385.89	1336.79
1387.50	1400.28	1394.08	1392.13	1393.66	1381.88
1362.61	1362.78	1339.15	1354.22	1336.43	1370.95
1336.07	1351.50	1365.19	1361.01	1365.96	1372.10
1379.00	1366.16	1368.59	1377.10	1368.34	1334.99
1360.99	1388.05	1390.82	1406.41	1409.98	1369.15
1397.80	1393.21	1407.58	1393.99	1384.14	1385.14
1389.88	1385.72	1385.95	1391.91	1379.08	1388.38
1356.37	1366.23	1321.71	1325.41	1322.08	1382.08
1308.28	1280.33	1277.31	1293.61	1317.83	1314.34
1332.30	1358.89				1330.27

Section 5.3, Figure 5.3(a)

Swedish Population Rates (in thousands); 100 annual observations, 1750–1849; source: Thomas (1940).

9	12	8	12	10	10	8	2	0	7	10	9
4	1	7	5	8	9	5	5	6	4	−9	−27
12	10	10	8	8	8	14	7	4	1	1	2
6	7	7	−2	−1	7	12	10	10	4	9	10
9	5	4	3	7	7	6	8	3	4	−5	−14
1	6	3	2	6	1	13	10	10	6	9	10
13	16	14	16	12	8	7	6	9	4	7	12
8	14	11	5	5	5	10	11	11	9	12	13
8	6	10	13								

Section 5.4, Figure 5.4(a)

Swedish Fertility Rates (in thousands); 100 annual observations, 1970–1849; source: Thomas (1940).

329	349	323	324	333	335	321	290	295	297	316
309	310	308	306	294	298	315	302	299	301	294
264	233	311	321	297	297	314	333	326	306	293
276	286	286	298	288	312	295	283	303	338	319
314	298	323	326	317	304	275	290	307	305	312
311	300	304	296	261	321	340	322	284	297	332
336	317	321	313	311	334	339	349	326	345	331
297	318	331	314	293	298	332	330	322	315	307
296	299	322	312	327	318	334	327	310	306	315
341										

BIBLIOGRAPHY

1978 Aaronson, D., C. T. Dienes and M. C. Musheno

Changing the public drunkenness laws: the impact of decriminalization. *Law and Society Review,* 12, 405–436.

1975 Anderson, T. W.

The Statistical Analysis of Time Series. New York: John Wiley and Sons.

1971 Blalock, H. M. Jr.

Causal Models in the Social Sciences. Skokie, IL: AVC.

1974 Bower, C. P., W. L. Padia and G. V. Glass

TMS: Two FORTRAN IV Programs for Analysis of Time Series Experiments. Boulder: University of Colorado Laboratory of Educational Research.

1964 Box, G. E. P. and D. R. Cox

An analysis of transformations. *Journal of the Royal Statistical Society,* B, 26, 211–243.

1976 Box, G. E. P. and G. M. Jenkins

Time Series Analysis: Forecasting and Control, Revised Edition. San Francisco: Holden-Day.

1975 Box, G. E. P. and G. C. Tiao

Intervention analysis with applications to economic and environmental problems. *Journal of the American Statistical Association,* 70–92.

1965 Box, G. E. P. and G. C. Tiao

A change in level of a nonstationary time series. *Biometrika,* 52, 181–192.

1962 Brown, R. G.

Smoothing, Forecasting and Prediction of Discrete Time Series. Englewood Cliffs, NJ: Prentice-Hall.

1963 Campbell, D. T.

From description to experimentation: interpreting trends as quasi-experiments. In C. W. Harris (ed.), *Problems of Measuring Change.* Madison: University of Wisconsin Press.

1966 Campbell, D. T. and J. C. Stanley

Experimental and Quasi-Experimental Designs for Research. Chicago: Rand-McNally.

1968 Campbell, D. T. and H. L. Ross

The Connecticut crackdown on speeding: time series data in quasi-experimental analysis. *Law and Society Review,* 3, 33–53.

1971 Caporaso, J. A. and A. L. Pelowski

Economic and political integration in Europe: a time series quasi-experimental analysis. *American Political Science Review,* 65, 418–433.

1979 Cook, T. D. and D. T. Campbell

Quasi-Experimentation: Design and Analysis Issues for Field Settings. Chicago: Rand-McNally.

1978 Deutsch, S. J

Deterrence effectiveness measurement. *Criminology,* 16, 115–131.

1979 Deutsch, S. J.

Lies, damn lies, and statistics: a rejoinder to the comment by Hay and McCleary. *Evaluation Quarterly,* 3, 315–328.

1977 Deutsch, S. J. and F. B. Alt

The effect of Massachusetts' gun control law on gun-related crimes in the city of Boston. *Evaluation Quarterly,* 1, 543–568.

1966 Draper, N. R. and H. Smith

Applied Regression Analysis. New York: John Wiley and Sons.

1974 Dhrymes, P. J.

Econometrics: Statistical Foundations and Applications. New York: Springer-Verlag.

1953 Doob, J. L.

Stochastic Processes. New York: John Wiley and Sons.

1960 Durbin, J.

The fitting of time series models. *Review of the International Institute of Statistics,* 28, 233–244.

1951 Durkheim, E.

Suicide. New York: Macmillan.

1968 Feller, W.

An Introduction to Probability Theory and Its Applications, Volume I, Third Edition. New York: John Wiley and Sons.

1971 Feller, W.

An Introduction to Probability Theory and Its Applications, Volume II, Second Edition. New York: John Wiley and Sons.

1969 Fried, M.

Deprivation and migration: dilemmas of causal interpretation. In D. P. Moynihan (ed.), *On Understanding Poverty*. New York: Basic Books.

1979 Friesema, H. P., J. Caporaso, G. Goldstein and R. McCleary

Aftermath: Communities After Natural Disasters. Beverly Hills, CA: Sage.

1968 Glass, G. V.

Analysis of data on the Connecticut speeding crackdown as a time series quasi-experiment. *Law and Society Review,* 3, 55–76.

1975 Glass, G. V., V. L. Willson and J. M. Gottman

Design and Analysis of Time Series Experiments. Boulder: Colorado Associated University Press.

1958 Goldberg, S.

Introduction to Difference Equations. New York: John Wiley and Sons.

1973 Goldberger, A. S. and O. D. Duncan

Structural Equation Models in the Social Sciences. New York: Seminar Press.

1972 Gottman, J. M. and R. M. McFall

Self-monitoring effects in a program for potential high school dropouts: a time series analysis. *Journal of Consulting and Clinical Psychology,* 39, 273–281.

1977 Granger, C. W. J. and P. Newbold

Forecasting Economic Time Series. New York: Academic Press.

1971 Hall, R. V., R. Fox, D. Willard, L. Goldsmith, M. Emerson, M. Owen, F. Davis and E. Porcia

The teacher as observer and experimenter in the modification of disputing and talking-out behaviors. *Journal of Applied Behavior Analysis,* 4, 141–149.

1979 Hay, R. A., Jr.

Interactive Analysis of Interrupted Time Series Models Using SCRUNCH. Evanston, IL: Northwestern University Department of Sociology and Vogelback Computing Center.

1979 Hay, R. A., Jr. and R. McCleary

Box-Tiao time series models for impact assessment: a comment on the recent work of Deutsch and Alt. *Evaluation Quarterly,* 3, 277–314.

1973 Hays, W. L.

Statistics for the Social Sciences, Second Edition. New York: Holt, Rinehart & Winston.

1975 Heise, D. R.

Causal Analysis. New York: John Wiley and Sons.

1977 Hibbs, D. A., Jr.

On analyzing the effects of policy interventions: Box-Tiao vs. structural equations models. In H. L. Costner (ed.), *Sociological Methodology 1977*. San Francisco: Jossey-Bass.

1974 Hibbs, D. A., Jr.

Problems of statistical estimation and causal inference in time series regression models. In H.L. Costner (ed.), *Sociological Methodology* 1973–74. San Francisco: Jossey-Bass.

1963 Holtzman, W.

Statistical models for the study of change in the single case. In C.W. Harris (ed.), *Problems in Measuring Change*. Madison: University of Wisconsin Press.

1980 Izenman, A. J. and S. L. Zabel

Babies and the blackout: the genesis of a misconception. *Social Science Research,* forthcoming.

1972 Johnston, J.

Econometric Methods, Second Edition. New York: McGraw-Hill.

1977 Keyfitz, N.

Applied Mathematical Demography. New York: Wiley-Interscience.

1971 Kmenta, J.

Elements of Econometrics. New York: Macmillan.

1959 Kuznets, S.

Six Lectures on Economic Growth. New York: Macmillan.

1980 Land, K. C.

Modeling Macro Social Change. In S. L. Leinhart (ed.), *Sociological Methodology 1980*. San Francisco: Jossey-Bass.

1976 Land, K. C. and M. Felson

A general framework for building dynamic macro social indicator models including an analysis of changes in crime rates and police expenditures. *American Journal of Sociology,* 82, 565–604.

1979 Lewis-Beck, M. S.

Some economic effects of revolution: models, measurement, and the Cuban evidence. *American Journal of Sociology,* 84, 1127–1149.

1976 Ljung, G. and G. E. P. Box

Studies in the modeling of discrete time series 3: a modification of the overall χ^2 test for lack of fit in time series model. University of Wisconsin, Department of Statistics, Technical Report 477.

1978 Makridakis, S. and S. C. Wheelwright

Interactive Forecasting, Second Edition. San Francisco: Holden-Day.

1970 Malinvaud, E.

Statistical Methods of Econometrics, Second Edition. Amsterdam: North-Holland.

1979 Mark, M. M.

The causal analysis of concomitancies in time series. Chapter 7 of T. D. Cook and D. T. Campbell, *Quasi-Experimentation: Design and Analysis Issues for Field Settings*. Chicago: Rand-McNally.

1963 Marquardt, D. W.

An algorithm for least squares estimation of nonlinear parameters. *Journal of the Society for Industrial and Applied Mathematics,* 2, 431–441.

1974 Mayer, T. F. and W. R. Arney

Spectral analysis and the study of social change. In H. L. Costner (ed.), *Sociological Methodology* 1973–San Francisco: Jossey-Bass.

1979 McCain, L. J. and R. McCleary

The statistical analysis of the simple interrupted time series quasi-experiment. Chapter 6 of T. D. Cook and D. T. Campbell, *Quasi-Experimentation: Design and Analysis Issues for Field Setttings*. Chicago: Rand-McNally.

1980 McCleary, R. and M. C. Musheno

Floor effects in the time series quasi-experiment. *Political Methodology,* 7(2).

1980 McDowall, D. and R. McCleary

Time series analysis: ARIMA models for the time series quasi-experiment. Sage University Paper Series on Quantitative Applications in the Social Sciences, forthcoming. Beverly Hills, CA: Sage.

1978 McSweeny, A. J.

The effects of response cost on the behavior of a million persons: charging for directory assistance in Cincinnati. *Journal of Applied Behavioral Analysis,* 11, 47–51.

1973 Nelson, C. R.

Applied Time Series Analysis for Managerial Forecasting. San Francisco: Holden-Day.

1964 Nerlove, M.

Spectral analysis of seasonal adjustment procedures. *Econometrica,* 32, 207–229.

1978 Ostrom, C. W., Jr.

Time series analysis: regression techniques. Sage University Paper Series on Quantitative Applications in the Social Sciences, 07–009. Beverly Hills, CA: Sage.

1977 Pack, D. J.

A Computer Program for the Analysis of Time Series Models Using the Box-Jenkins Philosophy. Columbus: Ohio State University Data Center.

1976 Pindyck, R. S. and D. L. Rubinfeld

Econometic Models and Economic Forecasts. New York: McGraw-Hill.

1978 Reed, D.

Whistlestop: A Community Alternative for Crime Prevention. Unpublished Ph.D. Dissertation. Evanston, IL: Department of Sociology, Northwestern University.

1970 Ross, H. L., D. T. Campbell and G. V. Glass.

Determining the effects of a legal reform: the British “breathalyzer” crackdown of 1967. *American Behavioral Scientist,* 13, 493–509.

1969 Smoker, P.

A time series analysis of Sino-Indian relations. *Journal of Conflict Resolution,* 13, 105–113.

1971 Theil, H.

Principles of Econometrics. New York: John Wiley and Sons.

1940 Thomas, D. S.

Social and Economic Aspects of Swedish Population Movements, 1750–1933. New York: Macmillan.

1968 Tyler, V. D. and G. D. Brown

Token reinforcement of academic performance with institutionalized delinquent boys. *Journal of Educational Psychology,* 59, 164–168.

1975 U.S. Department of Commerce

Historical Statistics of the United States: Colonial Times to 1970. Washington, D.C.: G.P.O.

1978 Vigderhous, G.

Forecasting sociological phenomena: application of Box-Jenkins methodology to suicide rates. In K. F. Schuessler (ed.), *Sociological Methodology 1978*. San Francisco: Jossey-Bass.

1975 Zimring, F. E.

Firearms and federal law: the gun control act of 1968. *Journal of Legal Studies,* 4, 133–198.

INDEX

Aaronson, D., 141, 175
ACF See Autocorrelation Function
Alt, F. B., 110, 174, 223
Anderson, T. W., 17
Arney, W. R., 17
Autocorrelation Function (ACF)
  Estimate, 66
  model identification, 94
  residual diagnosis, 98
  standard errors of, 94
Autoregressive process
  expected ACF, 72
  expected PACF, 76
  forecast profile, 215
  moving average identity, 57
  seasonal, 83
  stationarity bounds, 56, 135
  variance, 72

Backcasting, 225, 309
Backward Shift Operator, 45
Between-Series correlation, 229, 246
Bidirectional cause, 273
Blalock, H. M., Jr., 269
Boston armed robbery time series
  intervention model, 174
  univariate model, 120
Bower, C. P., 288, 307
Box, G.E.P., 18, 35, 91, 98, 100, 134, 135, 136, 141, 203, 225, 294, 307, 309
Brown, G. D., 141
Brown, R. G., 35

Campbell, D. T., 141, 142, 203
Caporaso, J. A., 141
CCF See Cross-Correlation Function
Chi square See Q statistic
Computer Programs, 305
Concomitant Variation, 227
Confidence intervals
  for ACF (k), 94
  for CCF (k), 232
  for forecasts, 211
  for PACF (k), 94
Cook, T. D., 142, 203
Cox, D. R., 135
Cross-Correlation function (CCF)
  estimate, 229
  expected value, 234
  residual diagnosis, 242
  standard errors, 232

Deseasonalization, 81
Detrending, 32
Deutsch, S. J., 110, 143, 174, 223
Dhrymes, P. J., 135, 272
Diagnosis
  multivariate, 242
  univariate, 98
  use of forecasting, 220
Difference equations, 36, 40
Differencing
  regular, 40
  seasonal, 83
Directory Assistance time series
  intervention model, 154
  ruling out alternatives, 169
  sum of squares function, 285
Doob, J., 17
Draper, N. R., 53, 294, 309
Drift, 35
Duncan, O. D., 269, 272
Durbin, J., 136
Durkheim, E., 20

Equilibrium state See Level
Estimation
  algorithms, 286
  least-squares, 281
  maximum likelihood, 276
  model-building strategy, 97
Event, 142
Expected value
  algebra, 136
  forecasts, 207

Factors, 84
Feller, W., 17, 134, 136
Felson, M., 225
Filters, 132
Fitting
  impacts, 197
  versus modeling, 36
Floor effects, 175
Forecast
  error, 211
  model (psi-weights), 209
  multivariate example, 236
  profiles, 213
Fried, M., 33
Friesema, H. P., 166

Gambler's ruin, 38
Gauss-Newton algorithm, 296
German immigration time series, 33
Glass, G. V, 81, 141, 160, 203, 227, 288, 307
Goldberg, S., 135
Goldberger, A. S., 269, 272
Gottman, J. M., 81, 141, 160, 203, 227, 307
Granger, C.W.J., 100, 121, 134, 224, 273, 309
Grid search algorithm, 288

Hall, R. V., 141
Harmonic analysis, 17
Hay, R. A., Jr., 104, 112, 174, 223, 225
Hays, W. L., 136
Heise, D. R., 269, 272
Hibbs, D. A., 111, 203, 271, 272
Holtzman, W., 160
Homoskedasticity, 53
Hyde Park Purse Snatchings time series
  forecast profile, 217
  intervention model, 200
  univariate model, 131

Identification
  bivariate, 229
  univariate, 93
Impact patterns
  abrupt, 146
  compound, 187
  gradual, 155
  temporary, 166
Infinite series
  autoregressive operator, 58
  differencing operator, 47
  forecast model, 209
  level change, 156
  temporary displacement, 168
Input-output models, 18
Integrated process
  differencing, 40
  expected ACF, 67
  forecast profile, 215
  input-output model, 45
Intercept, 32
Invertibility
  backward shift operator, 46
  bounds, 63
Izenman, A. J., 203

Jenkins, G. M., 18, 35, 91, 98, 100, 134, 136, 225, 294, 309
Johnston, J., 81, 269, 272

Keyfitz, N., 134, 273
Kmenta, J., 269, 272
Kuznets, S., 23

Land, K. C., 225, 273
Level
  abrupt change, 146
  asymptotic change, 157
  definition, 31
  nonstationary model, change, 147
Lewis-Beck, M. S., 141
Likelihood functions, 277
Ljung, G., 136
Log transformation
  interpreting parameters, 174
  variance nonstationarity, 51

Makridakis, S., 17, 81, 205, 225, 237, 294, 309
Malinvaud, E., 135, 272

Mark, M. M., 198
Mayer, T. F., 17
Marquardt, D. W., 293
Marquardt algorithm, 293, 298
McCain, L. J., 134, 203
McCleary, R., 112, 134, 174, 175, 203, 223, 225
McDowall, D., 203
McFall, R. M., 141
McSweeny, A. J., 147
Mean-square forecast error (MSFE), 209
Metadiagnosis, 100
MLE See Estimation, maximum likelihood
Model-building strategy
  multivariate, 251
  univariate, 91
Moving average process
  autoregressive identity, 63
  expected ACF, 69
  forecast profile, 218
  seasonal, 84
  variance, 70
MSFE See Mean-square forecast error
Multicollinearity, 305
Multiplicative models
  by log transformation, 173
  seasonal, 84
Multivariate time series analysis, 273
Musheno, M. C., 175

Nelson, C. R., 17, 134, 224, 294, 309
Nerlove, M., 81
Newbold, P., 100, 121, 134, 224, 273, 309
Nonfatal disabling mine injuries time series, 49
Normal equations, 283
Numerical solutions, 285

Ostrom, C. W., Jr., 272
Outlier effects
  estimation, 200
  identification, 131
Overdifferencing, 97
Overmodeling, 101

PACF See Partial autocorrelation function
Pack, D. J., 307
Padia, W. L., 288, 307
Parameter redundancy
  mixed models, 64
  estimation, 303
Paris-New York IBM time series, 236
Parsimony, 20, 171
Partial autocorrelation function (PACF)
  estimate of PACF (k), 75
  model identification, 93
  standard error, 94
  Yule-Walker equations, 135
Pelowski, A. L., 141
Perceptual speed time series
  abrupt intervention model, 163
  compound intervention model, 189
  gradual intervention model, 163
Pindyck, R. S., 17, 134, 205, 225, 309
Prewhitening, 243
Pulse function, 165

Q statistic, 99

Random shock, 40
Random walk, 36
Reed, D., 128, 201
Residual Mean Square (RMS), 101
RMS See Residual mean square
Ross, H. L., 141
Rubinfeld, D., 17, 134, 205, 225, 309

Seasonality, 81
Series B
  drift, 35
  forecast profile, 216
  sum of squares, 284
Slope
  change in, 159
  definition, 32
Smith, H., 53, 294, 309
Smoker, P., 141
Software, 305
Stanley, J. C., 141
Stationarity
  bounds, 56, 135
  homogeneous sense, 43
  variance (widest sense), 48
Steepest descent algorithm, 296
Step function, 142

Stochastic process
models, 17
realizations, 30
Sum of squares function
ANOVA, 34
ARIMA, 282
Sundbarg's hypothesis, 245
Sutter County workforce time series
forecast profile, 220
intervention model, 166
sum of squares function, 300
univariate model, 105
Swedish fertility rate time series, 257
Swedish harvest index time series
forecast profile, 219
multivariate model, 254
univariate model, 124
Swedish population growth time series, 244
System stability bounds, 155

Theil, H., 272
Thomas, D. W., 121, 244, 262, 273
Tiao, G. C., 18, 141, 203, 307
Time series, 30
Transfer functions
compound, 187
first order, 155
higher order, 185
relationships among, 169
zero order, 145
Trend
ANOVA model, 32
difference equation model, 42
Tyler, V. D., 141

U.S. Suicide time series
forecasts, 221
intervention model, 197

Vigderhous, G., 198, 225

Wheelwright, S., 17, 81, 205, 225, 237, 294, 309
White noise process
definition, 31
expected ACF, 67
forecast profile, 214
Willson, V. L., 81, 160, 203, 227, 307
Within-series correlation, 246

Yule-Walker equations, 134

Zabel, S., 203
Zimring, F. E., 143, 154

ABOUT THE AUTHORS

RICHARD McCLEARY is Associate Professor of Criminal Justice at Arizona State University. His Ph.D. is in sociology from Northwestern University. He has written several books and articles on applied time series analysis.

RICHARD A. HAY, JR., is on the staff of Vogelback Computing Center and is a Ph.D. candidate in sociology at Northwestern University. His research interests are in applied quantitative methodology and the political economy of socioeconomic development. He recently edited (with Janet Abu-Lughod) a volume of readings entitled *Third World Urbanization.* (Methuen).

ERROL E. MEIDINGER is Senior Fellow at the Natural Resources Law Center of Lewis and Clark University. He is currently a Ph.D. candidate in sociology at Northwestern University. His research interests and publications are in the social impacts of economic decisions, particularly those which allocate natural resources among alternative uses.

DAVID McDOWALL is Postdoctoral Fellow at the Center for Research on Social Organization, University of Michigan. His Ph.D. is in sociology from Northwestern University. His research interests and publications are in the areas of quantitative methodology and criminology.